STTO

Y0-BSF-843

Charts, Graphs & Stats Index

Charts, Graphs & Stats Index

1988–1991

Edited by Robert Skapura

Fort Atkinson, Wisconsin

Published by Highsmith Press
W5527 Highway 106
P. O. Box 800
Fort Atkinson, Wisconsin 53538-0800

∞
The paper used in this publication meets the minimum requirements of American National Standard for Information Science – Permanence of Paper for Printed Library Material, ANSI/NISO Z39.48-1984.

ISBN 0-917846-09-5
ISSN 1060-1465

Acknowledgments

I wish to thank Betty Bortz, Diablo Valley College, and Carol Bartlett, Contra Costa County Public Library, both in Pleasant Hill, California; and Ann M. Smith, John Swett Unified School District, Crockett, California. These librarians brought to the project many years of experience working at the reference desk. Each had a unique perspective from her own type of library. In addition, they all brought an enthusiasm that sustained us over the many months.

I am grateful also to Carol Shuey who was responsible for the data entry. Her rigorous proofreading and her many questions kept us all honest.

Thanks must also go to Monica Ertel, at Apple Computer Library, and Dennis Marshall, at Claris Corporation, for their advice and technical assistance.

Finally, I owe a special debt to Duncan Highsmith who encouraged and supported this project throughout, from the rough idea to the completed book.

Magazines Indexed

Black Enterprise

Bulletin of the Atomic Scientists

Business Week

Ms.

Newsweek

FDA Consumer

Scholastic Update

Time

U.S. News and World Report

Table of Contents

Preface *ix*

Explanatory Notes *1*

User's Guide *6*

Subject Index to Charts, Graphs & Stats *9*

List of Subject Headings *273*

Preface

While indexing articles for *The Cover Story Index*, all who worked on the project were struck by how frequently the stories contained statistics that we, as reference librarians, had struggled to find for our patrons: statistics on the homeless, guns in the United States, abortion, drug abuse, poverty, the trade balance. Not only were these statistics recent, they were usually presented graphically, so that at a glance trends could be seen and the numbers easily understood. And we all knew that once the magazines were put down, there was no easy access to those numbers. We then looked at a wide range of magazines covering the news, business, consumer and social issues, finally choosing nine that presented significant statistical information on current topics. We looked at all the articles, not just the cover stories, indexing the graphs, charts, diagrams, and maps that graphically presented quantitative information. We went back four years, 1988–1991, because beyond that period the numbers begin to appear in standard reference works. In order of the greatest number of citations, those magazines are:

U.S. News & World Report
Business Week
Newsweek
Time
Black Enterprise
Scholastic Update
FDA Consumer
Bulletin of the Atomic Scientists
Ms.

Most people, and younger students especially, want the most recent statistics. Generally, they don't realize how long it takes to gather and synthesize the data. To a student in his teens, three-year-old statistics seem very dated indeed. Magazines strive to be current, and the numbers that complement an article are frequently the result of the journalist contacting a government agency or a polling organization. So these statistics appear in a magazine long before they are printed in a reference book.

The fact is that most people, not just students, have a tendency to shy away from numbers. This is especially true when the numbers are presented in columns and rows. Typically this is the format of most almanacs and government publications like *Statistical Abstracts.*

In the last few years, two things have dramatically changed all this. *USA Today* (the newspaper) has demonstrated the appeal of visually illustrating the news. It could be argued that this type of reporting is to news what sound bites are to political campaigns, but it cannot be debated that the *USA Today* "look" has had an effect on other publications—both newspapers and magazines.

Technology has accelerated this trend. Desktop publishing has made it possible to add illustrations and graphics to an article even when there are only a few hours between the breaking story and its appearance in print. Books like *The Visual Display of Quantitative Information* and *Information Anxiety* have offered a theoretical basis for good graphs and charts.

The direction seems clear. More and more magazines are graphically illustrating statistical information. But does this trend justify an index of only graphs, charts, and statistics? Looking at a large number of the graphs, we discovered that many of them were only peripherally related to the story in which they appeared. Some were sidebars or background pieces. For example, in a major story on the Kuwait-Iraqi Invasion (1990), many graphs appeared showing American imports of crude oil, gasoline consumption in the United States, and the price of gasoline from 1960 to 1990. A student looking for information on oil imports or gasoline prices would never find these graphs because they are contained in articles indexed under the Iraqi invasion. They appeared as background information to the larger subject of the article.

Charts, Graphs & Stats Index provides easy access to the numbers behind the most important stories of the past four years: health care, the automobile industry, AIDS, the federal deficit, abortion, crime, the savings and loan failures, child care, drug abuse, imports and exports, unemployment, the recession, education, and the breakup of the Soviet Union. The numbers are the most recently available and they're graphically illustrated; it's like having the reporting staff of a major magazine at your finger tips The strength of *Charts, Graphs & Stats* is also its limitation. One can find out how many sexual harassment complaints have been filed with the EEOC, but not the top 100 magazines by circulation. The numbers cited are those behind today's most important stories. *Charts, Graph & Stats* complements an almanac; it does not replace it. The numbers it does cite generally reveal a range and detail not found in any other reference book. Anyone researching a topic that has been in the headlines during the past four years will find it useful.

Robert Skapura
February 1992

Explanatory Notes

The citations in this index are a departure from the traditional bibliographic entry. A typical citation to a magazine article includes the author and title as well as the publication information (magazine, date, page). It is the nature of a graph, a chart, and a diagram, however, that determines the format of these citations.

Structure of the Entries

Indexing charts and graphs contained in magazine articles is analogous to indexing the articles themselves, but the "parts" of the citation are somewhat different. A citation to a graph not only gives the publication information, it also attempts to describe something that is basically visual so that the reader will know what kind of information to expect. Therefore, in addition to the publication information, a citation to a graph usually consists of a title, a focus, and a date or date range. The graph below, although relatively simple, illustrates these three elements.

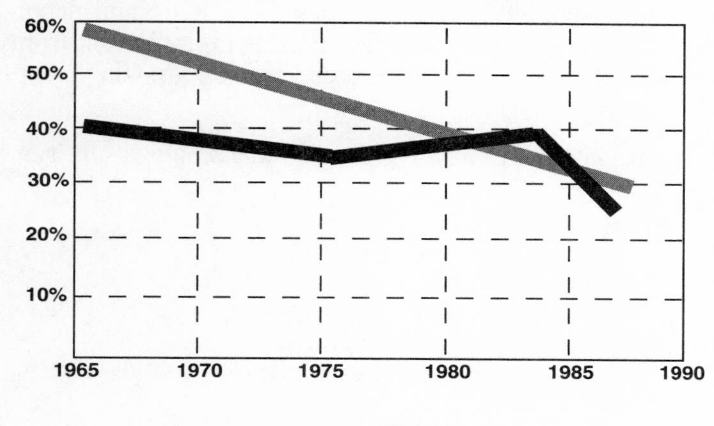

Title

A magazine article usually has a title that is not only descriptive but also attractive enough to capture the interest of the reader. A chart or graph does not have a "title" in quite the same way.

The title of a graph is usually an eye-catching "headline." It is not always descriptive, but is frequently in bold or large print at the top of the graph. The title in the example above is "Stubbing Out." Not all graphs have a title.

Focus

If the title is not very descriptive, what follows is usually a few phrases, sentences, or a short paragraph that describes what the graph is about or what the figures are trying to display. The focus in the example is "% of 20- to 24-year-olds who smoke. Male. Female." This is the subject of the numbers, the group that is being described. For example, another graph could have looked at "teenagers" or "only women." It is important for the citation to identify the focus, because it is this that will frequently determine whether the information is useful to the user. Where the focus was not explicit, additional information was included in brackets.

Date or Date Range

Very often the purpose of a graph is to compare, and frequently the comparison is made between different time periods. There is usually the implicit question, "Are things getting better or worse?" The date or date range is usually the X-axis of the graph. In the example, the date range is 1965–1990.

The Citation

The citation then includes these three elements: title, focus, and date or date range. Not every graph/chart has all three, just as not every bibliographic citation has an author. But these three items are combined to form the first part of the citation. These are followed by the magazine name, date of publication, and the page number. The citation does not try to convey the statistical information itself, but rather describes the type of information the user should expect to see. In the example, there is no indication from the citation whether more or fewer in that age group are now smoking compared to 1965.

Here are three typical citations, including the one that would result from the example:

SMOKING
Stubbing out. % of 20-to 24-year olds who smoke. Male. Female. 1965-1990.
Time Mar 5, 1990 p. 41

Leaders of the pack. Share of the U.S. cigarette market held [by various companies].
U.S. News Mar 5, 1990 p. 58

Who lights up? The demographics.
Proportion of smokers by education and
race. 1980-1987. Smokers by occupation.
Newsweek Feb 13, 1989 p. 20

Three kinds of statistical information were not indexed. We ignored what we considered to be trivial subjects. These were judgment calls. We did not reproduce public opinion surveys ("How do you think the president is handling domestic affairs?"). Also, we did not index projections ("The budget impact of the Democratic proposal on taxes").

In addition to the traditional graphs, we also indexed charts, diagrams and maps. Graphs display quantitative information which is illustrated by lines, bars and pies, in both two and three dimensions. Although "pie charts" contain the word "chart," in this context it was considered a graph. The distinction between charts and diagrams is usually in the material presented. Charts usually display relationships; diagrams try to explain how something works. For example, in February 1990, a chart appeared in *Time* [SOVIET UNION - POLITICS AND GOVERNMENT] showing Gorbachev as a tight-rope walker precariously balanced between the party and the government, with all their various ministries and committees. The chart illustrated his dual relationship to them as both general secretary and president.

In October 1990, in an article covering the Kuwait-Iraqi invasion, *U.S. News & World Report* displayed a diagram of how crude oil is refined into gasoline in four steps [PETROLEUM REFINING]. As used in this index, then, charts show relationships and diagrams show how something is done or how something works.

If a map displayed only geographic information, it was not indexed. But if it contained quantitative information (concentrations of acid rain, Israeli settlements in the West Bank), it was included.

Style Variations

A close reading of the citations will show an inconsistency of style as well as spelling. The reader will find inconsistencies in the use of punctuation, quotes, and dashes. For example, in some citations an expression will be placed within double quotation marks, in others the same expression will be in single quotes and in some with no quotes at all. This is a reflection of the various styles of the magazines themselves. It was decided to reproduce the words and phrases just as they appeared in the original graph.

Subject Index

OF CHARTS, GRAPHS & STATS

1988 - 1991

User's Guide

The Subject Index provides access to the statistical information behind the most important stories of the past four years. The numbers represented will be found in a graph, chart, diagram, or map that is generally part of a larger story in the magazine.

Each citation includes a title and/or description, magazine, date and page number on which the illustration appears.

The citations under each subject heading are arranged in reverse chronological order, i.e., the most recent citation on the subject is listed first. This puts the most recently published numbers first and provides an historical view of controversial subjects, such as abortion or the savings and loan failures. When more than one magazine covered the same subject on the same date, the citations are listed in alphabetical order by magazine. When preparing a bibliography, however, the graph title should not be used. Rather, the user should cite the title of the article in which the graph appeared.

CHARTS, GRAPHS, & STATS INDEX

Schools and transportation get less. Elementary and secondary education. Transportation. 1987-1989.
Business Week Jun 5, 1989 p. 88

More dollars, smaller classes. Average annual amount spent per student in public schools in public schools. Students per teacher in public schools. 1955–1985.
U.S. News Mar 20 1989 p.51

Educational spending by state. Map.
Scholastic Update Dec 2, 1988 p.9

U.S. Almanac: Comparing state stats. Publix school spending per student. Percent graduated from high school. [By state.]
Scholastic Update Dec 2, 1988 pp.18-19

What we spend on education and who pays. State government. Local government. Private sources. Federal government.
Black Enterprise Oct 1988 p.136

EDUCATION - TESTS AND MEASUREMENTS

A formula for failure. Math-achievement levels of high-school seniors. Below basic. Basic. Proficient. Advanced.
Newsweek Oct 14, 1991 p. 54

How the states rank. Rankings on the eighth-grade math-assessment test. [By state].
Newsweek Jun 17,1991 p. 65

Diplomas for dropouts. Here are the state-by-state results for high-school equivalency tests in 1989. Percentage who passed. Graduates over age 40.
U.S. News Jun 25, 1990 p. 66

Back to basics. Proficiency levels of students, age 13, in selected countries. Math. Science.
Time Sep 11,1989 p. 69

How students score. Average scores of 17-year-olds on proficiency tests. Reading. Math. Science. White. Hispanic. Black. 1970-1986.
U.S. News Mar20,1989 p. 51

An 'F' for the U.S. Among 18-to 24-year-olds, Americans scored dead last [out of nine countries] in identifying 16 geographic locations. Map.
Newsweek Aug 8, 1988 p. 31

EDUCATION, ELEMENTARY

Bookworms. Days spent in grade school per year. U.S. Japan. West Germany.
U.S. News Jul 16,1990 p. 26

It's arithmetic, in a landslide. Favorite subjects of children in school. Girls. Boys. Math. English. Art. Reading. Science.
U.S. News Sep 12,1988 p. 71

EDUCATION, HIGHER

Professorial shortfall. Faculty supply-and-demand projections in the social sciences and humanities. Supply. Demand. 1987-2012.
U.S. News Sep 25,1989 p.55

EDUCATION, PRESCHOOL

Preschool boom. Percentage of children enrolled in school. Ages 5 and 6. Ages 3 and 4. 1965-1985.
U.S. News Mar20,1989 p. 50

EDUCATION, SECONDARY

No pain, no gain. Time [spent] on homework [by] seventeen-year-old students. 1988. Chart.
Newsweek May27, 1991 p. 62

EGYPT- MILITARY STRENGTH

Quality versus quantity. Country. Population in millions. Armed forces: Regular, reserves. Tanks. Combat aircraft.
Newsweek May 1,1989 p. 44

ELDERLY
See OLDER AMERICANS

ELECTIONS - UNITED STATES

A bigger foothold for the G.O.P. Texas voters in primaries. Florida registered voters. California registered voters. 1980-1990.
Time Apr 23,1990 p. 22

The Democrats' electoral game plan [electoral votes by state]. Map.
Newsweek Ju125, 1988 p. 18

See Also PRESIDENTIAL ELECTIONS

ELECTIONS - UNITED STATES, 1988

Election results. State. Governor's name. U.S. Senators [political party]. U.S. Reps [number and political party].
Scholastic Update Dec 2,1988 pp. 16-17

SUBJECT HEADING

CONTENTS OF THE GRAPH, CHART OR STAT

MAGAZINE TITLE

PUBLICATION DATE

PAGE NUMBER

"SEE" REFERENCE
The correct subject heading for this subject is OLDER AMERICANS.

"SEE ALSO" REFERENCE
This reference directs you to related or more specific subject headings.

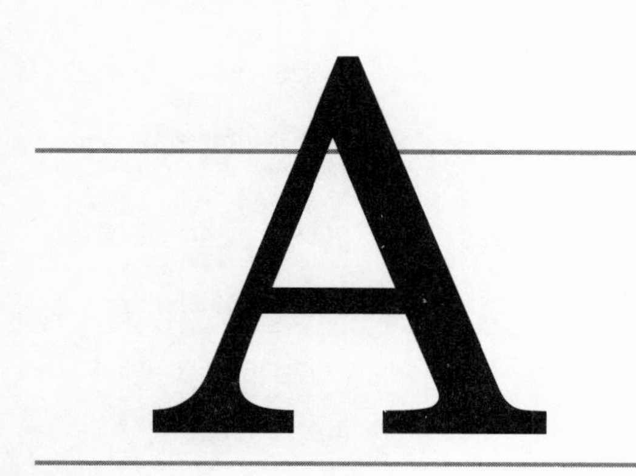

ABORTION

Abortion in America. How many women have abortions? Total number of live births and legal abortions. 1969-1985.
Scholastic Update Apr 20, 1990 p. 4

Women having an abortion by age.
Scholastic Update Apr 20, 1990 p. 4

Family income of women having an abortion, 1987.
Scholastic Update Apr 20, 1990 p. 5

Why women have abortions. The top six reasons given by women.
Scholastic Update Apr 20, 1990 p. 5

The who, when and where of abortion. Age. Race. Family income. Length of pregnancy. Locations.
Newsweek Jul 17, 1989 p. 18

Crucial timing [percent of abortions performed during various lengths of pregnancy].
Newsweek May 1, 1989 p. 30

Who has abortions.
Newsweek May 1, 1989 p. 31

Safe when legal. Comparative risk to U.S. women from abortion and childbirth, before and after *Roe v. Wade*. Number of deaths per 100,000. Childbirth mortality. Abortion mortality. 1970-1982.
Ms. Apr 1989 p. 90

Making the decision. The reasons women choose abortion.
Ms. Apr 1989 p. 93

ABORTION - LAWS AND REGULATIONS

Abortion-law scorecard. 1989-90 legislative sessions. Parental consent or notification. Sex selection. "Birth control" abortion. Number of states bill introduced. Number of states laws enacted.
U.S. News Jul 9, 1990 p. 21

The great abortion battle scorecard. Major wins and losses so far in 1990. Pro-life. Pro-choice.
Time Apr 23, 1990 p. 21

Countdown: The wars within the states [state-by-state standing on abortion issue].
Newsweek Jul 17, 1989 p. 24

Pro-choice. Pro-life. Battleground. Map.
Time Jul 17, 1989 p. 64

State by state. What might happen if the Supreme Court overturns Roe v. Wade.
Newsweek May 1, 1989 p. 38

State-by-state guide to parental consent and notification laws for minors' abortions.
Ms. Apr 1988 p. 51

ACCIDENTS

In harm's way. Home, office and playground mishaps that most often bring injured Americans to the emergency room. [Type and number of accidents].
U.S. News Oct 1, 1990 p. 95

Risky business. Deadly places to work [by selected occupation].
Newsweek Dec 11, 1989 p. 43

ACCIDENTS - MOTOR VEHICLE

Big cars are safest. Rankings reflect the frequency of injuries as reported in medical insurance claims after accidents involving 1987-89 models. The top 10. The bottom 10.
Business Week Nov 26, 1990 p. 130

An improved driving record. U.S. fatality rate. Deaths per 100 million miles. 1950-1989. Puts the U.S. among the safest countries. 1987 fatality rate. Deaths per 100 million miles [by country].
Business Week Nov 26, 1990 p. 130

In high drive. Percentage of injured drivers treated in emergency rooms who showed signs of: Alcohol and drugs. Drugs only. Alcohol only. No drugs or alcohol.
U.S. News Dec 25, 1989 p. 8

Teen tragedies. Teenager's motor-vehicle deaths in 1988. The percentage of teenagers who own a car.
U.S. News Aug 14, 1989 p. 70

Can a driver be too old? Accidents per 100 drivers in each age group. [Under 20 to 75 and older].
Time Jan 16, 1989 p. 28

On a collision course. Country [U.S. and various European countries]. Maximum speed limit. Traffic deaths per 100 million vehicle miles.
U.S. News Jan 9, 1989 p. 65

The 65-mph question. Motor-vehicle traffic fatalities in 1987 [by state]. Deaths per 100,000 population. Change since 1986.
U.S. News Aug 15, 1988 p. 78

More, but safer, miles. Average miles per driver. 1970-1987. Death rate per 100 million vehicle miles. 1970-1987.
U.S. News Aug 15, 1988 p. 78

Drivers in all accidents. Men. Women. Drivers in fatal accidents. Men. Women. 1977-1986.
Scholastic Update Apr 22, 1988 p. 21

ACID RAIN

Where the acid falls. Acidity of rainfall, pH units. Map.
U.S. News Jun 12, 1989 p. 52

Transformations in the air. Smog. Acid rain. Major sources of air pollutants. Diagram.
U.S. News Jun 12, 1989 p. 53

An environmental scorecard. Air. Where acid rain falls [by pH number]. Map.
Scholastic Update Apr 21, 1989 p. 16

Where the culprits are. U.S. power plants with highest sulfur-dioxide emissions. Areas of acid rain (pounds per acre). Map.
U.S. News Jul 25, 1988 p. 43

What causes acid rain. Sulfur emissions (trillions of grams per year). Power plants. Factories. Vehicles. 1940-1986.
U.S. News Jul 25, 1988 p. 44

What causes acid rain—and what the cleanup costs. Sulfur emissions. 1940-1986. Tons of emissions reduced. Cost.
U.S. News Jul 25, 1988 p. 44

Nowhere to hide. Air pollution. Tropical forests [annual loss]. Desertification. Map.
Newsweek Jul 11, 1988 p. 24

ACUPUNCTURE

How acupuncture works. Diagram.
U.S. News Sep 23, 1991 p. 70

ADDICTIVE BEHAVIOR

A consumer's guide to highs and lows. Alcohol. Cocaine. Nicotine. How it works. How it feels. How it hurts. How to get help.
Newsweek Feb 20, 1989 p. 56

ADOPTION

Cost of an independent adoption. Legal fees. Mother's medical. If Caesarean. Pediatrician. Mother's living costs.
Time Oct 9, 1989 p. 88

ADVANCED PLACEMENT EXAMINATIONS

Brain power. Most popular Advanced Placement subjects by number of exams given. Students taking Advanced Placement exams. 1965-1988.
U.S. News Mar 6, 1989 p. 66

ADVERTISING - EXPENDITURES

Miller is cranking up ad spending. Major media outlays on light beer brands. Miller Lite. Bud Lite. Coors Lite.
Business Week Jul 15, 1991 p. 29

Big corporate bankrolls. 1989's biggest sports advertisers on ABC, CBS, and NBC.
Business Week Dec 10, 1990 p. 221

Less growth in ad budgets. Total U.S. advertising spending. 1986-1990.
Business Week Oct 29, 1990 p. 78

A hedged bet. 1988 U.S. advertising expenditures [by media type].
Newsweek Apr 17, 1989 p. 43

Where the pitch is loudest. Countries where advertisers spend the most per person.
U.S. News Mar 20, 1989 p. 90

How ad budgets get hit after takeovers. Growth rates of combined ad spending before and after mergers in 1985 [by company]. Spending for seven major national media. 1983-1987.
Business Week Dec 5, 1988 p. 26

As Hispanic Americans grow wealthier, advertisers spend more to reach them. Total purchasing power. Ad spending aimed at Hispanics. 1983-1987.
Business Week Jun 6, 1988 p. 64

ADVERTISING, DIRECT MAIL

Third-class mail keeps growing despite big hikes in postage. Percentage of total pieces of mail. (Typical prices in cents). 1984-1988.
Business Week Mar 21, 1988 p. 53

ADVERTISING, MAGAZINE

Big magazines are losing ground. Annual growth rate in advertising revenue for 12 of the biggest magazines, compared to the rest of the industry, 1981-89 and 1989-91.
Business Week Sep 23, 1991 p. 71

ADVERTISING, TELEVISION

TV rivals are hurting the Big Three. Year-to-year percent change in advertising revenues. Cable TV. Syndicated TV. Three networks. 1981-1991.
Business Week Sep 23, 1991 p. 70

Revenues/billions. Three networks. Cable TV. Syndicated TV. 1981, 1991.
Business Week Sep 23, 1991 p. 70

Big corporate bankrolls. 1989's biggest sports advertisers on ABC, CBS, and NBC.
Business Week Dec 10, 1990 p. 221

Less market share for the networks. Television advertising dollars. Local and spot. Networks. 1983-1988.
Business Week Jan 11, 1988 p. 101

AEROBICS

What's your zone? [Aerobic heart rate formulas]. Heart rate (beats per minute). [Recommended lows and highs from] American Heart Association. American College of Sports Medicine. Institute for Aerobic Research.
U.S. News Oct 1, 1990 p. 86

AFFIRMATIVE ACTION

See DISCRIMINATION IN EMPLOYMENT

AFFIRMATIVE ACTION PROGRAMS

How companies are hiring. The following percentages of surveyed companies have written affirmative action plans or goals and timetables for hiring protected groups. Size of business. Type of business.
Black Enterprise Sep 1989 p. 46

AFGHANISTAN - POLITICS AND GOVERNMENT

The Afghan rebel factions: A house divided. 15 principal rebel groups. Map.
Newsweek Feb 13, 1989 p. 33

AFGHANISTAN-RUSSIAN INVASION, 1979-1989

The toll of a long war [effects of war on Afghanistan].
U.S. News Feb 13, 1989 p. 32

AFRICA

The three Africas. North Africa, sub-Saharan Africa, and Southern Africa. Main characteristics of each.
Scholastic Update Mar 23, 1990 pp. 4–5

AFRICA - ECONOMIC CONDITIONS

Africa fights back. Growth rates of real GDP per capita. 1965-2000.
U.S. News Dec 10, 1990 p. 60

Winds of change. Foreign investment. Export value index. 1981-1987.
Newsweek Jul 23, 1990 p. 27

Poverty abounds. GNP per capita (10 lowest countries).
Scholastic Update Mar 23, 1990 p. 13

The economics of despair. Per capita GDP. Change in per capita GDP 1980-85. Total external debt as share of GNP. Change in investment 1980-85.
U.S. News Jun 27, 1988 p. 29

The debt bomb. Total external debt of sub-Saharan Africa. Official development assistance to sub-Saharan Africa. 1981-1986.
U.S. News Jun 27, 1988 p. 31

AFRICA - HISTORY

The days of empire. Powerful cultures on African soil. 3000 B.C.-1800 A.D. Map.
Newsweek Sep 23, 1991 p. 44

AFRICA - SOCIAL CONDITIONS

Millions go hungry. People with "food insecurity" [by country].
Scholastic Update Mar 23, 1990 p. 13

Africa: A hungry continent. The hunger problems of six countries: Ethiopia, Sudan, Mozambique, Angola, Liberia and Zimbabwe. Map.
Scholastic Update Jan 27, 1989 pp. 4–5

Food production down. Index of African food production, per capita, from 1965-1986.
Scholastic Update Jan 27, 1989 p. 10

AFRICAN AMERICANS

See BLACKS

AGING

Long-term care. Annual premiums for long-term-care insurance, by age at the time of purchase. Policy A. Policy B. Policy C. 1989.
U.S. News Jul 30, 1990 p. 74

Who will care for us? Aging America. As a percent of total population. Children under age 17. Adults over age 65. 1980-2040.
Newsweek Jul 16, 1990 p. 53

Effects of human-growth hormone injections in men over 60.
Time Jul 16, 1990 p. 54

Gray power. Percent of population now over 65. U.S. Japan. West Germany. World average.
U.S. News Jul 16, 1990 p. 25

Age 25. Age 45. Age 85. Max. heart rate. Lung capacity. Cholesterol level. Muscle strength. Kidney function.
Newsweek Mar 5, 1990 p. 44

Projected life expectancy. [People] born in 1990. Your life expectancy at birth. Your life expectancy today. Men. Women.
Newsweek Mar 5, 1990 p. 46

The aging of America. People 50 or over. Percent of U.S. populaton. 1988-2025.
Business Week Apr 3, 1989 p. 66

As Europe's population ages [by country]. 1950-2040.
Business Week Mar 13, 1989 p. 54

The aging of America. Percent of the population 65 or older. 1940. 1990. 2040. 2090.
FDA Consumer Oct 1988 p. 22

Common ailments in older Americans. Percent of American men and women aged 65 or older who suffer from one or more of six common ailments.
FDA Consumer Oct 1988 p. 23

The war on aging. No two people age alike. The brain. Skin. Muscle. The heart. Bones. Immune system.
Business Week Feb 8, 1988 p. 59

AGRICULTURAL SUBSIDIES

The high cost of subsidies. Total 1989 cost of agricultural protectionism and subsidies. Cost to taxpayers. Cost to consumers. European community. Japan. United States.
Business Week Oct 29, 1990 p. 45

Pay as you grow. Following a sharp increase, net expenditures for farm subsidies have declined 59 percent between 1986 and 1989. 1980-1989.
U.S. News Apr 30, 1990 p. 39

AGRICULTURE

Love those traditional crops. Acres planted in: Corn. Soybeans. 1987-1991.
Business Week May 20, 1991 p. 46

Leaner days down on the farm. Net cash farm income. 1986-1991.
Business Week Jan 14, 1991 p. 110

America abandons the farm. Counties where farming accounts for 20% or more of income. 1950, 1986.
Business Week Sep 25, 1989 p. 93

Food from abroad. American agricultural imports total $21 billion. Percentages coming from [various countries].
U.S. News Mar 27, 1989 p. 56

The farm economy. Total farm income. Total farm debt. 1981-1988.
Business Week May 23, 1988 p. 53

Grain sales are growing faster. Consumption of American grain. Corn. Wheat. Soybeans. U.S. Export. 1985-1988.
Business Week Mar 28, 1988 p. 81

Less surplus in the farm belt. U.S. inventories of major grain crops. 1983-1988.
Business Week Jan 11, 1988 p. 123

See Also FARMS AND FARMING

AGRICULTURE - CALIFORNIA

Arid farmlands. Rainfall from Oct. 1, 1990 to Feb. 1, 1991. Hardest hit [crops].
Time Feb 18, 1991 p. 55

AGRICULTURE - SOVIET UNION

How Soviet farms stack up. Average annual wheat yields. Average annual livestock productivity. Average annual yield of major Soviet and comparable U.S. crops.
U.S. News Oct 7, 1991 p. 40

Agriculture. Fixed capital investment in agriculture. Metric tons of grain produced. 1970-1988. What market prices would mean. Estimated market price. State price.
U.S. News Apr 3, 1989 p. 44

AGRICULTURE, COOPERATIVE

Some top farm co-ops. Co-op. 1987 sales.
Business Week Mar 21, 1988 p. 96

AIDS (DISEASE)

New AIDS cases. Number reported among U.S. adults annually and the breakdown by percent. Heterosexuals. Male homosexuals. IV drug users. Male-homosexual IV drug users. Others. 1983-1991.
Time Nov 25, 1991 p. 73

Who has AIDS? Homosexual males. Heterosexual intravenous drug users. Heterosexuals. Transfusions. Hemophiliacs. Undetermined. Chart.
Newsweek Nov 18, 1991 p. 59

Where the cases are. A breakdown, by profession, of all reported cases of AIDS in health-care workers since the epidemic began in the early 1980s.
Newsweek Jul 1, 1991 p. 50

AIDS moves in many ways. Number of people who have contracted AIDS from blood transfusions. Transplants. Health-care workers.
Time Jun 3, 1991 p. 56

Among women, minorities, and young people. Percentage increase in number of AIDS cases reported, 1988-89.
Scholastic Update Feb 22, 1991 p. 3

AIDS: Female age distribution. Total of 15,665 U.S. cases through September 1990.
Ms. Jan 1991 p. 27

AIDS toll. Deaths for different age groups. Age. 1988, 1989.
U.S. News Oct 22, 1990 p. 82

The AIDS case load will surge in the early 1990s. Persons in the U.S. under treatment for AIDS or HIV infection. High estimate. Low estimate. 1989-1996.
U.S. News Aug 13, 1990 p. 56

Changing profile. % of reported cases [of AIDS]. Heterosexuals. Male-homosexual IV drug users. All other IV drug users. Male homosexuals and bisexuals. Other. 1981, 1989.
Time Jul 2, 1990 p. 42

How the virus attacks, and how to attack the virus. Chart.
Newsweek Jun 25, 1990 p. 22

The changing profile of AIDS. Most AIDS patients are still men and IV drug users. But the rate of increase among heterosexuals and newborns proves that the virus knows no boundaries.
Newsweek Jun 25, 1990 p. 26

The steep cost of AIDS home care. Item. Monthly usage. Approximate cost. Billing rate.
Business Week Jun 11, 1990 p. 20

How many people have the AIDS virus? Number of people with the virus in the U.S. [various sources]. Projections of the cost of medical treatment in 1991 [various sources].
U.S. News Jan 29, 1990 p. 28

AIDS. Cases reported between November 1988 and October 1989. [By state].
Scholastic Update Jan 12, 1990 p. 29

Women with AIDS. Race. Age. Transmission. Women infected through heterosexual contact.
Ms. May 1989 p. 82

Homosexual males: Number of cases of AIDS reported (includes bisexuals). 1985-1988.
Time Jan 30, 1989 p. 60

Intravenous drug abusers: Number of cases of AIDS reported. 1985-1988.
Time Jan 30, 1989 p. 61

The deadly tally. White. Black. Hispanic. Asian and others. Percent of cases.
Newsweek Dec 5, 1988 p. 24

AIDS and blood. Pints of blood donated by surgery patients for self-transfusion. 1982-1987. Pints of blood donated through the American Red Cross. 1976-1988.
U.S. News Oct 24, 1988 p. 81

A doubled dose of dollars? Federal spending on AIDS. 1982-1990.
U.S. News Jun 13, 1988 p. 7

Three approaches to blocking the AIDS virus. Vaccine. Decoy receptors. "Antigene." Diagram.
U.S. News Feb 29, 1988 p. 57

Where the risk lies [various groups].
U.S. News Feb 29, 1988 p. 58

Inpatient hospital costs. Outpatient costs. Cost. Sample. Study.
U.S. News Jan 18, 1988 p. 22

AIDS (DISEASE) - AFRICA

Estimated AIDS cases in thousands. Women and children in sub-Saharan Africa. Women. Children. 1983-1992.
Ms. Jan 1991 p. 21

The African picture. Evidence of HIV-1. Evidence of HIV-2. Evidence of both HIV-1 and HIV-2. Map.
Newsweek Jan 4, 1988 p. 60

AIDS (DISEASE) - CHILDREN

Estimated AIDS cases in thousands. Women and children in sub-Saharan Africa. Women. Children. 1983-1992.
Ms. Jan 1991 p. 21

Children's crusade. AIDS. Distribution of HIV-infected females. [By region.]
Newsweek Oct 8, 1990 p. 48

Pediatric AIDS cases reported through May. Hemophilia. Undetermined. Blood transfusion. Mothers with or at risk for HIV infection.
FDA Consumer Oct 1990 p. 11

AIDS young victims. Total AIDS cases among children under 13, reported from June, 1981, through September, 1988.
U.S. News Jul 3, 1989 p. 62

Children: Number of cases of AIDS reported 13 years or younger. 1985-1988.
Time Jan 30, 1989 p. 61

The youngest victims. Children born with AIDS. 1982-1987.
U.S. News Feb 1, 1988 p. 49

AIDS (DISEASE) - RESEARCH

AIDS ranks fourth in research dollars. Cancer. Genetic diseases. Heart disease. AIDS. Infectious diseases.
Business Week Sep 17, 1990 p. 97

Closing the gap. AIDS funding has soared. Cancer grants haven't kept pace. N.C.I. spending on cancer research. All federal spending on AIDS. 1984-1989.
Newsweek Mar 6, 1989 p. 47

AIDS (DISEASE) - TESTING

Where the virus strikes. Population group. Infection rate. Number screened.
U.S. News Nov 28, 1988 p. 79

AIDS (DISEASE) - VACCINES

Ways to trigger immunity. Four different kinds of vaccines.
Time Apr 1, 1991 p. 64

Vaccine strategies. AIDS virus (cut away). Diagram.
U.S. News Nov 19, 1990 p. 65

AIDS (DISEASE) - WOMEN

Among women, minorities, and young people. Percentage increase in number of AIDS cases reported, 1988-89.
Scholastic Update Feb 22, 1991 p. 3

AIDS cases per 100,000 women. [World map].
Ms. Jan 1991 p. 19

Estimated AIDS cases in thousands. Women and children in sub-Saharan Africa. Women. Children. 1983-1992.
Ms. Jan 1991 p. 21

Women with AIDS. Race. Age. Transmission. Women infected through heterosexual contact.
Ms. May 1989 p. 82

AIDS (DISEASE) - WORLD

The growing epidemic. Number of reported AIDS cases worldwide, 1981-1990.
Scholastic Update Feb 22, 1991 p. 3

Total AIDS cases reported in 1989, by continent. Africa. Asia. Europe. Australia and Oceanic countries. North and South America (minus the U.S.). United States.
Scholastic Update Feb 22, 1991 p. 3

AIDS cases per 100,000 women. [World map].
Ms. Jan 1991 p. 19

Global epidemic. U.S. AIDS cases. World AIDS cases. 1982-2000.
Newsweek Jun 25, 1990 p. 20

Cumulative AIDS cases reported to WHO [World Health Organization]. 1979-1988.
Newsweek Jun 27, 1988 p. 47

Global grip of AIDS. Reported AIDS cases. Map.
U.S. News Feb 8, 1988 p. 8

AIR TRANSPORT - MARKET SHARE

Up, up and away. Manufacturers, retailers and individuals are shipping more things by air. International shipments are growing fastest.
U.S. News Oct 2, 1989 p. 50

AIR TRAVEL

Meals made to order—by request only. Airline. Types of special meals offered. Most popular specialty meal. Total meals served daily. % specialty meals served. Notice required.
Newsweek Nov 26, 1990 p. 70

Painful airports. U.S. News ranked the 10 busiest U.S. airports on the inconvenience each inflicts on travelers.
U.S. News Nov 26, 1990 p. 68

The biggest carriers. How they treat you. Airline. Bags lost. Bumps. On-time arrivals. Death or illness. Unaccompanied children. Frequent-flyer miles.
U.S. News Nov 26, 1990 p. 70

Too many seats, too few passengers. Growth in filled seats. Growth in available seats. Oct.'89 - Aug.'90.
Business Week Oct 8, 1990 p. 42

How the bumpers stack up. [Airlines] differ in how they compensate people who voluntarily give up their seats. Airlines. Passengers bumped per 10,000. Typical compensation.
U.S. News Apr 16, 1990 p. 55

Miles in the bank. You could fly to Pluto and back 90 times and not use up all the unclaimed frequent-flier miles. Here is the total still unused. 1981-1989.
U.S. News Apr 2, 1990 p. 68

Improving. % of on-time arrivals [by airline]. 1987, 1988.
Newsweek Feb 20, 1989 p. 8

How deregulation has changed the U.S. airline industry. Many more people are flying. Millions of passengers. 1978-1988.
Business Week Dec 19, 1988 p. 71

Traffic growth has slowed. Percent increase in passenger traffic. 1983-1988.
Business Week Aug 8, 1988 p. 24

AIR TRAVEL - ACCIDENTS

Crash payment. Average awards to victims, in thousands of dollars.
Time Aug 7, 1989 p. 42

What causes accidents. Share of aircraft accidents involving [various factors].
U.S. News May 16, 1988 p. 18

See Also AIR TRAVEL - SAFETY

AIR TRAVEL - FARES

Senior discount coupons. Airline. Price.
U.S. News Dec 9, 1991 p. 84

Same seat, several prices. Round-trip international airline tickets. Route. Airline fare. Advance purchase. Unrestricted. Discounted fare.
U.S. News Jul 16, 1990 pp. 58–59

Rising fares. Consumer price index. Airline travel prices. 1978-1988.
Time May 15, 1989 p. 52

How deregulation has changed the U.S. airline industry. Fares may have nowhere to go but up. Average ticket revenue per passenger mile. 1978, 1987, 1988.
Business Week Dec 19, 1988 p. 71

Air fares: A sampler [by route and city].
Business Week Dec 12, 1988 p. 31

AIR TRAVEL - SAFETY

How deregulation has changed the U.S. airline industry. Service problems may have peaked. Consumer complaints. Near collisions in midair. 1978, 1987, 1988.
Business Week Dec 19, 1988 p. 71

White-knuckle flights. Near collisions by aircraft. 1980-1987.
U.S. News Aug 29, 1988 p. 102

See Also AIR TRAVEL - ACCIDENTS

AIRLINE INDUSTRY

Crash landing? Net profit, major U.S. airlines. Aircraft delivered to major U.S. airlines. 1985-1991.
U.S. News Nov 18, 1991 p. 70

American. United. Delta. Northwest. Purchases. Cost in millions. 1985-1991.
Time Jul 22, 1991 p. 49

Unfriendly skies. Operating profits of U.S. scheduled airlines. 1988, 1989, 1990.
Newsweek Feb 11, 1991 p. 52

The nosedive in airline earnings. Operating profits of U.S. scheduled airlines. 1986-1991.
Business Week Jan 14, 1991 p. 91

Where Aeroflot leads the pack. Passenger-miles flown, 1989 (millions). Freight-tons carried, 1989 (thousands). [By airline].
Business Week Dec 17, 1990 p. 46

As airline costs soar to new heights. Operating expenses. Losses may sink to a record level. Operating profits/losses. 1985-1990.
Business Week Dec 10, 1990 p. 31

The biggest carriers. How they treat you. Airline. Bags lost. Bumps. On-time arrivals. Death or illness. Unaccompanied children. Frequent-flyer miles.
U.S. News Nov 26, 1990 p. 70

Too many seats, too few passengers. Growth in filled seats. Growth in available seats. Oct.'89 - Aug.'90.
Business Week Oct 8, 1990 p. 42

A crash landing for airlines? Operating profits for U.S. scheduled airlines. Pre-Mideast-crisis forecast. Post-Mideast-crisis forecast. 1988-1990.
U.S. News Sep 24, 1990 p. 61

American's pilots' pay: Flying low. American. Northwest. Delta. USAir. Typical monthly pay. Captains. First officers.
Business Week Jul 2, 1990 p. 33

How the bumpers stack up. [Airlines] differ in how they compensate people who voluntarily give up their seats. Airlines. Passengers bumped per 10,000. Typical compensation.
U.S. News Apr 16, 1990 p. 55

Airlines hit a downdraft. Operating profits for the 18 largest U.S. scheduled airlines. 1985-1990.
Business Week Jan 8, 1990 p. 88

How old are the aircraft? Airline. Average age of fleet, in years. Maintenance expenses per hour of operation. 1988.
Time Mar 13, 1989 p. 41

Friendly skies ahead for the airlines. Average operating margin of 15 carriers. 1984-1989.
Business Week Jan 9, 1989 p. 88

Airlines stay in the black—but barely. Operating earnings or losses, after interest expense. 1983-1988.
Business Week Jan 11, 1988 p. 108

AIRLINE INDUSTRY - MARKET SHARE

Airlines. Key acquisitions. 1990, 1991. Revenue passenger miles (August). Market share. 1985 ranking. 1990 ranking.
Business Week Oct 14, 1991 p. 87

The business is in fewer hands. Market share of the top five U.S. airlines. Total operating expenses. Traffic growth. Net profit margins.
Business Week Jan 21, 1991 pp. 58–59

Wide wings. Carriers with the largest percentages of total revenue passenger miles. 1978, 1988.
U.S. News Aug 21, 1989 p. 16

Bigger shares. Number of revenue passenger miles in billions. 1984, 1989.
Time May 15, 1989 p. 53

American: Flying high. Share of passenger traffic. Market value of stock. 1988 operating margin.
Business Week Feb 20, 1989 p. 54

How deregulation has changed the U.S. airline industry. The biggest carriers are gaining dominance. Five largest airlines' percentage of total passenger traffic. 1978-1988.
Business Week Dec 19, 1988 p. 71

Airlines: The top players have changed. Airlines ranked by passenger traffic. Billions of miles. 1983, 1987.
Business Week Aug 8, 1988 p. 24

AIRPLANES, MILITARY

Tomorrow's top gun [fighter maneuverability]. Diagram.
U.S. News Feb 20, 1989 pp. 56–57

U.S. and Soviet nuclear-capable aircraft (1988). Country and type. Year introduced. Number. Comments.
Bull Atomic Sci Nov 1988 p. 48

See Also PERSIAN GULF WAR, 1991 - WEAPONS

ALASKA

Portrait of a state. How the population has followed Alaska's fortunes. Damage from the Exxon Valdez. Map.
Time Apr 17, 1989 p. 59

See Also EXXON VALDEZ (SHIP) OIL SPILL, 1989

ALCOHOL AND YOUTH

An early start. Teen drinking habits, 1989. (High school seniors).
Scholastic Update Nov 16, 1990 p. 4

ALCOHOLIC BEVERAGES

Bourbon streak. Exports of bourbon and Tennessee whisky. 1985-1990.
Business Week Jul 1, 1991 p. 62

An expensive habit. 1988 U.S. retail expenditures for alcohol. Beer. Spirits. Wine. Total.
Scholastic Update Nov 16, 1990 p. 4

Where liquor sells best. States with the highest annual liquor purchases per capita. States with the lowest annual liquor purchases per capita. Gallons purchased per person. Total gallons sold.
U.S. News Feb 27, 1989 p. 75

States with the lowest annual beer purchases per capita. Gallons purchased per person. Total gallons sold.
U.S. News Feb 27, 1989 p. 75

High spirits in December. Monthly share of annual liquor consumption in the U.S.
U.S. News Dec 12, 1988 p. 96

ALCOHOLIC BEVERAGES, DRINKING OF

The spirits U.S. drinkers are pouring. Category [type of alcohol]. Percent of total market. Total consumption. 1988, 1987, 1978.
Business Week Jun 12, 1989 p. 54

Smoking and drinking: Taxes vs. the social cost. Taxes per pack or ounce. Social costs per pack or ounce (medical care, property loss, deaths, etc., not borne by smoker or drinker).
Business Week Jun 5, 1989 p. 27

A consumer's guide to highs and lows. Alcohol. Cocaine. Nicotine. How it works. How it feels. How it hurts. How to get help.
Newsweek Feb 20, 1989 p. 56

Smoking, drinking and oral cancer. The risk of getting oral cancer for smokers and drinkers. Nondrinkers. Nonsmokers. Moderate smokers and drinkers. Heavy smokers and drinkers.
FDA Consumer Feb 1989 p. 40

The changing tastes of U.S. drinkers. Category [type of alcoholic beverage]. 1987, 1986, 1977.
Business Week Jun 27, 1988 p. 54

ALCOHOLISM

Alcoholism remains dangerously widespread. Its annual costs are staggering.
Business Week Mar 25, 1991 pp. 76–77

Substance abuse: The toll on blacks. Smoking-attributable deaths. Cirrhosis of the liver caused by alcohol abuse. Deaths caused by drug abuse.
Black Enterprise Jul 1990 p. 39

Like father. Evidence that genetic factors can contribute to alcoholism.
Time Apr 30, 1990 p. 88

ALIENS, ILLEGAL

The high cost of leaving. Prices for smuggling one person into the U.S. Mexicans, Dominicans. Central and South Americans. Yugoslavs. Portuguese. Indians. Pakistanis. Bangladeshis. Filipinos. Koreans. Poles. Iranians.
Time May 14, 1990 p. 71

The U.S. immigration story: Today and yesterday. Illegal immigrants arrested in U.S. 1965-1987.
Scholastic Update May 6, 1988 p. 4

Those who wish to stay: Amnesty requests by illegals. 1987-1988.
Scholastic Update May 6, 1988 p. 4

ALLERGIES

The complete guide to allergy relief. Respiratory allergies: Pollen, dust, mold, animals. Poison ivy, oak and sumac. Chemicals, metal, cosmetics. Insect stings.
U.S. News Feb 20, 1989 p. 71

Why you feel miserable [allergic reaction]. Diagram.
U.S. News Feb 20, 1989 p. 72

Achoo! Regions of U.S. where pollen-producing plants are most active. When pollen-producing plants are most active.
U.S. News Apr 11, 1988 p. 68

ALZHEIMER'S DISEASE

The brain under siege. Area affected. Alzheimer's ruinous pathways. The patient's symptoms. Diagram.
U.S. News Aug 12, 1991 pp. 42–43

Facing the facts [various statistics].
Newsweek Dec 18, 1989 p. 55

AMERICA - DISCOVERY AND EXPLORATION

The Columbian Exchange. New world portrait. New world to old [animals and plants brought]. Old world portrait. Old world to new. Diagram.
U.S. News Jul 8, 1991 pp. 29–31

AMERICAN FLAG

See FLAGS - UNITED STATES

AMERICANS

[Two] centuries of census. We've become older. American's average age. 1790-1980.
Scholastic Update Jan 12, 1990 p. 6

Americans: Who are we? By age. Percent of U.S. population in each age group. Male. Female. Chart.
Scholastic Update Dec 2, 1988 p. 5

Americans: Who are we? How much we earn. Yearly household income. Percent of households earning [by various yearly incomes]. Chart.
Scholastic Update Dec 2, 1988 p. 5

Americans: Who are we? Religion. Religious affiliation, by percent of population. Chart.
Scholastic Update Dec 2, 1988 p. 5

Americans: Who are we? Race. Percent of population. White. Black. Hispanic. Native American. Other.
Scholastic Update Dec 2, 1988 p. 5

AMUSEMENT PARKS

Are we having fun? Amusement-park attendance.
U.S. News Mar 7, 1988 p. 77

ANDES

Parallel worlds. Andes region. Other regions. 13,000 B.C. - 1532 A.D. Map. Chart.
U.S. News Apr 2, 1990 pp. 48–49

ANIMAL EXPERIMENTATION

Gene farming. Giving animals human DNA is becoming routine. [Three-step process.] Diagram.
Newsweek Sep 9, 1991 p. 55

The toll on the animal kingdom [used in laboratory experiments].
Newsweek Dec 26, 1988 p. 51

The price of doing business. The cost of lab animals varies widely. The more exotic or complex the genetic makeup, the more expensive the animal.
Newsweek Dec 26, 1988 p. 57

The ark is overpopulated [dogs and cats in laboratory experiments].
Newsweek Dec 26, 1988 p. 59

Instead of men, guinea pigs. Researchers test new drugs and procedures on animals whose biological characteristics resemble those of humans. Animal. Some systems and structures similar to humans.
Newsweek May 23, 1988 p. 60

See Also LABORATORY ANIMALS

ANIMALS

The hunting toll on U.S. wildlife. Animals taken by licensed hunters in the 1988-89 season. Mammals. Birds.
U.S. News Feb 5, 1990 p. 33

The government's kill. The wildlife taken by federal Animal Damage Control program in 1988. Mammals. Intentionally killed. Inadvertently killed. Birds. 1988.
U.S. News Feb 5, 1990 p. 37

Pelt pile-up. Number of animals killed to make an average full-length fur coat [by animal].
U.S. News Dec 5, 1988 p. 93

ANOREXIA NERVOSA

See EATING DISORDERS

ANTARCTICA

570 million years ago. Map.
Time Apr 8, 1991 p. 66

Antarctica. Year-round stations. Territorial claims.
Time Jan 15, 1990 p. 59

Antarctica's untapped mineral wealth. Research station. Known hard-mineral deposits. Possible petroleum deposits. Map.
U.S. News Oct 24, 1988 p. 66

Antarctica. Polar riches. Potential resources. Oil & gas. Coal. Map.
Time Jun 20, 1988 p. 38

ANTHROPOLOGY

All in the family. A family tree for the human species. Genetic groups. Major language families.
U.S. News Nov 5, 1990 p. 65

APARTHEID

Scrambling for seats. If negotiations do begin, it will not be easy to settle who sits at the table. Some possible players.
Time Feb 5, 1990 p. 28

APPLIANCES, HOUSEHOLD

Appliances. Key acquisitions. 1986, 1989, 1991. Major appliances. Market share. 1985 ranking. 1990 ranking.
Business Week Oct 14, 1991 p. 86

The leaders in major appliances. Estimated market share [by company]. U.S. Europe.
Business Week Sep 5, 1988 p. 70

Appliance anniversaries. Life expectancy of appliances, in years.
U.S. News Apr 4, 1988 p. 73

APPORTIONMENT (ELECTION LAW)

Redistricting: Who calls the shots [by state]. Democratic governor & legislature. Republican governor & legislature. Split governor & legislature. Map.
Business Week Nov 19, 1990 p. 48

ARAB NATIONS

A vast battleground. The Arab world. [Type of government]. Population. Per capita income. Religion. [By Arab country]. Map.
U.S. News Jan 21, 1991 pp. 26–27

To have and have not. The Arab world is sharply divided between wealthy nations that have oil and poor ones that don't. Haves. Have-nots. GNP per capita. Population.
U.S. News Aug 20, 1990 p. 27

Israel and the Middle East. [Profiles of] Israel. Egypt. Lebanon. Jordan. Syria. Saudi Arabia. Map.
Scholastic Update Apr 22, 1988 pp. 2–3

ARAB-ISRAELI RELATIONS

See ISRAELI-ARAB RELATIONS

ARMED FORCES - PERSONNEL

See MILITARY PERSONNEL

ARMED FORCES - UNITED STATES

U.S. nonnuclear defense missions. Worldwide commitments. U.S. defense spending—and missions. Map.
U.S. News Oct 14, 1991 pp. 28–29

U.S. nonnuclear defense missions. The U.S. stockpile.
U.S. News Oct 14, 1991 p. 29

The shrinking U.S. military. Budget. 1990, 1995. Personnel. 1987, 1995.
U.S. News Apr 15, 1991 p. 42

Caught with their guard down. Total Armed Forces. Active Armed Forces. Reserves. National Guard.
Time Feb 4, 1991 p. 15

Quick reaction. Gulf. Vietnam.
Newsweek Nov 5, 1990 p. 33

Military balance in the Pacific and Asia. U.S. Pacific forces. U.S.S.R. Pacific forces. Map.
U.S. News Apr 23, 1990 p. 33

Strength in numbers. Active duty forces. World War I. World War II. Korean War. Vietnam War. 1990-1995.
Newsweek Mar 19, 1990 p. 21

U.S. forces in the Pacific. U.S. bases. Soviet bases. South Korea. Japan. Philippines. Guam.
Time Mar 5, 1990 p. 16

The soldier of the future. He's armed with high-tech electronics and state-of-the-art materials. And he—or she—will cost less than a fleet of B-2 bombers. Diagram.
Time Feb 12, 1990 p. 21

See Also MILITARY PERSONNEL

ARMED FORCES - WORLD

Concentration of forces. Seven of the world's 20 largest military establishments are in the Mideast. Country. Manpower. Tanks. Artillery. Combat aircraft.
Newsweek Jul 2, 1990 p. 29

ARMS CONTROL

Shrinking nuclear arsenals. ICBM. SLBM. Bombs, SRAMs, ALCMs.
U.S. News Oct 21, 1991 p. 54

The mushrooming race. Three decades of agreements. [Timeline of nuclear weapons treaties]. United States. Soviet Union. 1945-1990.
U.S. News Oct 7, 1991 pp. 24–25

The lingering threat. Nuclear warheads. Armed forces. Nuclear missiles. Bombers. Fighters. Tanks. U.S. U.S.S.R.
Newsweek Sep 16, 1991 p. 43

New arsenals. United States. Soviet Union. ICBMs. SLBMs. Bombers. 1982, 1990, 1999.
Newsweek Aug 12, 1991 p. 36

Unequal sacrifices. Strategic nuclear delivery vehicles. Nuclear warheads. Soviet. U.S. Current. Under START.
Time Aug 5, 1991 p. 22

The last arms deal? Number of warheads. United States. Soviet Union.
U.S. News Aug 5, 1991 p. 16

A real cut. Total number of warheads. U.S. U.S.S.R. 1960, 1991, after START.
Newsweek Jul 29, 1991 p. 15

Before and after [CFE - Treaty on Conventional Armed Forces in Europe]. Weapon. Alliance/country. Current holdings. Holdings after CFE.
Bull Atomic Sci Jan 1991 pp. 33–34

Countries that have not signed the Non-Proliferation Treaty. Nuclear parties. Non-nuclear parties. Non-parties. Map.
Bull Atomic Sci Jul 1990 pp. 24–25

Jump start. Nuclear warheads. ICBMs. SLBMs. Bombers. 1981, 1989, 1998.
Newsweek Jun 11, 1990 p. 19

The numbers game. START was billed as cutting long-range nuclear forces by 50 percent. But loopholes exempt many warheads. Current. After START. Missile warheads. Bomber. Sea-launched cruise. Total.
U.S. News Jun 11, 1990 p. 31

Nuclear balance. Warheads on submarines. On ICBMs. On bombers. U.S.S.R. U.S.
Time Dec 18, 1989 p. 37

Counting down. Existing forces. NATO's count of its own forces. NATO's count of Warsaw Pact. Warsaw Pact's count of NATO. Warsaw Pact's count of its own forces.
Time Jun 12, 1989 pp. 30–31

Arms control treaties and agreements since World War II. Name. Parties. Effective date. Description.
Bull Atomic Sci May 1989 p. 36

Drawing down. While estimates of force levels vary from source to source, NATO is demanding far greater cuts in Warsaw Pact weaponry than Gorbachev proposed unilaterally. NATO proposal. Gorbachev's plan.
Newsweek Dec 19, 1988 p. 31

ARMS DEALERS

B.C.C.I. arms deals. [By country.] Diagram.
Time Sep 2, 1991 p. 56

Arms merchants to the world. Interlocking government ministries and Chinese firms depend on foreign sales to finance arms purchases and research efforts.
U.S. News Jul 22, 1991 p. 37

The sellers. The top 10 arms exporters from 1985 to 1989. The buyers. The top five arms importers.
Newsweek Apr 8, 1991 p. 24

Anatomy of an arms deal. Sale of cluster bombs to Iraq involved some odd routes. Map.
Newsweek Apr 8, 1991 p. 28

Who built the arsenal? [War materials supplied to Iraq by] France. Soviet Union. Italy. West Germany. United States.
Newsweek Feb 4, 1991 p. 57

Arms suppliers, 1983-87. [By country].
Bull Atomic Sci May 1990 p. 19

Look homeward, Mikhail. Per capita GNP. Soviet arms transfers. Soviet troops, advisers [for six countries receiving Soviet Military Aid].
U.S. News May 9, 1988 pp. 33–35

How top 10 stack up. Arms suppliers to the Third World [by country].
U.S. News Apr 11, 1988 p. 45

ARMS RACE

See NUCLEAR WEAPONS

ARMS SALES

Arming up. Who's buying up the most artillery and where they're getting it. Country. Arms imported (in billions). Main supplier.
Newsweek Jul 16, 1990 p. 8

Top 10 Third World arms importers, 1983-87. [By country].
Bull Atomic Sci May 1990 p. 15

ARMY - UNITED STATES

Measuring up. U.S. Army physical fitness standards for people age 22 to 26. Push-ups. Sit-ups. Two-mile run. Women. Men.
Newsweek Aug 5, 1991 p. 29

Who packs the most punch? Marine Battalion [vs.] Army Division. Personnel. Tanks. Armored vehicles. Artillery. Attack helicopters. Attack aircraft.
Time May 21, 1990 p. 28

They're in the Army now: Portrait of one branch of the Armed Forces. [Percent] by race. By sex. By rank. Chart.
Scholastic Update Oct 6, 1989 p. 21

ART - PRICES

Art prices: The rise and fall. Contemporary. Impressionist. Modern. 1988-1990.
Business Week Dec 31, 1990 p. 141

ART MUSEUMS

Museum pieces take a great fall. Value of objects donated. 1985-1990.
Business Week Feb 4, 1991 p. 102

Art attack. 1988 attendance [at various art museums].
U.S. News Jun 5, 1989 p. 69

ARTIFICIAL BODY PARTS

The body shop. A partial list of the nearly five dozen artificial body parts now available. Diagram.
U.S. News Nov 12, 1990 pp. 76–77

ASIA - ECONOMIC CONDITIONS

Asia's hot new growth zones. Gross national product. Population. Per-capita GNP. [By country].
Business Week Nov 11, 1991 p. 57

ASIAN AMERICANS

A tempting market. Asian-Americans: A small but fast-growing group with upscale demographics. 1990 U.S population [by ethnic group]. Asian-Americans. Total U.S. Median household income. Percent with college degree. Percent in management and professions.
Business Week Jun 17, 1991 pp. 54–55

By 2056, whites may be a minority group. Non-Hispanic whites as a % of total population. 1920-2056.
Time Apr 9, 1990 p. 30

[Percent] increase in each population group (1980-1988). Asians and others. Hispanics. Blacks. Whites.
Time Apr 9, 1990 p. 30

Birth rates and immigration. Whites. Blacks. Hispanics. Asians and others.
Time Apr 9, 1990 p. 31

Growing Asian enrollment. Percentage of American university students who are of Asian or Pacific-islander descent. 1976-1986.
U.S. News Mar 28, 1988 p. 53

ASPIRIN

The wonder drug. [Effects on the body]. Potential benefits.
Time Dec 16, 1991 p. 66

The battle for the painkiller market. Ibuprofen. Acetaminophen. Aspirin. Market share. Percent. 1987 advertising expenditures.
Business Week Aug 29, 1988 p. 60

ASSAULT RIFLES

Playing with firepower. Semiautomatic rifles range from military weapons to guns used for hunting. Assault rifle (banned). Assault rifle (legal). Hunting rifle (legal). Diagrams.
Newsweek Mar 27, 1989 p. 28

Street favorites. Assault weapons available over the counter.
Time Feb 6, 1989 p. 25

ASSETS, PERSONAL

Assets. Amount and types of assets, excluding real-estate equity, per household. [By] age of head of household.
U.S. News Jul 30, 1990 p. 72

ASSOCIATIONS

A nation of joiners. Here are some of the most popular associations that Americans join. Membership.
U.S. News May 21, 1990 p. 78

ASTEROIDS

Death from the sky. An asteroid or a comet perhaps 8 km (5 miles) in diameter may have hit the earth 65 million years ago. Diagram.
Time Jul 1, 1991 p. 60

ASTHMA

Anatomy of an asthma attack. A healthy airway. In an attack. The aftermath. Diagram.
U.S. News Mar 4, 1991 p. 60

The cases rise along with the death toll. Asthma sufferers. Annual deaths. 1980-1989.
U.S. News Mar 4, 1991 p. 61

When breathing breaks down. Asthma cases by age [1977-1986]. Asthma deaths by race [1977-1986].
Newsweek Sep 4, 1989 p. 61

ASTROLOGY

Daily newspapers running horoscope columns. Books in print about Nostradamus. [Other selected statistics.]
U.S. News Dec 16, 1991 p. 31

ASTRONOMY

The blackness of day. Hawaii's total eclipse. Longest blackout. Graying of the continent. Percentage of sun eclipsed. Map.
U.S. News Jul 15, 1991 pp. 44–45

Death from the sky. An asteroid or a comet perhaps 8 km (5 miles) in diameter may have hit the earth 65 million years ago. Diagram.
Time Jul 1, 1991 p. 60

A journey across the spectrum. Equipment to view energy emitted by heavenly bodies. Gamma rays. Visible light. Near infrared. Far infrared.
Newsweek Jun 3, 1991 p. 48

Wavelengths (in meters). X-ray. Ultraviolet. Microwaves. Television.
Newsweek Jun 3, 1991 pp. 48–49

Eye on the sky. Keck telescope in Hawaii. Diagram.
Newsweek Jun 3, 1991 p. 50

Evolution of the universe. New satellite measurements of "fossil" radiation from the big-bang explosion at the origin of the universe are causing theoreticians to scratch their heads. Diagram.
U.S. News Jan 29, 1990 pp. 48–49

Orbiting observer [Hubble Space Telescope]. Operations. Maintenance. Diagram
U.S. News May 15, 1989 pp. 54–55

ASYLUM, POLITICAL

Huddled masses. Applications for political asylum by Hispanics and Haitians now in the U.S. Country. Cases pending. As of Nov. 1, 1988.
Business Week Feb 6, 1989 p. 52

ATHLETIC SHOES - MARKET SHARE

Foot race. Adidas. Nike. Reebok. Revenues in Europe. 1988-1990.
Business Week Mar 11, 1991 p. 56

ATLANTIC CITY (NEW JERSEY)

Atlantic City [various statistics].
Newsweek Oct 29, 1990 p. 82

ATLANTIC STATES - ECONOMIC CONDITIONS

The Southeast. South Central. Employment. Population. Median home price. Annual rate of change from previous year. 1989-1990.
U.S. News Nov 13, 1989 p. 62

New England. Mid-Atlantic. Employment. Population. Median home price. Annual rate of change from previous year. 1989-1990.
U.S. News Nov 13, 1989 p. 64

ATOMIC BOMBS
See NUCLEAR WEAPONS

ATTORNEYS
See LAWYERS

AUTOGRAPHS

Celebrity signings. Current value of movie stars' autographs.
U.S. News Apr 3, 1989 p. 76

Washington's John Hancock. Value of a presidental signature, as most commonly signed.
U.S. News Oct 24, 1988 p. 81

AUTOMOBILE ACCIDENTS
See ACCIDENTS - MOTOR VEHICLE

AUTOMOBILE DRIVING

Tooting your own horn. When 500 drivers were asked to compare their driving habits with those of other motorists, here is the percentage of drivers who gave excellent or good marks to [selected driving skills for] themselves. Other drivers.
U.S. News Dec 18, 1989 p. 82

Which driver would you choose? Skills of drivers at different ages. Male. Female. Risk of fatality for unrestrained passengers relative to risk for an unrestrained driver.
U.S. News Nov 14, 1988 p. 82

The 65-mph question. Motor-vehicle traffic fatalities in 1987 [by state]. Deaths per 100,000 population. Change since 1986.
U.S. News Aug 15, 1988 p. 78

More, but safer, miles. Average miles per driver. 1970-1987. Death rate per 100 million vehicle miles. 1970-1987.
U.S. News Aug 15, 1988 p. 78

AUTOMOBILE INDUSTRY

Car crash. Earnings, 20 large U.S. auto suppliers. 1986-1991.
U.S. News Aug 12, 1991 p. 35

Riding the economy. Big Three's profits and losses. 1981-1991.
Time May 13, 1991 p. 41

Planting U.S. roots. Assembly capacity at foreign-owned North American automotive plants. 1983-1993.
Business Week Dec 10, 1990 p. 29

Japanese plants in North America. U.S. plants. [Comparison]. Assembly time per car. Worker training. Defects per 100 cars. Inventory supply. Work teams.
Business Week Oct 8, 1990 p. 83

Profits are skidding. Quarterly net profits. 1988, 1989.
U.S. News Dec 4, 1989 p. 50

Japanese auto plants in the U.S. and Canada. Parent. Production started. Planned yearly capacity. Planned employment. Map.
Business Week Aug 14, 1989 pp. 74–75

Japanese plants take a bigger slice. Cars made in America. Percent of domestically produced cars. 1984-1989.
Business Week Aug 14, 1989 p. 78

Continuing car glut. Passenger cars produced in North America. Supply. Demand. 1988,1989. 1990-1991 [projection].
U.S. News Apr 17, 1989 p. 47

The big three automakers: How they stack up. Number of black-owned dealerships, 1988. Amount of minority business, 1987. Ford. GM. Chrysler.
Black Enterprise Mar 1989 p. 18

Honda's success breeds imitation. Major foreign-owned auto plants in the U.S. and Canada. Company. Locations. Open. Capacity. Comments.
Business Week Apr 25, 1988 p. 96

Unrestrained competitors. Cars and small trucks. Imports. Produced in U.S. by foreign manufacturers. 1979-1987.
U.S. News Mar 21, 1988 p. 52

A setback for U.S. car makers. Aftertax profits of Detroit's auto companies. 1983-1988.
Business Week Jan 11, 1988 p. 75

AUTOMOBILE INDUSTRY - ADVERTISING

It's getting noisier. Ad spending. Billions of dollars. 1984-1988.
Business Week Jun 12, 1989 p. 81

AUTOMOBILE INDUSTRY - MARKET SHARE

GM's total profits and losses in billions. GM's share of the U.S. car market. 1980-1991.
Time Dec 31, 1991 p. 56

The selling race: An update. Big Three U.S. car companies and their top Japanese rivals. Market share. 1981, 1990, 1991.
Business Week Oct 21, 1991 p. 42

Detroit is losing ground. Market share. Chrysler. Ford. General Motors. Asians. Europeans. 1990, 1991.
Business Week Aug 19, 1991 p. 25

Gaining fast. Japan's share of the U.S. car market. Transplants. Imports. 1986-1990.
Newsweek Apr 8, 1991 p. 42

Chrysler is losing market share. Percent of total U.S. car and truck sales. Chrysler. Ford. General Motors. Foreign rivals. 1988-1990.
Business Week Mar 25, 1991 p. 94

GM has halted its market share slide. GM's share of U.S. auto and light truck market. Total sales of autos and light trucks in the U.S. Total net earnings [GM]. 1986-1991.
Business Week Feb 4, 1991 p. 95

Who's hot and who's not. 1990 U.S. market share [by various Japanese companies]. Percent change from previous year.
Business Week Jan 21, 1991 p. 37

Share of car sales in the U.S. GM. Japan. 1980-1990.
Time Oct 29, 1990 p. 76

How Japan's top three are faring. Toyota. Nissan. Honda. U.S. market share. 1985-1989.
Business Week Mar 12, 1990 p. 99

Japanese plants take a bigger slice. Cars made in America. Percent of domestically produced cars. 1984-1989.
Business Week Aug 14, 1989 p. 78

Bad news for the Big Three. Market share of cars sold in the U.S. 1988-1990.
U.S. News Apr 17, 1989 p. 47

Battle for position. GM's market share. 1980-1985.
Newsweek Feb 20, 1989 p. 39

Honda's growing stake in the U.S. U.S. car sales as percent of worldwide car sales. Company. Cars sold in U.S. Share of total. 1980, 1987.
Business Week Apr 25, 1988 p. 92

Unrestrained competitors. Cars and small trucks. Imports. Produced in U.S. by foreign manufacturers. 1979-1987.
U.S. News Mar 21, 1988 p. 52

Driving hard bargins. Foreign manufacturers' share of U.S. auto market. 1979-1987.
U.S. News Mar 21, 1988 p. 52

Even with import quotas, Detroit has lost market share. Percent of autos sold in the U.S. GM, Ford, Chrysler. Japanese imports. Other imports. Transplants. 1981-1987.
Business Week Mar 7, 1988 p. 61

AUTOMOBILE THEFT

Car-nabbers picks and pans. These 1988 cars were the most and least popular with thieves that year. Cars stolen per 1,000. Total thefts in 1988.
U.S. News Jan 29, 1990 p. 62

Havens for car snatchers. Reported motor-vehicle thefts in 1987 [by state]. Per 100,000 registered vehicles. Total thefts.
U.S. News Jan 16, 1989 p. 70

Cars that get stolen. Average theft payments for 1987 cars. Highest claims. Lowest claims.
U.S. News Feb 1, 1988 p. 73

AUTOMOBILES

Needs clutch, runs great. A new car's value can drop as much as 40 percent in the first two years. Here are the models that are best and worst at holding their value after five years.
U.S. News Apr 16, 1990 p. 62

America's car fleet is showing its age. Average age of U.S. passenger cars. 1958-1988.
Business Week Mar 26, 1990 p. 26

New superpower. Registered autos per 1,000 people. United States. United Germany. Japan. Soviet Union.
Newsweek Feb 26, 1990 p. 23

The top ten: Now and then. Best-selling cars in the United States. 1978, 1989.
Newsweek Jan 22, 1990 p. 42

The market is more crowded. Car and light truck models on the U.S. market. Number of models. 1984-1988.
Business Week Jun 12, 1989 p. 81

What do you drive? Auto ownership by blacks.
Black Enterprise Nov 1988 p. 47

What car dealers really think. Dealers who rated the models they sell "excellent." Reliablility and dependability. Delivery condition.
U.S. News Jun 27, 1988 p. 64

Cars they make here. Automobiles and pickup trucks manufactured in U.S. by Japanese companies. 1983-1987.
U.S. News May 9, 1988 p. 59

Picking car colors. Most popular. Least popular.
U.S. News Apr 18, 1988 p. 83

AUTOMOBILES - AIR CONDITIONING

Cool cars for the future. How CFC's cool. Automobile air conditioners work on the same principle as home refrigerators, by circulating a fluid, now CFC-12. Here's how it works. Diagram.
U.S. News Nov 6, 1989 p. 82

AUTOMOBILES - CRASHWORTHINESS

Bumper bashing. A crash test of 1991 mid-size cars. The range of repair costs [for lowest and highest models].
U.S. News Apr 8, 1991 p. 69

Question of survival. These cars had the highest and lowest death rates per 10,000 cars over the 1986-88 period.
U.S. News Jan 8, 1990 p. 66

AUTOMOBILES - DEFECTS

A race to be glitch-free. Europe. U.S. Japan. Problems per 100 vehicles. 1987-1991.
Business Week Oct 25, 1991 p. 70

Ford narrows the quality gap. Defects per 100 vehicles. Ford. Japanese. 1986-1989.
Business Week Apr 30, 1990 p. 116

Detroit is narrowing the quality gap. Overall customer satisfaction index. Asian models. Domestic models. 1985-1988.
Business Week Jun 12, 1989 p. 81

AUTOMOBILES - FUEL EFFICIENCY

The drop in fuel economy. Corporate average fuel economy for passenger cars sold in the U.S. All cars sold in the U.S. U.S. nameplates only. 1987-1990.
Business Week Jun 3, 1991 p. 26

Choosing a route. [Various automobile fuels]. Advantages. Disadvantages.
U.S. News Feb 4, 1991 p. 64

Stalled performance. Average fuel economy of cars sold in U.S. Federal minimum. Actual. 1978-1990.
Time Oct 8, 1990 p. 64

Fueling pollution. These gas guzzlers emit the most carbon dioxide. Carbon dioxide emitted per 100,000 miles.
U.S. News Jul 2, 1990 p. 62

Only Chrysler is home free. Estimated 1989 fuel economy. 1989 standard. 1990 standard. Chrysler. GM. Ford.
Business Week Apr 10, 1989 p. 41

AUTOMOBILES - IMPORTS AND EXPORTS

Detroit tries again. The Big Three's auto exports to Japan. GM. Ford. Chrysler. 1979-1990.
Business Week Jul 22, 1991 p. 82

Sticker shock in South Korea. Why a $10,000 imported car costs so much in Korea.
Business Week Feb 15, 1988 p. 47

AUTOMOBILES - LEASING AND RENTING

The leasing binge. Cars leased by individuals. 1975-1995.
Business Week May 20, 1991 p. 104

Buying vs. leasing. Vehicle: 1988 Dodge Daytona. Conventional loan. Lease.
Black Enterprise May 1988 p. 49

Don't get taken for a rent-a-car ride. How rates vary. Company. Cost.
Business Week Mar 14, 1988 p. 146

AUTOMOBILES - PURCHASING

Trendy transport. American makes. Foreign makes. [By manufacturer]. Black consumers. White consumers.
Black Enterprise Nov 1990 p. 49

Sticker price of new Saturn. Average weeks worked to buy car in 1980, 1989. Costs to operate a passenger car. Best-selling cars, 1990.
U.S. News Oct 15, 1990 p. 18

Motor trends. The car-buying woman. Proportion of new cars bought by women. Profile.
Ms. Jan 1989 p. 137

AUTOMOBILES - SAFETY DEVICES

Saved by a [car] seat. Number of children's lives saved. Ages 0-4. 1979, 1984, 1989.
Newsweek Jan 14, 1991 p. 43

Safety, 1991 style. This exclusive guide to virtually all of the 1991 [passenger-car] models shows which of six important safety features are available.
U.S. News Dec 3, 1990 pp. 72–78

AUTOMOBILES - SALES

Auto sales off. In millions. 1988-1991.
Time Dec 9, 1991 p. 22

Domestic auto sales. 1989-1991.
Time Nov 25, 1991 p. 64

Recession? Maybe it's not history after all. Auto sales. 1980-1991.
U.S. News Sep 9, 1991 p. 13

Auto sales show signs of life. U.S. dealer sales of cars and light trucks. 1990, 1991.
Business Week Aug 19, 1991 p. 25

U.S. auto sales (10-day selling periods). 1990-1991. Companies with the biggest sales declines. Japanese car makers' share of the American market. 1990 average new car prices.
U.S. News Jun 3, 1991 p. 15

A pain meter. U.S. car and light-truck unit sales. Big three. Japanese. European.
Business Week Apr 1, 1991 p. 29

Another setback for Detroit. U.S. new-car and light-truck sales. 1986-1991.
Business Week Jan 14, 1991 p. 67

Battling to hold on as No. 1. U.S. retail sales of passenger cars. Imports. GM. Ford. Chrysler. 1980-1989.
Newsweek Oct 22, 1990 p. 51

Solid sales. Auto sales in U.S. 1989, 1990.
U.S. News Oct 22, 1990 p. 59

Auto makers slip into reverse. Total U.S. sales. Big three U.S. sales. 1985-1990.
Business Week Jan 8, 1990 p. 69

Is Motown ready for a drop in demand? Domestic unit sales. Cars and light trucks, including imports. 1983-1989.
Business Week Oct 3, 1988 p. 30

GM's response to a changing world. GM sales as a percent of total domestic car sales. 1987, 1992.
Business Week May 9, 1988 pp. 114–115

A model year for Ford. The 20 best-selling cars of 1987. The 20 slowest sellers.
U.S. News Apr 18, 1988 p. 83

Will the auto glut choke Detroit? U.S. auto sales. Imports. Production capacity. Excess capacity.
Business Week Mar 7, 1988 pp. 54–55

AUTOMOBILES, COST OF OPERATION

Sticker price of new Saturn. Average weeks worked to buy car in 1980, 1989. Costs to operate a passenger car. Best-selling cars, 1990.
U.S. News Oct 15, 1990 p. 18

Cars. Average annual cost, per household, of owning and operating cars. Income. Car costs. Cars per household. [By income].
U.S. News Jul 30, 1990 p. 72

Affording the car you want. Here's a five-year cost comparison for someone in the 28% tax bracket on an Oldsmobile Cutlass Supreme with a sticker price of $18,000, bargained down to $16,500.
Newsweek May 22, 1989 p. 65

Car care. Your driving costs (1988 average). Operating cost (per mile). Ownership cost (per year).
Black Enterprise Nov 1988 p. 47

Buying vs. leasing. Vehicle: 1988 Dodge Daytona. Conventional loan. Lease.
Black Enterprise May 1988 p. 49

AUTOMOBILES, EXPERIMENTAL

Three ways to go. Electric. Compressed natural gas. Methanol. Diagrams.
Business Week Apr 8, 1991 pp. 56–57

Choosing a route. [Various automobile fuels]. Advantages. Disadvantages.
U.S. News Feb 4, 1991 p. 64

AUTOMOBILES IN EASTERN EUROPE

Why Boris takes the bus. Here's the car count in Eastern Europe. People per car. Cars needed to equal Western Europe's average.
U.S. News Mar 26, 1990 p. 73

AUTOMOBILES IN JAPAN

Foreign cars Japan likes. Number of cars imported to Japan [by country and make of car].
U.S. News Aug 22, 1988 p. 65

AUTOMOBILES, USED

More and more U.S. cars are showing their age. Passenger cars that are at least nine years old. 1970, 1980, 1989.
Business Week Mar 4, 1991 p. 12

Needs clutch, runs great. A new car's value can drop as much as 40 percent in the first two years. Here are the models that are best and worst at holding their value after five years.
U.S. News Apr 16, 1990 p. 62

AWARDS

Pulitzers awarded. Top journalism winners. Most awards. [Other selected statistics.]
U.S. News Sep 16, 1991 p. 11

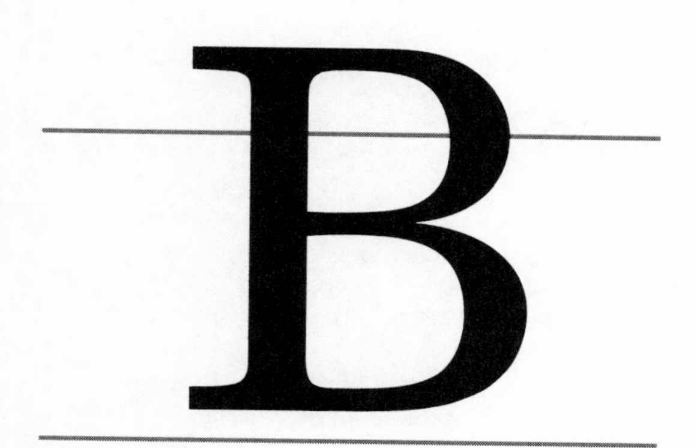

BABIES, PREMATURE

See INFANTS, PREMATURE

BABY BOOM GENERATION

Middle-age bulge. Vital statistics on the generation born between 1946 and 1964. Percent of the population. Education. Marital status. Birth rate. Wealth.
Business Week May 20, 1991 p. 108

A wealth gap divides the generations. Average real net worth of families. Baby boomers. Parents.
Business Week Jan 28, 1991 p. 22

A baby-boomer future. People who will reach age 65, year by year. 1990-2050.
U.S. News Aug 15, 1988 p. 65

BACK

Back. Ruptured disk. The usual way: Disk excision (laminectomy). The minimal way: Percutaneous diskectomy. Procedures per year. Average cost. Hospital stay. Recovery time. Risk.
U.S. News May 20, 1991 p. 77

BALANCE OF TRADE

See INTERNATIONAL TRADE

BALDNESS

Men with "pattern baldness" in United States. Women with same problem. Hair transplants. Number of wigs. [Other selected statistics.]
U.S. News Aug 26, 1991 p. 13

BALTIC STATES

Estonia. Latvia. Lithuania. Population. Per capita GNP. [Ethnic makeup]. Map.
Time Sep 9, 1991 p. 36

BANK FAILURES

The number of bank failures is declining, but the assets involved are growing, depleting the bank insurance fund. Total assets of failed banks. Bank insurance fund balance. 1987-1991.
Business Week Dec 9, 1991 p. 30

Flood tide. Bank failures. 1950-1992.
U.S. News Oct 14, 1991 p. 55

Bank profits bumping along the bottom. Troubled commercial real estate loans. FDIC bailout fund. [Other selected statistics.] 1985-1991.
U.S. News Sep 23, 1991 p. 17

Bank runs and bailouts: A short and costly history lesson. Crisis. Solution. FDR's bank holiday. Penn Central. New York City. Chrysler Corporation.
U.S. News Feb 20, 1989 pp. 50–51

Cracks in the system. Costs to the FDIC from bank failures. Bank failures. Costs to the FSLIC from savings and loan failures. Savings and loan liquidations, mergers and acquisitions. 1981-1987.
Time Aug 29, 1988 pp. 54–55

Shaky S&L's. A dying breed. Bank failures resolved by: Mergers & acquisitions. Liquidations. Changes in management. 1978-1987.
U.S. News Aug 29, 1988 p. 84

Busting the bank. Bank failures. 1978-1987.
U.S. News Aug 15, 1988 p. 42

Busting the bank. Pricey bailouts. 1978-1987.
U.S. News Aug 15, 1988 p. 42

Hit-or-miss economy. See-sawing stock markets. Rising bank failures. The diving dollar.
U.S. News Apr 11, 1988 p. 47

The surge in bank failures. Number of bank failures. 1982-1987.
Business Week Apr 4, 1988 p. 30

BANK OF CREDIT AND COMMERCE INTERNATIONAL

B.C.C.I. arms deals. [By country.] Diagram.
Time Sep 2, 1991 p. 56

The connections. The bank's global web was designed to mystify. B.C.C.I. holdings. Ownership. Customer. Dubious loans. Charity donations. Diagram.
Time Jul 29, 1991 pp. 42–43

BANKRUPTCY

Corporate undertaking. Commercial bankruptcies are growing in both number and size. Debts of failed businesses. 1984-1990.
U.S. News Apr 8, 1991 p. 51

A sick economy. Bankruptcies per year, 1985-1990.
Scholastic Update Mar 8, 1991 p. 3

The bankruptcy boom. Total liabilities in business failures. 1984-1991.
Business Week Jan 21, 1991 p. 29

Total bankruptcies filed. Three largest bankruptcies. Ratio of consumer bankruptcies to business bankruptcies.
U.S. News Jan 21, 1991 p. 8

Personal bankruptcies are climbing. Filings. 1987-1990.
Business Week Dec 10, 1990 p. 205

Why merchants are miserable. Bankruptcies are on the rise. Number of retailers filing for bankruptcy protection. 1985-1990.
Business Week Nov 26, 1990 p. 135

Corporate dropouts. The 10 most significant corporate bankruptcies of 1990. Gross revenues. Date of filing.
U.S. News Nov 19, 1990 p. 58

Personal bankruptcies are rocketing. Bankruptcies per thousand persons. 1980-1990.
Business Week Aug 20, 1990 p. 18

Plenty of nothing. The bankruptcy rate in the 1960s, '70s and '80s swamped the rate during the Depression years. 1900-1989.
U.S. News Jun 4, 1990 p. 73

A run for cover. Personal bankruptcies. 1981-1989.
Newsweek Apr 2, 1990 p. 40

More people are becoming deadbeats. Personal bankruptcy filings. 1980-1988.
Business Week Oct 17, 1988 p. 22

The bankruptcy boom drags in more bonds. Bonds newly in default. Companies defaulting. 1982-1988.
Business Week Sep 5, 1988 p. 85

Bankruptcy bonds. Public debt of bankrupt companies. 1984-1987.
U.S. News Jan 11, 1988 p. 58

See Also BUSINESS FAILURES

BANKS AND BANKING

Bank profits bumping along the bottom. Troubled commercial real estate loans. FDIC bailout fund. [Other selected statistics.] 1985-1991.
U.S. News Sep 23, 1991 p. 17

American vault doors are closing. Number of insured U.S. commercial banks. 1980-1991.
Time Jul 15, 1991 p. 48

The U.S. is still overbanked by comparison. [Number of] commercial banks, latest figures. [By country].
Time Jul 15, 1991 p. 48

The banks' eroding market share. Portion of total assets held by financial institutions. 1960, 1989.
Business Week Apr 22, 1991 p. 75

Who's who among big card issuers and how banks are losing share. 1986-1990.
Business Week Apr 15, 1991 p. 29

As bad bank loans explode, the bank insurance fund shrinks. Net charge-offs by insured commercial banks. Yearend balance. 1980-1991.
Business Week Jan 21, 1991 p. 25

Banks make a smaller share of loans and depositors have other choices. Lending by commercial banks as a share of total new debt. 1978-1990. Money held in money-market funds as a share of bank deposits. 1983-1990.
Business Week Jan 21, 1991 p. 30

A steady slide. Return on assets for all FDIC-insured banks. 1973-1990.
Newsweek Jan 21, 1991 p. 42

A weakening safety net. Amount in FDIC fund for every $100 in insured deposits, year-end. 1934-1991.
Time Jan 21, 1991 p. 54

Tightening credit. Rate of change year to year. Treasury securities held by banks. Commercial and industrial loans. 1985-1990.
U.S. News Dec 24, 1990 p. 51

Estimated cost of FDIC rescues [various banks].
Time Sep 24, 1990 pp. 66–67

Can't get out of the red. Net income or deficit of the Federal Deposit Insurance Corporation in billions of dollars. 1983-1990.
Time Sep 24, 1990 p. 67

Adding up the columns. Nonperforming loans for all U.S. banks. Federal deposit insurance per insured $100 of U.S. bank deposits. 1985-1989.
U.S. News Aug 20, 1990 p. 43

Banks. Is big trouble brewing? Loan losses for federally insured banks, after recoveries. 1976-1990. Noncurrent loans as a percent of all real estate loans. Change in real estate loans. Map.
Business Week Jul 16, 1990 pp. 146–147

Sagging profits. Real-estate values helped erode commercial-bank profits. Net income of commercial banks. 1985-1989.
U.S. News Jul 2, 1990 p. 38

Bottom lines. Banks with the highest total in deadbeat real-estate loans. Bank. Problem real-estate loans [dollar amount]. Percentage of all real-estate loans.
U.S. News Jul 2, 1990 pp. 38–39

Less lending. Banks are reducing their real-estate exposure. Growth in real-estate loans. Commercial real-estate loans. Construction and land development loans. 1985-1990.
U.S. News Jul 2, 1990 p. 39

The next crisis. The government is already suffering losses on many other guarantees. What it backs in all. Program. Taxpayer liability (in billions).
Newsweek May 21, 1990 p. 24

Lending by commercial banks. Percentage change in growth of loans for commercial and industrial projects. 1988-1990.
U.S. News Apr 23, 1990 p. 52

Lending by commercial banks. Percentage change in growth of real-estate loans. 1988-1990.
U.S. News Apr 23, 1990 p. 52

Lending by commercial banks. Percentage change in growth of personal loans. 1988-1990.
U.S. News Apr 23, 1990 p. 52

Where credit is cheapest. The average interest rate on Visa and MasterCard purchases is 18.5 percent, but many banks offer better rates. These banks give the best deals.
U.S. News Oct 30, 1989 p. 82

Biggest home bankers. Program. Monthly Fee. Special features. 1982-1988.
Time Jan 9, 1989 p. 49

BANKS AND BANKING - MARKET SHARE

Banking. Key acquisitions. 1991. Assets at yearend. Billions of dollars. 1985 ranking. 1990 ranking.
Business Week Oct 14, 1991 p. 86

New top 5 U.S. banks following mergers. Assets in billions (as of June 30).
Time Aug 26, 1991 p. 42

Who's got the money in the vault? Approximate shares of world banking assets. Net 1989 income for the top ten banks. Japan. Europe. U.S. 1984-1988.
Time Jul 30, 1990 p. 48

Biggest banks. Home base of the world's top 100 banks. U.S. Japan. West Germany.
U.S. News Jul 16, 1990 p. 24

BARBECUE COOKERY

Summer kitchen. Number of times an average family barbecues these items per month.
U.S. News Jul 4, 1988 p. 63

BASEBALL CARDS

Save that box. At Sotheby's first auction of sports memorabilia, bidders paid these prices for vintage food packages imprinted with baseball cards.
U.S. News Apr 8, 1991 p. 69

It's in the cards. Experts who keep track of baseball cards say [here are] the 15 most valuable.
U.S. News Jul 3, 1989 p. 62

BASEBALL, PROFESSIONAL

Are salaries soaring out of the ballpark? Salaries as percent of league revenues. Including penalties. As paid. 1986-1990.
Business Week Dec 30, 1991 p. 40

Minority hiring in baseball: A scorecard. Major league club front-office employees. Blacks. Hispanics. Women. Major league players. Blacks. Hispanics. 1988, 1991.
Business Week Dec 30, 1991 p. 48

Percentage of overall blacks vs. blacks in front-office management. NBA. NFL. MLB.
Black Enterprise Dec 1991 p. 47

Who holds the ball. (White and black head coaches and managers). NBA. NFL. MLB.
Black Enterprise Dec 1991 p. 47

The money is bigger than ever but it doesn't last long. Average salary. Average length of professional-sports career. Baseball. Basketball. Football. Hockey. 1981, 1991.
Business Week Jun 3, 1991 p. 55

[Retired] National Football League players [and] major league baseball players. [Other selected statistics.]
Business Week Jun 3, 1991 p. 58

Ticket lineup. The teams with the biggest [ticket price] increases.
U.S. News Apr 22, 1991 p. 71

Franchise price tag. Player salaries. Attendance at major-league games.
U.S. News Oct 1, 1990 p. 14

Baseball [television] contract. Increase from 1984.
Time Mar 26, 1990 p. 67

Salary scorecard [selected baseball players].
Newsweek Apr 10, 1989 p. 44

Where the cash flows in baseball. Broadcasting contracts. Local and national revenues. Revenues. Expenses.
Business Week Apr 3, 1989 pp. 86–87

Stakes in the October [World series] game. World Series winner's share. Actual prize money per player. Adjusted for inflation. 1908-1987.
U.S. News Oct 24, 1988 p. 81

BASKETBALL, COLLEGE

Setting hurdles. Average SAT scores at [various universities]. NCAA cutoff.
U.S. News Jan 30, 1989 p. 69

BASKETBALL, PROFESSIONAL

Percentage of overall blacks vs. blacks in front-office management. NBA. NFL. MLB.
Black Enterprise Dec 1991 p. 47

Who holds the ball. (White and black head coaches and managers). NBA. NFL. MLB.
Black Enterprise Dec 1991 p. 47

NBA revenue. Average NBA salaries. College referee's pay per game. Graduation rate of college players. [Other selected statistics.]
U.S. News Nov 25, 1991 p. 9

The money is bigger than ever but it doesn't last long. Average salary. Average length of professional-sports career. Baseball. Basketball. Football. Hockey. 1981, 1991.
Business Week Jun 3, 1991 p. 55

N.B.A. [television] contract. Increase from 1986.
Time Mar 26, 1990 p. 67

BATTERIES

Batteries not included. Almost 1.2 billion batteries will be sold this holiday season in the U.S. Here is where that power goes. Items in use. Percentage of battery dollar volume.
U.S. News Dec 18, 1989 p. 82

BEACHES

Trashy shores. Here are the pounds of debris collected per mile in [various states].
U.S. News Jan 15, 1990 p. 67

BEER CONSUMPTION

States with the lowest annual beer purchases per capita. Gallons purchased per person. Total gallons sold.
U.S. News Feb 27, 1989 p. 75

BEER INDUSTRY - MARKET SHARE

Top competitors are gaining. Share of U.S. beer market. Miller Lite. Bud Lite. Coors Lite.
Business Week Jul 15, 1991 p. 29

Miller is cranking up ad spending. Major media outlays on light beer brands. Miller Lite. Bud Light. Coors Light.
Business Week Jul 15, 1991 p. 29

How the brewers stack up in Japan. Market share.
Business Week Mar 4, 1991 p. 41

The top imported beers. Company. Country. 1987 share of U.S. imports.
Business Week Apr 18, 1988 p. 83

BENEFITS

See EMPLOYEE BENEFITS

BETTING

See GAMBLING

BEVERAGES

Cola competition. Market share. Coke. Coca-Cola Classic. Pepsi. 1985-1989.
Newsweek Mar 19, 1990 p. 38

Way behind with Citrus Hill. Share of U.S. orange juice market. Minute Maid. Tropicana. Citrus Hill.
Business Week Jan 23, 1989 p. 38

Caffeine content of beverages and foods.
FDA Consumer Jan 1988 p. 24

Consumption of coffee and other beverages. 1962, 1987.
FDA Consumer Jan 1988 p. 25

Caffeine content of soft drinks. Regular. Diet drinks.
FDA Consumer Jan 1988 p. 25

BICYCLES AND BICYCLING

Racers 40,000. Commuters. Regular riders. 1983, 1990.
Time Aug 19, 1991 p. 43

Adults who bicycled in 1990. Rode bicycles or walked to help save oil. Number of Americans who ride bicycles to work.
U.S. News May 13, 1991 p. 14

BILINGUALISM

Which second language? Languages likely to be helpful in the '90s.
U.S. News Dec 25, 1989 p. 66

BILLIONAIRES

Billionaires on America's Main Street. Where this year's *U.S. News* 100 live or work.
U.S. News Aug 1, 1988 pp. 38–39

The *U.S. News* 100. Individuals or families [who] own at least 5 percent of the shares in publicly traded companies. Rank. Name. Holdings. Value of holdings. % of shares owned. 1987 rank.
U.S. News Aug 1, 1988 pp. 40–41

BIOFEEDBACK

How biofeedback works. Diagram.
U.S. News Sep 23, 1991 p. 73

BIOLOGICAL RHYTHMS

A circadian 'Believe It or Not.' Studies of circadian rhythms have turned up the following sometimes hard-to-believe information.
FDA Consumer Jul 1990 p. 21

BIOLOGICAL WARFARE

See CHEMICAL AND BIOLOGICAL WARFARE

BIRMINGHAM (ALABAMA)

Birmingham. Population. Median family income. Median home. Unemployment.
Newsweek Feb 6, 1989 p. 44

BIRTH CONTROL

Effectiveness rates of contraceptive methods. Method. Lowest expected. Typical. Number of pregnancies for every 100 women during the first year of use.
FDA Consumer May 1991 p. 10

In search of alternatives. Of America's 58 million women of childbearing age, 60 percent practice some form of contraception. What they use [by type].
Newsweek Dec 24, 1990 p. 68

The choices [of birth control methods]. Couples who use [various methods]. Effectiveness.
U.S. News Dec 24, 1990 pp. 58–59

How Norplant works. Diagram.
U.S. News Dec 24, 1990 p. 60

A user's guide to 14 methods. Pregnancy/failure rates. Cost. Advantages. Disadvantages.
U.S. News Dec 24, 1990 pp. 62–63

The surgical solution. Sterilizations. Women. Men. 1971-1987.
U.S. News Dec 24, 1990 p. 64

Two techniques on the drawing board. For women. For men. Diagram.
U.S. News Dec 24, 1990 p. 64

Estrogen levels of pill drop. Amount of estrogen in oral contraceptives. 1960s - 1990s.
FDA Consumer Dec 1990 p. 10

Preventing pregnancy. Actual effectiveness rates of the various forms of contraception.
FDA Consumer Dec 1990 p. 11

Methods [of birth control] used in the U.S. Methods used abroad.
Time Feb 26, 1990 p. 44

While the world's population has climbed steadily in the past ten years, real U.S. spending on family planning has declined. 1979-1989.
Time Dec 18, 1989 p. 62

Comparing contraceptives. Oral contraceptive. Intrauterine. Diaphragm. Cervical cap. Overall effectiveness rate.
FDA Consumer Sep 1988 p. 34

Birth control. What contraceptives do women choose? Method. Age.
Ms. Mar 1988 p. 76

BIRTH DEFECTS

Birth defects. Disability. Projected cases, 1990.
Newsweek Oct 22, 1990 p. 78

BIRTH ORDER

Sibling smarts. When the Mensa society, a group open to anyone who scores in the top 2 percent on a standardized IQ test, surveyed its members, here is how they fell in their family's lineup.
U.S. News May 22, 1989 p. 81

BIRTHS AND BIRTH RATE

Declining birth rates. United States. France. Sweden. United Kingdom. Switzerland. Germany. Italy. 1989.
Newsweek Dec 16, 1991 p. 43

Some [consumers] have more family responsibilities. Number of births per 1,000 women ages 30 to 34. 1975-1988.
Business Week Nov 11, 1991 p. 134

Birthrates are up. Average number of children per woman under age 50. 1970-1990.
Business Week Aug 5, 1991 p. 49

The over-40 parent boom. First births. Women age 40 to 44. All births. 1947-1988. All births. Men age 40 to 44. 1945-1988.
U.S. News Oct 29, 1990 p. 105

Baby boomlet. Birth rate. Births. Number per 1,000 population. 1960-1989.
U.S. News Sep 24, 1990 p. 99

Women and children. More women are working—and fewer are having babies. Female workers as a percentage of the total labor force. 1979-1989. Birthrate. Children per family. 1965-1990.
Newsweek Jul 16, 1990 p. 39

Birth rates and immigration. Whites. Blacks. Hispanics. Asians and others.
Time Apr 9, 1990 p. 31

More women, fewer babies. Total fertility rate. Average number of children born to women in: Less developed countries. More developed countries. Sub-Saharan Africa. Northern Africa. Mexico. 1965-2025.
U.S. News Dec 18, 1989 p. 23

Birth rates per 1,000 unmarried women age 15-44 per year. Blacks. Whites. 1970-1986.
U.S. News Jul 3, 1989 p. 29

Many births, many deaths. Countries with highest birth rates. Fertility rate. Infant-mortality rate. Countries with lowest birth rates.
U.S. News May 23, 1988 p. 76

BLACK ACCOUNTANTS

Accounting for black accountants. Where black accountants are employed.
Black Enterprise Nov 1990 p. 22

BLACK ATHLETES

Percentage of overall blacks vs. blacks in front-office management. NBA. NFL. MLB.
Black Enterprise Dec 1991 p. 47

Who holds the ball. (White and black head coaches and managers). NBA. NFL. MLB.
Black Enterprise Dec 1991 p. 47

BLACK CHILDREN

Deep-rooted poverty. Percentage of persons below the poverty line 1990. White. Black. Black children.
U.S. News Nov 4, 1991 p. 63

Poverty rates for black children: 1969-1989.
Black Enterprise May 1991 p. 36

A mixed picture. Rate of black children living with mother only. 1960-1988.
U.S. News Jan 22, 1990 p. 28

Poverty rate of black families: 1977 to 1987. Poverty rate of black children under 18 years old: 1977 to 1987. Poverty thresholds, by size of family unit: 1987.
Black Enterprise Nov 1989 p. 49

BLACK COLLEGE STUDENTS

Degrees of separation. College enrollment among Blacks. 1970, 1989. Master's degrees. Professional degrees. Doctorate degrees.
Newsweek May 6, 1991 p. 27

Integrating the ivory tower. African-American representation among students and full-time faculty at undergraduate and selected graduate institutions, 1989-90.
Black Enterprise May 1991 p. 74

Students rate themselves above average in [various characteristics]. Freshmen at black colleges. Freshmen overall.
Black Enterprise Sep 1990 p. 45

BLACK COLLEGES AND UNIVERSITIES

The largest historically black public colleges. University. Main location. Enrollment.
Time Nov 11, 1991 p. 81

BLACK CONSUMERS

Black vs. white consumers. Planned major purchases. Blacks. Whites.
Black Enterprise Oct 1990 p. 45

Black vs. white consumers. Types of financial vehicles owned. Blacks. Whites.
Black Enterprise Oct 1990 p. 45

BLACK FAMILIES

The continued breakup of the black family. Distribution of black and white families by type, 1940-1985. Other male head. Female head. Husband-wife. Black families. White families.
Black Enterprise Jan 1989 p. 52

BLACK FARMERS

The declining number of black farmers. 1900-1982.
Black Enterprise Jul 1988 p. 21

BLACK TEACHERS

Number and percent of bachelor's degrees in education conferred on blacks, 1977-1985.
Black Enterprise Dec 1990 p. 18

BLACKS

In 20 years Hispanics will be the largest minority group. Blacks. Hispanics. 1980-2020.
Time Jul 29, 1991 p. 15

A generation of change? Then [1960s]. Now [1980s]. White. Black. Income. Life expectancy. Infant mortality. Death rate. Education. Arrested. Crime victims. 1960-1990.
U.S. News Jul 22, 1991 pp. 20–21

Homicide rates. (Gun-related). Black males. Black females. White males. White females. 1984-1988.
Newsweek Dec 17, 1990 p. 33

Growing gap. Life expectancy at birth. Whites. Blacks. 1980-1988.
Time Dec 10, 1990 p. 78

By 2056, whites may be a minority group. Non-Hispanic whites as a % of total population. 1920-2056.
Time Apr 9, 1990 p. 30

[Percent] increase in each population group (1980-1988). Asians and others. Hispanics. Blacks. Whites.
Time Apr 9, 1990 p. 30

Birth rates and immigration. Whites. Blacks. Hispanics. Asians and others.
Time Apr 9, 1990 p. 31

A mixed picture. Rate of black children living with mother only. 1960-1988.
U.S. News Jan 22, 1990 p. 28

Tale of two cities. The northwest corner of D.C. is white and well off. The rest of the city is poorer and mostly black. Average household income. Location of homicides.
Newsweek Mar 13, 1989 p. 17

The statistics: Separate and unequal [blacks and whites compared].
Newsweek Mar 13, 1989 p. 19

BLACKS - ECONOMIC CONDITIONS

The mortgage squeeze on minorities. Mortgage denial rates.
U.S. News Nov 4, 1991 p. 21

Deep-rooted poverty. Percentage of persons below the poverty line 1990. White. Black. Black children.
U.S. News Nov 4, 1991 p. 63

Living above our means? Black consumer buying habits. How we invest. How we spend. Hooked on plastic. White. Black.
Black Enterprise Oct 1991 p. 41

The poverty line. White. Black. Hispanic. Americans living below it.
Time Sep 30, 1991 p. 30

Devastating minorities. Median annual family income. By type of family head under 30. White. Black. Hispanic. 1973, 1989.
Business Week Aug 19, 1991 p. 81

Median annual earnings. White males. Black males. White females. Black females. 1969-1989.
Business Week Jul 15, 1991 p. 10

Money income, by race, 1988-1991. White. Black. Other races.
Black Enterprise Jan 1991 p. 48

Not much of a cushion. Blacks earning $24,000 to $48,000. Median assets. Blacks. Whites. Net worth.
Newsweek Dec 31, 1990 p. 41

A persistent black income deficit. Received. Parity. 1988, 1989, 1990.
Black Enterprise Jun 1990 p. 224

A mixed picture. Family income. Per capita income. White. Black. 1979-1988.
U.S. News Jan 22, 1990 p. 28

Selected characteristics of black households. Black households. All households.
Black Enterprise Jan 1990 p. 45

Median income by race. 1960-1987. White family income. Black family income.
Black Enterprise Jan 1990 p. 58

Catching up, sort of. Percentage of households with incomes of $50,000 or more. Whites. Blacks. Hispanics. 1972-1988.
U.S. News Dec 11, 1989 p. 74

Poverty rate of black families: 1977 to 1987. Poverty rate of black children under 18 years old: 1977 to 1987. Poverty thresholds, by size of family unit: 1987.
Black Enterprise Nov 1989 p. 49

Median family incomes, by race and family type, 1987. Black. White. Black/white ratio.
Black Enterprise Nov 1989 p. 59

Black per capita income rises, still below that of whites. Overall. Whites. Blacks. 1967, 1987.
Black Enterprise Nov 1989 p. 59

Black gains slow. Black men. Black women. Percent of average full-time income. 1955-1985.
Business Week Sep 25, 1989 p. 95

Rich vs. poor: The gap widens. Percent change in household income from 1979 to 1987. Married couples with children. Single mothers with children. Heads of household under 25. Heads of household over 65. Blacks.
Business Week Apr 17, 1989 pp. 78–79

Income. Living in poverty. Unemployment. Median income. Blacks. Whites. 1966, 1987.
Scholastic Update Apr 7, 1989 p. 21

Housing. Percent of blacks and whites who own their own homes. Whites. Blacks. 1974, 1985.
Scholastic Update Apr 7, 1989 p. 21

Median family income in 1987 dollars. Black. White. 1967, 1987.
Time Mar 13, 1989 p. 62

The widening gap between the rich and poor. Percentage of black families receiving income in each range. 1970, 1986.
Black Enterprise Jan 1989 p. 51

Playing the money management game. Black profile: Major purchase contemplated. Black consumer spending by category. Black profile: Financial investment held.
Black Enterprise Oct 1988 p. 43

The times are not a-changin'. Median family income. White. Black. 1968-1986. Unemployment rate for those 16 years and older. Black. White. 1968-1987.
Black Enterprise May 1988 p. 35

Upward mobility: Blacks make progress but gaps persist. Household income. Home ownership. Net worth. College graduates. Managers.
Business Week Mar 14, 1988 p. 68

Race gap then and now. Living in poverty. Unemployment. Median income. Blacks. Whites. 1986. Latest.
U.S. News Mar 14, 1988 p. 11

The salary gap. Salaries. Median usual weekly earnings of full-time wage and salary workers, annual averages. White. Black. 1979-1987.
Newsweek Mar 7, 1988 p. 24

Ghetto woes. The underclass. Growth, in millions. 1970, 1980.
Newsweek Mar 7, 1988 p. 43

Black stock ownership, 1982-1987. Total corporate stock outstanding. Amount. Percent of total.
Black Enterprise Feb 1988 p. 73

BLACKS - EDUCATION

Degrees of separation. College enrollment among Blacks. 1970, 1989. Master's degrees. Professional degrees. Doctorate degrees.
Newsweek May 6, 1991 p. 27

Against the odds. College enrollment (1988). Black males. Black females. White males. White females.
Black Enterprise Feb 1991 p. 41

An American dilemma. Who goes to college. Black males. Black females. White males. White females. As of 1988.
Newsweek Oct 15, 1990 p. 67

The big ten. Number of black students entering medical school in 1989 from [ten large universities].
U.S. News May 28, 1990 p. 61

A mixed picture. High school graduation rate (for those 25-34 years old). Black. White. 1970-1988.
U.S. News Jan 22, 1990 p. 28

Education. Percent of 18-to 24-year-old high school graduates enrolled in college. Whites. Blacks. 1976, 1985.
Scholastic Update Apr 7, 1989 p. 21

The rising drop-in rate. Percentage of 25-to-29-year-olds who completed: Less than 4 years of high school. 4 or more years of college. White. Black. 1950-1987.
U.S. News Mar 20, 1989 p. 50

[Percent] of blacks who are college graduates. 1967, 1987.
Time Mar 13, 1989 p. 68

Largest individual cash gifts to selected United Negro College Fund schools. Contributor. Amount. Year.
Black Enterprise Feb 1989 p. 27

Doctorate degrees by race. Selected years. Black. White. 1975-1976. 1980-1981. 1984-1985.
Black Enterprise Sep 1988 p. 23

Federal school desegregation lawsuits, since 1963. States with ten or more districts that are under permanent injunction to remain desegregated.
Black Enterprise Sep 1988 p. 25

High school graduates as a percentage of total 18-24 year olds, selected years. Black. White. Total. 1976, 1981, 1985.
Black Enterprise Sep 1988 p. 43

Percentage of 18-24 year old high school graduates enrolled in college, selected years. Black. White. Total. 1976, 1981, 1985.
Black Enterprise Sep 1988 p. 43

Upward mobility: Blacks make progress but gaps persist. Household income. Home ownership. Net worth. College graduates. Managers.
Business Week Mar 14, 1988 p. 68

See Also BLACK COLLEGE STUDENTS

BLACKS - EMPLOYMENT

Little change for black males. Since 1975, despite Affirmative Action, black male employment has barely grown. Male. Female. 1975-1988.
Black Enterprise Oct 1991 p. 37

Breakdown of U.S. employees in executive, administrative and managerial positions, 1990. [Men, women, white, black].
Black Enterprise Aug 1991 p. 42

Blacks have made solid gains. Share of jobs at companies with 100 or more employees. All jobs. Officials/managers. Professionals. Technicians. Skilled craft. Clerical. 1966-1989.
Business Week Jul 8, 1991 p. 52

Whites are still faring much better. Working as: Managers/professionals. Administrators. Service personnel. Laborers. Whites. Blacks.
Business Week Jul 8, 1991 p. 53

Civilian labor force and employment/unemployment, by race, 1989-1991.
Black Enterprise Jan 1991 p. 48

Where blacks are in selected occupations. Occupation. Total Americans employed in field (in thousands). Percentage black. 1989.
Black Enterprise Jun 1990 p. 91

U.S. population, civilian labor force, employment, and unemployment; By race, 1987-1990.
Black Enterprise Jan 1990 p. 60

Selected white-collar occupations filled by blacks, 1940-1980. Managerial. Professional and technical. Number (percent).
Black Enterprise Nov 1989 p. 58

Formal training by sex and race of trainee. Share of training. Share of the work force. Males. Females. Whites. Blacks. Hispanics. [By] age.
Black Enterprise Aug 1989 p. 41

Percentage of women, blacks and Hispanics in technical occupations. All workers. All technical workers. Technicians and related support.
Black Enterprise Aug 1989 p. 41

Jobs. Careers of black workers. White collar. Farm. Blue collar. Household help. Other. 1965, 1985.
Schoiastic Update Apr 7, 1989 p. 21

Income. Living in poverty. Unemployment. Median income. Blacks. Whites. 1966, 1987.
Scholastic Update Apr 7, 1989 p. 21

Number of black managers and professionals in millions. 1967, 1987.
Time Mar 13, 1989 p. 67

The black employment chasm. 1987-percentage of jobs held by blacks age 16 and over. Percentage change in demand for jobs. 1986-2000. Where we are. Where we need to be [by occupation].
Black Enterprise Feb 1989 p. 68

Unemployment rates for black workers, age 16 to 19 years. Year. Annual average (in percent). 1985-1987.
Black Enterprise Jan 1989 p. 33

Metro areas with fewest black males working full time. Metro areas with most black males working full time.
Black Enterprise Jan 1989 p. 33

The times are not a-changin'. Median family income. White. Black. 1968-1986. Unemployment rate for those 16 years and older. Black. White. 1968-1987.
Black Enterprise May 1988 p. 35

BLACKS - HEALTH

Causes of death. Heart disease. Cancer. Stroke. Diabetes. Liver disease. Asthma. Tuberculosis. Whites. Blacks. 1988.
Time Sep 16, 1991 p. 51

Life expectancy. White females. Black females. White males. Black males. 1980-1990.
Time Sep 16, 1991 p. 51

The state of black health. Tracking the silent killers. Mortality statistics by race. Cardiovascular disease. Heart disease. Stroke. Black. White. Men. Women.
Black Enterprise Jul 1991 p. 43

Growing up against the odds [black infants health and mortality].
Newsweek Sep 11, 1989 p. 31

Hospitals in black communities. Hospitals serving black communities (with more than 500 beds). Number of hospitals serving black communities, selected years.
Black Enterprise Jul 1988 p. 41

BLACKS - INTEGRATION

[Percent] of black families living in the suburbs. 1967, 1987.
Time Mar 13, 1989 p. 61

Living separately and unequally. Of 60 major metropolitan areas in America, the 20 most segregated are.
U.S. News Aug 15, 1988 p. 20

BLACKS - POLITICAL ACTIVITY

Black vote for GOP presidental candidates. 1956-1988.
Black Enterprise May 1989 p. 35

Officeholders. Number of black elected officials. 1964-1988.
Scholastic Update Apr 7, 1989 p. 20

The gap is narrowing in voter turnout. Whites. Blacks. Percent reported voting. 1968-1984.
Business Week Jul 4, 1988 p. 97

Black voters play a bigger role. Percent of total vote. 1968-1984.
Business Week Jul 4, 1988 p. 97

Who votes for the Democrats? Percent of total black vote. Percent of total white vote. 1968-1984.
Business Week Jul 4, 1988 p. 97

Voter turnout by gender and race, presidential elections. Black women. Black men. 1972-1984.
Ms. Apr 1988 p. 77

Distribution of black elected officials by census region, January 1987.
Black Enterprise Mar 1988 p. 37

Black mayors of cities with populations over 100,000. Name. Term expires. City. Percentage of blacks.
Black Enterprise Mar 1988 p. 37

Annual rate of increase of black elected officials, 1977-1987.
Black Enterprise Mar 1988 p. 37

Percentage of black elected officials who are women [by region].
Black Enterprise Jan 1988 p. 20

BLACKS - RELIGION

Black Catholics in U.S. and black bishops. 1975, 1985, 1989.
Black Enterprise Oct 1989 p. 28

BLACKS IN BUSINESS

Ten largest major industry groups in receipts for black-owned firms: 1987. Major industry group. Firms (number). Receipts (in millions).
Black Enterprise Apr 1991 p. 45

Black-owned firms in ten largest U.S. cities: 1987. Firms (number). City. Receipts (in millions).
Black Enterprise Apr 1991 p. 45

Distribution of black-owned firms by state: 1987.
Black Enterprise Feb 1991 p. 158

Percent distribution of black-owned companies and all U.S. companies by major group: 1987.
Black Enterprise Jan 1991 p. 23

Types of franchised businesses owned by blacks, 1986.
Black Enterprise Sep 1989 p. 39

Population, money income and business sales. 1988-2000. Total. Black. Black business receipts. Percent of black income.
Black Enterprise Jun 1989 p. 156

The big three automakers: How they stack up. Number of black-owned dealerships, 1988. Amount of minority business, 1987. Ford. GM. Chrysler.
Black Enterprise Mar 1989 p. 18

That negative feeling. The annual growth of private capital in the minority venture business, 1980-1987.
Black Enterprise Jun 1988 p. 42

Sales of black-owned business by major industry division, 1982.
Black Enterprise Jun 1988 p. 95

Selected characteristics of black and nonminority businesses owners, 1983.
Black Enterprise Jun 1988 p. 95

Black business receipts: 1982-1986.
Black Enterprise Jun 1988 p. 264

Plugging into the airwaves. Minority ownership of radio and television stations, 1979-1986.
Black Enterprise May 1988 p. 36

Dollar value of federal contracts awarded to 8(a) [socially and economically disadvantaged business persons] firms, 1983-87. Number of firms. Contracts awarded.
Black Enterprise Mar 1988 p. 19

BLACKS IN NEWSPAPER PUBLISHING

Good news/bad news. Black representation in the departments of the nation's newspapers.
Black Enterprise Sep 1990 p. 20

BLIND

Blind and visually impaired people. Number of guide-dog schools. Percentage of U.S. population that wears prescription glasses. [Other selected statistics.]
U.S. News Nov 18, 1991 p. 22

BLOOD

Blood tests doctors often order. Complete blood count. Fats. Other common tests.
U.S. News Jun 18, 1990 p. 72

AIDS and blood. Pints of blood donated by surgery patients for self-transfusion. 1982-1987. Pints of blood donated through the American Red Cross. 1976-1988.
U.S. News Oct 24, 1988 p. 81

BLOOD BANKS

The rise of personal blood banking. Personal blood bank deposits. Thousand of units. 1980-1989.
Business Week Oct 22, 1990 p. 114

BODY FAT

Fatty risks: How to estimate health hazards. Body-mass index (BMI) is a standard method doctors use to estimate degree of body fat.
U.S. News May 14, 1990 p. 58

BODY IMAGE

Fanciful fat. About 24 percent of men and 27 percent of women are truly overweight. Here is the percentage who said they are [various weights]. Male. Female.
U.S. News Feb 19, 1990 p. 70

BONDS

Increasing rates. Yields on government bonds. France. U.S. Germany. June 1989 - Nov. 1990.
U.S. News Dec 17, 1990 p. 73

America needs more cash. Net issuance of treasury debt. 1987-1991. Foreigners are bailing out. Net foreign purchases of treasury debt. 1987-1990.
Business Week Oct 22, 1990 p. 30

The bankruptcy boom drags in more bonds. Bonds newly in default. Companies defaulting. 1982-1988.
Business Week Sep 5, 1988 p. 85

T-bond yields have exceeded their ceiling for decades. Legal ceiling on new issues. Average annual yields on 10-year treasury bonds. 1960-1987.
Business Week Sep 5, 1988 p. 89

BOOK PUBLISHERS AND PUBLISHING

Total books sold (billions). Percentage that were fiction last year. Average hardback price. Breakdown of fiction sold. [By type].
U.S. News Nov 5, 1990 p. 14

Campus readers. Robert Fulghum's volume of inspirational essays has topped college bestseller lists for 16 months. Here are the books that university students like best.
U.S. News May 7, 1990 p. 75

Voracious readers. Book sales rose 18 percent between 1988 and 1989. 1982-1989.
U.S. News Apr 16, 1990 p. 41

Lightweight literature. The best-selling paperback books of all times [by title].
U.S. News Mar 5, 1990 p. 72

Reader's block. The percentage of [college] students who could name the authors of the following literary works.
U.S. News Nov 13, 1989 p. 89

Mysterious bestsellers. Mysteries are the most popular. Based on estimates of the number of books sold weekly, here's how the other kinds of novels stack up.
U.S. News Aug 14, 1989 p. 70

Hot topics. Top eight subjects of books published in 1987. Category. Books published in U.S. Books imported. Average U.S. price.
U.S. News Aug 1, 1988 p. 62

Awash in an ever-deepening sea of print. New listings. 1949-1988.
Newsweek Jul 11, 1988 p. 66

Best sellers. Proportion of books [by category] bought by women and men.
Ms. Jun 1988 p. 68

BOSTON (MASSACHUSETTS)

How Massachusetts plans to clean up the Harbor. Diagram and map.
U.S. News Sep 24, 1990 pp. 58–59

BOTANY - ECOLOGY

What old-growth trees do for the ecosystem and for the economy. Diagram.
Time Jun 25, 1990 p. 62

BRAIN

A new window on the mind. PET scan shows the area of the brain that became active when a test taker tried to [perform various tasks].
Newsweek Nov 25, 1991 p. 67

How long-term memories are formed. Diagram.
Time Oct 28, 1991 p. 86

Brain circuits. The brain appears "programmed" to size up the emotional importance of certain stimuli much more quickly than scientists once thought. Diagram.
U.S. News Jun 24, 1991 pp. 58–59

How the brain really works its wonders. The things memories are made of. Diagram.
U.S. News Jun 27, 1988 pp. 48–49

BRAZIL - ECONOMIC CONDITIONS

Brazil's fatter trade surplus. Billions of dollars. 1984-1988.
Business Week Nov 14, 1988 p. 84

BREAST EXAMINATION

[Six-step] breast self-examination (BSE). Diagram.
FDA Consumer Jul 1991 p. 12

BUDGET - BUSH ADMINISTRATION

With the budget growing more slowly Bush is relying on heftier tax receipts to lighten the deficit's impact. Outlays. Receipts. Deficit. Deficit as percentage of Gross National Product. 1982-1991.
Business Week Mar 13, 1989 p. 101

BUDGET - UNITED STATES

Red-ink syndrome. The federal budget. 1976-1989
U.S. News Feb 29, 1988 p. 50

BULGARIA - ECONOMIC CONDITIONS

Adding up the damage. Bulgaria. Czechoslovakia. Hungary. Poland. Romania. Real economic growth. Hard-currency balance. Oil imports.
U.S. News Nov 5, 1990 pp. 55–56

BULIMIA NERVOSA

See EATING DISORDERS

BUSH, GEORGE

Bush's box score. Bush campaigned for 63 candidates since May. Here's how they fared.
Time Nov 19, 1990 p. 31

The President's environmental report card. *U.S. News* asked leaders of 10 conservation groups and 10 business and trade associations to grade the Bush administration's environmental efforts in 12 key areas. Issue. Grade.
U.S. News Mar 19, 1990 p. 22

BUSINESS

See CORPORATIONS

BUSINESS ETHICS

Living up to high standards. Countries with high business ethics. Most ethical region of U.S. The most ethical professions.
U.S. News Mar 14, 1988 p. 76

See Also CORPORATE SOCIAL RESPONSIBILITY

BUSINESS FAILURES

Unemployment is climbing. Civilian unemployment rate. Jan. 90 - Jan. 91. So are business failures. 1990-1991.
Business Week Mar 18, 1991 p. 31

Least likely to succeed. In 1988, the following percentages of firms 5 years old or younger were projected to fail. Industry. Percentage of firms projected to fail.
U.S. News Oct 23, 1989 p. 80

Failure factors. Common reasons companies went under in 1988. Cause of failure. Percentage of failures attributed primarily to this problem.
U.S. News Oct 23, 1989 p. 80

Starts and stops. The number of small businesses that opened and closed their doors in 1987, by U.S. region.
U.S. News Oct 23, 1989 p. 80

Business failures have been heading down. 1984-1988.
Business Week Jan 9, 1989 p. 26

See Also BANKRUPTCY

BUSINESS SCHOOLS

Where the money is. Estimated 1991 starting salaries for graduates of top business schools. Consulting. Investment banking. Marketing.
Business Week May 6, 1991 p. 82

Top 25 business schools. 25 graduate schools of business with the highest scores in the *U.S. News* survey.
U.S. News Apr 29, 1991 p. 68

[Top 20 business schools] BW rank. School. BW 1988 rank. Corporate poll rank. Graduates' poll rank. Full-time enrollment. Annual tuition. Applicants accepted.
Business Week Oct 29, 1990 p. 54

How the schools stack up: *Business Week* rates the top 20. School. Corporate poll rank. Graduates' poll rank. Highlights. 1988 graduates. Annual tuition. Applicants accepted. Average starting pay.
Business Week Nov 28, 1988 p. 78

Women in business school. 1988-1990.
Business Week Mar 14, 1988 p. 39

A scorecard on the B-schools. School. Special characteristics. Student/faculty ratio. Annual tuition. Percentage of applicants accepted. 1987 grads. Avg. starting salary. Ratings.
Business Week Jan 11, 1988 pp. 164–165

BUSINESSES, SMALL

See SMALL BUSINESSES

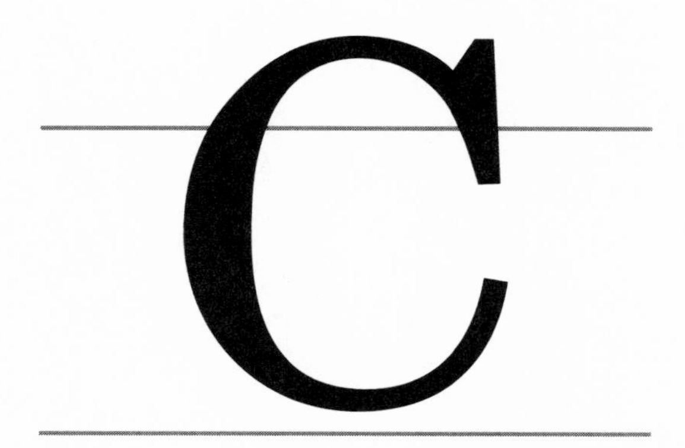

CAFFEINE

Unexpected pick-me-up. Some prescription and over-the-counter medications contain caffeine. Here is the caffeine in [various medications].
U.S. News Sep 11, 1989 p. 87

The caffeine kick. Average milligrams of caffeine per serving. Coffee. Tea. Soft drinks.
U.S. News May 23, 1988 p. 76

Caffeine content of beverages and foods.
FDA Consumer Jan 1988 p. 24

Caffeine content of soft drinks. Regular. Diet drinks.
FDA Consumer Jan 1988 p. 25

Caffeine content of drugs. Prescription drugs. Nonprescription drugs. Alertness tablets. Pain relief. Diuretics. Cold/allergy remedies.
FDA Consumer Jan 1988 p. 26

CALIFORNIA

Big spenders. Per capita spending, 1988. Education. Welfare. Highways. Health. Capital improvements. New York. California.
U.S. News Dec 23, 1991 p. 53

The golden state. [Characteristics of the population]. If California were a country [comparison of the state economy to countries in the world]. Map.
Time Nov 18, 1991 pp. 50–51

One year's new arrivals [in California]. Where they came from in 1989.
Time Nov 18, 1991 p. 68

Rain, rain, come again [California drought]. Little rainfall. Reservoir levels down. Snowpack down. Map.
Newsweek Feb 18, 1991 p. 56

The long drought is draining California's reservoirs. Reservoir storage. Precipitation. 1986-1991.
Business Week Feb 11, 1991 p. 30

The good life [in California].
Newsweek Jul 31, 1989 p. 24

The states of California. A guide to California's state of mind. The coast. The emerging California. The Central Valley. The state of Los Angeles. The south counties. Map.
Newsweek Jul 31, 1989 p. 25

The tough life [in California].
Newsweek Jul 31, 1989 p. 27

CALIFORNIA - ECONOMIC CONDITIONS

The nation of California. Migration to California. Economic growth. Jobs. State tax net increase. 1989 economic output. Office vacancy [by selected cities]. Defense industry employment. Homebuilding.
Business Week Dec 30, 1991 pp. 32–33

Bicoastal pain. This recession has hit New York and California harder than the rest of the country. Civilian unemployment rate. 1989, 1990, 1991.
U.S. News Dec 23, 1991 p. 52

Bad karma. 1987-1991.
Newsweek Oct 28, 1991 p. 50

Economic outlook. Will California hold back the recovery? Civilian unemployment rate. [Employment]. Percentage of California labor force in defense-related jobs. Median sales price, existing homes. California. U.S.
U.S. News Apr 8, 1991 p. 56

CALORIES

Summer coolers. Despite their natural sources, fruit juices and drinks are surprisingly high in sugar. Here's how various 8 oz. servings stack up.
U.S. News Jul 10, 1989 p. 62

Paying the piper. Time needed to burn the calories consumed in [various foods and by various exercises].
U.S. News Jul 18, 1988 p. 54

CAMBODIA

Bare facts of economic desperation. Vietnam. Cambodia. Land area. Population. Life expectancy at birth. Per capita GNP. Value of exports. Value of imports.
U.S. News Aug 1, 1988 p. 34

CAMPAIGN FINANCING

Cash and carry. Senate campaign costs. House campaign costs. [Number of] political action committees. Biggest PAC givers. 1974-1988.
U.S. News Oct 22, 1990 p. 31

The high cost of winning [a seat in the] House. Senate. 1976-1988.
Scholastic Update Oct 19, 1990 p. 11

Total spending for congressional races in millions of dollars. House. Senate. 1972-1989.
Time May 7, 1990 p. 32

Money to burn. Even when running unopposed, congressmen raise big campaign chests. [Here] is this year's Top 10 list. Name, state. Amount raised.
Newsweek Nov 14, 1988 p. 23

The cost of winning a congressional seat has almost tripled since 1978. The Senate. The House. 1978-1986.
Time Oct 31, 1988 p. 45

Campaign costs. Total spending for congressional races. House of Representatives. Senate. 1972-1986.
Time Mar 7, 1988 p. 23

[Campaign spending]. Republicans. Democrats. Contributions in 1987. Cash on hand as of 1-1-88.
Time Feb 15, 1988 p. 19

Campaign costs. Primary spending in 1984. Democrats. Republicans.
Scholastic Update Jan 29, 1988 p. 2

CANADA

One country, two cultures. Canada. Quebec. Language. GDP per capita. Area. Population.
U.S. News Jun 18, 1990 p. 30

CANADA - ECONOMIC CONDITIONS

A tale of three nations. Comparative economic and social indicators. Population. GNP. Unemployment rate. Average hourly earnings. U.S. Mexico. Canada.
Black Enterprise Oct 1991 p. 27

Why Canada is costlier. Unit labor costs in manufacturing. Canada. U.S. 1985-1990.
Business Week Apr 8, 1991 p. 47

Canada's economy has slipped into recession. Gross domestic product. Percent change at annual rate. 1987-1990.
Business Week Oct 15, 1990 p. 20

CANADA AND THE UNITED STATES

The shape of North America Inc. The new North America. Population. GNP. Total trade. Canada. U.S. Mexico.
Business Week Nov 12, 1990 p. 103

The free-trade agreement. [U.S. and Canada]. Merchandise exports. Direct investment.
Time Dec 5, 1988 p. 40

CANCER

The world's most common forms of cancer.
Time Oct 28, 1991 p. 88

A method to their madness. [How] mutations in cancer cells enable them to metastasize or spread. Diagram.
U.S. News Sep 24, 1990 p. 71

Landmark experiments in the genetics of cancer.
U.S. News Dec 11, 1989 p. 58

Fitness pays. Deaths per 10,000 people in a year. All causes. Cardiovascular. Cancer. Men. Women.
Time Nov 13, 1989 p. 13

How light-activated drugs kill cancer. Diagram.
Business Week May 29, 1989 p. 106

CANCER - CAUSES

Some of the surprising sources of cancer risk. HERP scale measures carcinogen hazards from natural and man-made chemicals. Exposure. Carcinogen. Risk.
Business Week Oct 15, 1990 pp. 58–59

Proportions of cancer deaths attributed to various different factors.
FDA Consumer Jun 1990 p. 8

Smoking, drinking and oral cancer. The risk of getting oral cancer for smokers and drinkers. Nondrinkers. Nonsmokers. Moderate smokers and drinkers. Heavy smokers and drinkers.
FDA Consumer Feb 1989 p. 40

Minimizing the risks from nature's bounty. Food. Predicted cancers per 1 million people. What to do.
Newsweek Jan 30, 1989 p. 75

Where it [radon] gets in. When to take steps. Exposure. Lung-cancer deaths per 1,000 people exposed. Comparable exposure level. Comparable risk. Action recommended. How to get it out. Diagram.
U.S. News Sep 26, 1988 pp. 62–63

The big 7. Cancer risks you can control. Risk factors. Cancers. 1987 deaths. Survival rates. How to reduce risk.
FDA Consumer Apr 1988 pp. 27–29

CANCER - NUTRITIONAL ASPECTS

Danger in the diet. Fat intake. Breast-cancer rate. [By country].
Time Jan 14, 1991 p. 52

CANCER - RESEARCH

Dollars for diseases. Disease [type of cancer]. Deaths per year. Annual U.S. govt. research funding.
Newsweek Dec 10, 1990 p. 65

Closing the gap. AIDS funding has soared. Cancer grants haven't kept pace. N.C.I. spending on cancer research. All federal spending on AIDS. 1984-1989.
Newsweek Mar 6, 1989 p. 47

CANCER - SURVIVAL RATES

More cancer survivors. People alive five years after cancer diagnosis. 1955-1990.
U.S. News Nov 6, 1989 p. 97

Five-year cancer survival rates for whites. Five year cancer survival rates for blacks. [By type of cancer].
Black Enterprise Apr 1989 p. 43

Ups and downs in survival rates. Percentage of cancer victims [by type of cancer] who survived five years after diagnosis of their disease in 1979-84 compared with five years earlier. 1974-79. 1979-84.
U.S. News Feb 15, 1988 p. 12

CANCER (BREAST)

Can sunshine save your life? Deaths from breast cancer [by U.S. cities]. Amount of solar radiation received.
Newsweek Dec 30, 1991 p. 56

[Six-step] breast self-examination (BSE). Diagram.
FDA Consumer Jul 1991 p. 12

Incidence of breast cancer per 100,000 U.S. women. 1977-1987.
Time Jan 14, 1991 p. 49

How the risk shapes up. Apple shaped. Pear shaped. Where cancer is most likely to be found.
Time Jan 14, 1991 p. 51

Danger in the diet. Fat intake. Breast-cancer rate. [By country].
Time Jan 14, 1991 p. 52

The outlook. Incidence vs. mortality. Incidence. Mortality. 1973-1987.
Newsweek Dec 10, 1990 p. 66

Prostate vs. breast cancer. Cases. Deaths. Federal funding. 1989.
U.S. News Jul 10, 1989 p. 57

Cancer watch. Breast cancers per 100,000 women. Incidence. Deaths. 1973-1987.
U.S. News Jul 11, 1988 p. 53

Flirting with trouble. Share of all women surveyed who have had a mammogram. Reasons cited by women age 45 and older for not having had a mammogram during the last two years.
U.S. News Jul 11, 1988 p. 54

Catching it early. The five stages of breast cancer. Five-year-survival rate.
U.S. News Jul 11, 1988 p. 57

CANCER (CERVICAL)

U.S. cervical cancer mortality rate. 1950-1985.
FDA Consumer Sep 1989 p. 22

CANCER (COLON)

The steps to colon cancer. Mutations in specific types of genes are the underlying factor that causes normal cells to become cancerous. Chart.
Time Nov 12, 1990 p. 97

CANCER (LUNG)

The hunt for killer DNA. Smokers with lung cancer are more likely to have a deadly enzyme in their blood. Diagram.
Newsweek Oct 21, 1991 pp. 56–57

A smoking gun for smokers. Four of 5 people who die of lung disease each year are smokers. The number of deaths from lung disease by state. Rate of deaths per 100,000. Deaths.
U.S. News Nov 13, 1989 p. 89

From killer to lifesaver. Doubling the odds of survival. Percentage of lung-cancer patients still alive—with and without vaccine. Diagram.
U.S. News Apr 4, 1988 p. 61

CANCER (PROSTATE)

Unmasking a stealthy cancer. Cases of prostate cancer per 100,000 U.S. men per year. 1973-1987.
Time May 6, 1991 p. 45

The man-killing cancer. Prostate-cancer cases and deaths. 1973-1988.
U.S. News Jul 10, 1989 p. 55

Two prostate solutions. The surgery option. The radiation option.
U.S. News Jul 10, 1989 pp. 56–57

Prostate vs. breast cancer. Cases. Deaths. Federal funding. 1989.
U.S. News Jul 10, 1989 p. 57

CANCER (SKIN)

The plan of attack. In one of the experiments that received approval last week, researchers will use genetically engineered blood cells to fight skin cancer. Diagram.
Time Aug 13, 1990 p. 61

Assault on the skin. Ultraviolet A. Ultraviolet B. Melanoma. Basal-cell carcinoma. Squamous-cell carcinoma.
Time Jul 23, 1990 p. 69

CANCER (TESTICULAR)

Hopeful trend in testicular cancer survival. Five-year survival rates, by race, for cases of testicular cancer diagnosed during the following years. [Various years 1960-1984].
FDA Consumer Jan 1989 p. 18

CAPITAL GAINS

Capital-gains history [changes in capital gains tax]. 1942-1988.
U.S. News Feb 13, 1989 p. 83

CAPITAL PUNISHMENT

Dead end. Countries that have abolished the death penalty. Countries that retain the death penalty.
Time Mar 4, 1991 p. 52

Who's on death row. Awaiting execution. Number executed. Racial breakdown. 1980-1989. Executed since 1976; on death row as of July, '89 [by selected states].
Time Apr 2, 1990 p. 19

Slow go on the row. State. Inmates on death row. Executions since 1976.
Newsweek Feb 19, 1990 p. 73

A decade of death—and interminable life. Inmates on death row. Executions. 1980-1989.
Newsweek Jan 8, 1990 p. 60

Updating the death penalty. Following are states that have executed the most people since the penalty was reinstated in 1976, and their death-row populations.
Newsweek Nov 6, 1989 p. 8

Death toll. Executions. 1935-1988.
U.S. News Jan 30, 1989 p. 81

Death sentences. Prisoners on death row. 1955-1988.
U.S. News Jan 30, 1989 p. 81

CARBON

The Earth's thermostat. Carbon. A complex cycle regulates its distribution around the planet. Diagram.
U.S. News Oct 31, 1988 p. 58

The global carbon budget. Carbon added to atmosphere (in metric tons per year). Carbon removed from atmosphere (in metric tons per year).
U.S. News Oct 31, 1988 p. 58

CAREER PLATEAUS

Measuring the glass ceiling. Why white males monopolize the top of the corporate ladder. Men. Women. Minorities. Employees. Managers. Top execs.
U.S. News Aug 19, 1991 p. 14

CARIBBEAN REGION

Caribbean tourism. How much they spent. Caribbean tourist arrivals. Tourist arrivals by origin. U.S. tourists main destinations. 1986.
Black Enterprise May 1988 p. 61

The rewards of rest and relaxation. [Caribbean country]. Travel receipts. Other exports. Total. Travel receipts as % of total.
Black Enterprise May 1988 p. 66

Cruising toward increased revenue. Estimates of visitor expenditure: 1982-1986 [by Caribbean country].
Black Enterprise May 1988 p. 68

Conflict and change in the Caribbean. El Salvador. Honduras. Nicaragua. Cuba. Haiti. Panama. Map.
Scholastic Update Mar 11, 1988 pp. 6–7

CARPAL TUNNEL SYNDROME

Repetitive motion injuries. 1983-1987. Diagram.
Business Week Jan 30, 1989 p. 92

CATHOLIC CHURCH

Catholics give less to the church. Share of income contributed by U.S. Catholics. Protestants. 1960-1989.
Business Week Jun 10, 1991 p. 60

Where the Jesuits are [by geographic area]. Map.
Time Dec 3, 1990 p. 89

Black Catholics in U.S. and black bishops. 1975, 1985, 1989.
Black Enterprise Oct 1989 p. 28

CELLULAR PHONES

Cellular's surging revenue growth hasn't produced lower prices. Total California retail cellular service revenues. Cellular service rates. 1984-1989.
Business Week Mar 26, 1990 p. 36

Ringing up sales. Number of subscribers. Costly infrastructure. Total capital investment. Top operators. Leading cellular phone companies. Subscribers.
U.S. News Jul 24, 1989 p. 41

CENSORSHIP

Book bashing. Most frequently challenged books in public schools, 1982-88. Number of times challenged.
U.S. News Dec 12, 1988 p. 96

CENSUS

Just a traditional census. A half century of census undercounts. 1940-1990.
U.S. News Jul 29, 1991 p. 10

Americans the Census Bureau estimates it missed in 1990. Percentage of Americans who never returned their 1990 census form. Percentage of blacks not counted in 1990. [Other selected statistics.]
U.S. News Jul 15, 1991 p. 12

[Two] centuries of the census. Our population has grown. Population (in millions). 1790-1980.
Scholastic Update Jan 12, 1990 p. 6

[Two] centuries of the census. [Number of] places of 50,000 or more. 1790-1980.
Scholastic Update Jan 12, 1990 p. 6

[Two] centuries of census. We've become older. American's average age. 1790-1980.
Scholastic Update Jan 12, 1990 p. 6

[Two] centuries of census. We've moved west. U.S. center of population. 1790-1980. Map.
Scholastic Update Jan 12, 1990 p. 6

CENTRAL AMERICA

How the Sandinista defeat affects Nicaragua's neighbors. Guatemala. El Salvador. Cuba. Honduras. Costa Rica. Panama. Map.
Time Mar 12, 1990 p. 14

Conflict and change in the Caribbean. El Salvador. Honduras. Nicaragua. Cuba. Haiti. Panama. Map.
Scholastic Update Mar 11, 1988 pp. 6–7

CENTRAL AMERICA - ECONOMIC CONDITIONS

The legacy of poverty. A few reap all the riches. Income distribution by country. Poorest 20% of population. Richest 20% of population.
Scholastic Update Feb 9, 1990 p. 5

Economies plagued by debt. National income vs. foreign debt (per capita) [by country].
Scholastic Update Feb 9, 1990 p. 5

CENTRAL AMERICA - SOCIAL CONDITIONS

A poor standard of living. Literacy rate and average life expectancy [by country].
Scholastic Update Feb 9, 1990 p. 5

CENTRAL AMERICA AND THE UNITED STATES

The Yankee dollar. Money is a main ingredient of American policy in the region. U.S. aid 1980-88 [in Central America].
Newsweek Apr 4, 1988 p. 19

CHARITY

Charitable giving per capita. Personal income Americans gave to charity last year. Estimated value of time volunteers gave in 1989. [Other selected statistics.]
U.S. News Dec 9, 1991 p. 16

The rich are giving less and less to charity. Average charitable contributions. Taxpayers with income over $1 million. Taxpayers with income from $500,000 to $1 million. 1980-1988.
Business Week Nov 5, 1990 p. 29

Untithed. Religious denomination. % of income members give to charity.
Newsweek Jan 22, 1990 p. 8

CHEMICAL AND BIOLOGICAL WEAPONS

An anti-gas uniform. Despite the intense heat, U.S. soldiers may have to wear added layers of heavy gear for protection against Iraq's chemical weapons. Diagram.
Newsweek Aug 27, 1990 p. 25

Chemical warfare: Name your poison. Type. Physical effect. Used to. Time to take effect.
Newsweek Sep 18, 1989 p. 29

Biological weapons: The future of war? If they ever make it out of the laboratory, biological weapons could be the next weapons of choice for some armies, or even for terrorists.
Newsweek Jan 16, 1989 p. 23

The tools of chemical war. Poison gas comes in many forms.
Newsweek Jan 16, 1989 p. 23

Bad chemical reactions. How they are made. Who has them. What they can do.
U.S. News Jan 16, 1989 pp. 30–31

Department of Defense support for chemical and biological warfare research and development.
Bull Atomic Sci Jan 1989 p. 53

Fatal fumes. Countries known to have chemical weapons. Countries reported to have chemical weapons.
Newsweek Sep 19, 1988 p. 30

Catalog of death. Chemical weapon. Blood agents. Choking agents. Blistering agents. Nerve agents. Effects.
Time Aug 22, 1988 p. 48

CHEMICAL INDUSTRIES

Chemical exports rise again. The value of products sold abroad by U.S. chemical processing companies. 1983-1988.
Business Week Jan 11, 1988 p. 78

CHICAGO (ILLINOIS)

A grim second city. Chicago built housing projects to provide safe, low-cost family shelter. Living comparisons. Violent crimes per 1,000 residents.
Newsweek Jan 2, 1989 p. 25

CHILD ABUSE

The grim tally. Cases of sexual child abuse. 1976-1985.
U.S. News Jun 13, 1988 p. 22

CHILD CARE

What a child care savings plan saves. Net saving above tax credit. One child. Two children. Adjusted gross income (thousands of dollars).
Business Week Nov 19, 1990 p. 177

The high costs of child care.
Newsweek Aug 7, 1989 p. 26

Who's minding the kids. Primary type of care for preschool kids with working mothers.
Business Week Jul 10, 1989 p. 68

How young children are cared for [by provider]. Paying for child care. Amounts paid by parents per year. Parent satisfaction with today's system.
Black Enterprise Jul 1989 p. 35

Taking care of baby. The weekly cost of day-care centers operating for profit and the percentage of family income that represents in these metropolitan areas. Highest-cost areas. Lowest-cost areas.
U.S. News Jun 5, 1989 p. 69

As the pool of young workers shrinks, women will fill the gap, and more working mothers will increase the demand for child care. Population 16-24 [years]. 1979-2000. Female share of the work force. 1950-2000.
Business Week Sep 19, 1988 pp. 112–113

On-the-job care. [Numbers of] employers who provide child-care options. 1978-1988.
U.S. News Sep 19, 1988 p. 45

Who's minding the kids? Care of preschool children in families with mothers in the workforce. Relative. Sitter. Family daycare. Daycare center. 1965-1985.
Ms. Aug 1988 p. 86

How fathers feel about child care. Male Du Pont employes who want [leaves and benefits]. 1985, 1988.
U.S. News Jun 20, 1988 p. 68

Bringing up baby. The cost of child care in the United States varies widely. Family day care. Day-care center. Full-time babysitter. [Selected cities].
Newsweek Feb 15, 1988 p. 57

CHILD CARE - EUROPE

Europe pays to tend its children. Percentage of children in publicly funded child-care facilities. Percentage of mothers with children (age 4 and younger) who are in labor force [by country].
U.S. News Aug 22, 1988 p. 36

CHILD REARING

A bringing up baby budget. The yearly cost of raising a child varies by the child's age. As a rough national average, here is the annual expense for various ages.
U.S. News Feb 19, 1990 p. 70

Bringing up baby. Here's how much the women surveyed report their husbands help with the baby.
U.S. News Oct 2, 1989 p. 70

Bringing up baby. Average time parents spend in activities devoted solely to child care. Age of children. Employment status. Household income. Men. Women. Hours per week.
Ms. Sep 1989 p. 86

Bringing up baby. National estimates for the cost of raising one child to age 18 in 1987 dollars for a two-parent, two-child urban household with earnings between $25,000 and $50,000 per year. 1960-1987.
Black Enterprise Nov 1988 p. 35

How fathers feel about child care. Male Du Pont employes who want [leaves and benefits]. 1985, 1988.
U.S. News Jun 20, 1988 p. 68

Who does the work? Time spent on housework and childcare for married couples with children. 1969-1983.
Ms. Feb 1988 p. 19

CHILDBIRTH

Caesarean peak? Caesareans per 100 deliveries. 1965-1989.
U.S. News Jul 29, 1991 p. 63

CHILDREN

Death. Deaths of children under age 5 (per 1,000 live births, 1988) [by country].
Scholastic Update Jan 25, 1991 pp. 4–5

Children's crusade. Causes of death.
Newsweek Oct 8, 1990 p. 48

Children's crusade. Education. Illiteracy. Developed countries. Nondeveloped countries. Global total.
Newsweek Oct 8, 1990 p. 48

Children's crusade. AIDS. Distribution of HIV-infected females. [By region.]
Newsweek Oct 8, 1990 p. 48

Dumpling decade. Body mass index ages 10 and 11. Boys. Girls. 1980-1988
Newsweek Aug 27, 1990 p. 63

How social investment in children pays off. $1 invested in [a social program] saves [remedial care].
Business Week Sep 19, 1988 p. 123

The dark and other fears. Percentage of children afraid of [various fears].
U.S. News Jun 20, 1988 p. 78

Favorites of the lunch bunch. Share of children age 6-12 citing these foods as their favorites for lunch.
U.S. News May 2, 1988 p. 77

Child labor. Percentage of children age 6 to 11 who clean and vacuum. Fix own breakfast. Fix own lunch. Shop for groceries. Cook own dinner.
U.S. News Mar 21, 1988 p. 73

Average weekly income, expenditures and savings of children.
U.S. News Mar 21, 1988 p. 73

CHILDREN - DISEASES

Disease. Child deaths attributable to [various] diseases (1988).
Scholastic Update Jan 25, 1991 pp. 4–5

Immunization requirements. Grades K-12. [By state]. Diphtheria. Tetanus. Pertussis. Measles. Mumps. Rubella. Polio.
FDA Consumer Sep 1990 p. 24

Childhood disease comes creeping back. Number of cases. Mumps. Measles. Whooping cough. 1981-1986.
Business Week Jan 25, 1988 p. 67

CHILDREN - FITNESS

Fitness test. These short tests for strength, aerobic fitness and flexibility can give parents a general idea of a child's fitness level.
U.S. News May 20, 1991 pp. 90–91

CHILDREN - HEALTH

Hunger. Percentage of infants with low birth weight (1982-88) [by country].
Scholastic Update Jan 25, 1991 pp. 4–5

Prevention. The cost of helping a child. Many life-saving measures cost nothing more than nickels and dimes.
Scholastic Update Jan 25, 1991 pp. 4–5

Children's crusade. Malnutrition of children under age 5 [by region].
Newsweek Oct 8, 1990 p. 48

Recommended immunization schedule. Immunization schedule recommended for infants and children. Recommended age. Vaccines.
FDA Consumer Sep 1990 p. 22

CHILDREN - RECREATION

Kids at play. Percentage of children who after school [do certain activities].
U.S. News May 9, 1988 p. 84

CHILDREN, GIFTED

Children in the United States who are gifted. Average SAT score of college-bound high-school seniors. Who dropped out of high school. [Other selected statistics.]
U.S. News Nov 11, 1991 p. 14

CHILDREN OF DIVORCED PARENTS

Children of divorce. Children under 18. 1950-1985.
Business Week Sep 25, 1989 p. 102

CHILDREN'S LITERATURE

Flooding the market. Number of children's book titles published each year. 1980-1986.
U.S. News Aug 1, 1988 p. 50

Oldies but goodies. Top-selling old favorites. For babies and toddlers. Picture books. For younger readers. For middle readers. For young adults.
U.S. News Aug 1, 1988 p. 51

CHILE - ECONOMIC CONDITIONS

Chile breaks from the pack. Percentage change in—Chile, Latin America. Exports (1980-1989). Inflation (1988, 1989). Foreign debt outstanding (1986-1989). Per capita gross domestic production (1981-1989).
U.S. News Mar 19, 1990 p. 44

CHINA

Technologies of freedom. Letters delivered. Telephone subscribers. Long-distance calls. Long-distance lines. Television sets. Facsimile machines.
Newsweek Jun 19, 1989 p. 29

Two giants. Soviet Union and the People's Republic of China [vital statistics]. Map.
Scholastic Update May 5, 1989 pp. 4–5

CHINA - ECONOMIC CONDITIONS

Asia ascending. Japan. U.S.A. China. Per capita GNP. 1991, 2041.
Newsweek Dec 9, 1991 p. 6

Rise in China's gross national product. Annual inflation rate. Average annual income. Persons per telephone, per automobile.
U.S. News Dec 10, 1990 p. 12

Slower growth, higher prices. Political unrest has hurt the Chinese economy. Real growth in GDP. Consumer price index (percent change). 1981-1990.
U.S. News Mar 12, 1990 p. 44

Plowing for profit. Gross value of Chinese agricultural output. Per capita annual income for rural households, adjusted for inflation (in yuan). 1978-1988.
U.S. News Jul 10, 1989 p. 35

China's 10-year economic explosion. Fueling growth but boosting prices. GNP. Inflation. Percent change from prior year. 1979-1988.
Business Week Jun 19, 1989 p. 33

End of an era? Total trade with China. 1980-1988.
Newsweek Jun 19, 1989 p. 26

Where the two Chinas are drifting apart. Declining China. Emerging China. Share of total industrial output. Annual rate of industrial growth. Prices and wages. Management. Trade. Financing.
Business Week Jun 5, 1989 p. 38

The long march ahead. U.S. China. People per TV set in use. Telephone. Hospital bed. Physician.
U.S. News Jun 5, 1989 p. 22

Annual inflation rate [China]. 1977-1989.
U.S. News Jun 5, 1989 p. 23

Value of Chinese foreign-exchange holdings. 1977-1988.
U.S. News Jun 5, 1989 p. 23

Value of Chinese foreign-exchange holdings. 1977-1988.
U.S. News Jun 5, 1989 p. 23

Percentage of the labor force employed in agriculture, industry, services. China. U.S.
U.S. News Jun 5, 1989 p. 23

China overheats. Economic growth. Factory output. Retail prices. 1987, 1988.
Business Week Apr 3, 1989 p. 54

China's robust economy. Increase in GNP. Industrial output. Per capita income. Retail price index. 1986-1988.
Business Week May 9, 1988 p. 61

China's hunger for consumer products. Annual growth in per capita consumption. Total goods. 1981-1987.
Business Week Jan 25, 1988 p. 69

CHINA - FOREIGN INVESTMENTS

Is the party over? Total foreign investments in China. U.S. investments in China. 1979-1988.
U.S. News Jul 3, 1989 p. 31

CHINA - MILITARY STRENGTH

Where the troops are. The People's Liberation Army [China]. Missile divisions. Armored divisions. Infantry divisions. Airborne brigades. Major air bases. Naval bases. Map.
U.S. News Jun 19, 1989 p. 22

CHINA - POLITICS AND GOVERNMENT

Forced labor. Prison camps in China. Number of camps in each region. Map.
Newsweek Sep 23, 1991 p. 28

CHINA AND THE SOVIET UNION

Trading the China card. Total Soviet exports. Total Soviet imports. Soviet exports to China. Soviet imports from China. 1980-1990.
U.S. News Aug 19, 1991 p. 30

U.S.S.R. China. [Locations of armed forces]. Strategic missile sites. Military headquarters. Map.
Time Jul 18, 1988 p. 32

CHINA AND THE UNITED STATES

What China ships to the U.S. Toys and games. Clothing. Footwear. Petroleum. Export value 1990.
Business Week Jul 29, 1991 p. 38

China widens the gap. Trade surplus with the U.S. 1985-1990.
Business Week Apr 22, 1991 p. 46

Some U.S.-China joint ventures. Company. Product. U.S. investment. Millions of dollars.
Business Week Jun 5, 1989 p. 51

CHLORINATION OF WATER

The chlorination quandary. Solutions. Diagram.
U.S. News Jul 29, 1991 p. 50

CHLOROFLUOROCARBONS

Cool cars for the future. How CFC's cool. Automobile air conditioners work on the same principle as home refrigerators, by circulating a fluid, now CFC-12. Here's how it works. Diagram.
U.S. News Nov 6, 1989 p. 82

CHOCOLATE

Nation of chocoholics. Annual per-person consumption of chocolate candy [by country].
U.S. News Feb 27, 1989 p. 75

CHOLERA

Number of cholera cases in the continental United States and Latin America. Map.
FDA Consumer Sep 1991 p. 15

The epidemic. Epicenter, widespread infection. Fewer cases. At risk. Map.
Newsweek May 6, 1991 p. 44

Death in the time of cholera. Countries with cholera, date of first case and number of reported cases. Spread of cholera. Map.
Time May 6, 1991 p. 58

CHOLESTEROL

Tale of the tube. Cholesterol guidelines. Children. Adults.
Time Apr 22, 1991 p. 76

Cholesterol count. Food. Saturated fat. Cholesterol.
U.S. News Apr 24, 1989 p. 74

The arteries' heros and villains. Diagram.
Time Dec 12, 1988 pp. 62–63

Healthy artery. Diseased artery. Diagram.
Time Dec 12, 1988 pp. 64–65

Facts on fat. 3.5 oz. of [various foods]. Saturated fat. Cholesterol. Calories.
Time Dec 12, 1988 p. 68

Battling fat and cholesterol. Be careful and enjoy. Grams fat. Milligrams cholesterol. [Selected foods].
Newsweek Feb 8, 1988 p. 58

CHOLESTEROL TESTING

Sloppy screening. A U.S. study of public cholesterol testing found widespread unacceptable procedures. Problem. % of people citing problem.
Time Mar 12, 1990 p. 79

CHRISTMAS - ECONOMIC ASPECTS

Letters sent to Santa at North Pole, Alaska, last year. Toy guns as a percentage of the U.S. toy market. Amount toy companies spent per child on advertising last year. [Other selected statistics.]
U.S. News Dec 2, 1991 p. 8

Clearly Christmas. Percentage of Christmas-tree light sets sold that are multicolored [and] clear. 1975-1988.
U.S. News Dec 12, 1988 p. 96

CHRISTMAS CARDS

Average number of Christmas cards received per household last year. Busiest mail day of the year. Greeting cards sent by President Bush in 1990. [Other selected statistics.]
U.S. News Dec 23, 1991 p. 12

CIGARETTE INDUSTRY

See TOBACCO INDUSTRY

CIGARETTE SMOKING

See SMOKING

CITIES AND TOWNS

Changing cities. Minorities are majorities in 15 of the 28 U.S. cities with a population of 400,000 or more. The newcomers to the list.
U.S. News Oct 28, 1991 p. 90

New York. Population. Unemployment rate. Poverty rate. Bond rating. Outstanding debt per capita. Budget surplus or deficit.
U.S. News Oct 29, 1990 p. 81

Philadelphia. Population. Unemployment rate. Poverty rate. Bond rating. Outstanding debt per capita. Budget surplus or deficit.
U.S. News Oct 29, 1990 p. 81

The newest boom towns. Even in the midst of a real-estate recession housing values in a number of cities are soaring. The five hottest metropolitan areas in the last year.
Newsweek Oct 1, 1990 p. 49

Senior savings. Looking for a retirement home at a good price? City. Average condo price.
Newsweek Mar 5, 1990 p. 8

The lure of city lights. Here's how today's 50 biggest cities will change in population by the year 2000.
U.S. News Feb 19, 1990 p. 70

[Two] centuries of the census. [Number of] places of 50,000 or more. 1790-1980.
Scholastic Update Jan 12, 1990 p. 6

Where the stars are rising. Technology, population shifts, foreign competition and a booming service sector have created a string of new economic hubs that one day may rival the country's traditional centers of commerce. [By U.S. city].
U.S. News Nov 13, 1989 pp. 54–55

Suburbs outearn cities. Suburbs. Central cities. Real family income. 1959-1989.
Business Week Sep 25, 1989 p. 95

City crime: A matter of perception. Perceived safety. Violent crimes per 100,000 population [by city].
Newsweek Aug 14, 1989 p. 6

Cities of students. These cities have the most four-year colleges and universities within their boundaries.
U.S. News Aug 14, 1989 p. 70

'I need more money!' Here's the salary needed to match the lifestyle of a family of four with a $55,000 income living in an average suburban town.
Newsweek Jul 3, 1989 p. 40

Plenty of everything, plenty of nothing. The richest. The poorest. Suburb. Per capita income. Metropolitan area.
Newsweek Jun 26, 1989 p. 23

Coming and going. Most popular cities to move to. Cities people leave most often.
U.S. News Apr 3, 1989 p. 76

The top 10: Having it all. Population. Median family income. Median home. Unemployment.
Newsweek Feb 6, 1989 pp. 42–44

St. Paul. Population. Median family income. Median home. Unemployment.
Newsweek Feb 6, 1989 p. 43

Birmingham. Population. Median family income. Median home. Unemployment.
Newsweek Feb 6, 1989 p. 44

Portland. Population. Median family income. Median home. Unemployment.
Newsweek Feb 6, 1989 p. 44

A tour of the urban killing fields. Cities with the highest per capita murder rates. The biggest increases. 1988.
Newsweek Jan 16, 1989 p. 45

A grim second city. Chicago built housing projects to provide safe, low-cost family shelter. Living comparisons. Violent crimes per 1,000 residents.
Newsweek Jan 2, 1989 p. 25

Hot cities. Where Americans will spend their vacations this summer.
U.S. News Jun 13, 1988 p. 71

Elbowroom in Anchorage. U.S. cities with largest land areas. Square miles. Population.
U.S. News May 16, 1988 p. 79

Turning downtowns into ghost towns. Vacancy rates for center-city offices in the five worst-hit U.S. cities.
Business Week Jan 11, 1988 p. 131

See Also MUNICIPAL FINANCE

CITY GOVERNMENT

See MUNICIPAL GOVERNMENT

CIVIL RIGHTS

Civil rights cases in 1990 [by type]. Voting. Accommodation. Employment. Welfare. Other.
Black Enterprise Aug 1991 p. 21

CIVIL RIGHTS - INTERNATIONAL ASPECTS

Cases investigated by Amnesty International. Documented cases of political imprisonment. Percentage of countries with death penalty.
U.S. News May 20, 1991 p. 12

CIVILIZATION, ANCIENT

Parallel worlds. Andes region. Other regions. 13,000 B.C. - 1532 A.D. Map. Chart.
U.S. News Apr 2, 1990 pp. 48–49

CLIFFS NOTES

Study shortcuts. The percentage of Cliffs Notes users by grade. Of the 210 titles published by Cliffs Notes, here are their 10 bestsellers.
U.S. News Aug 21, 1989 p. 66

CLIMATE AND WEATHER

1990 warm spots. World map.
Time Jan 21, 1991 p. 65

[Weather conditions and forecasting in the U.S.]. Limit for a reasonably accurate daily-weather prediction. Percentage of storms the National Weather Service fails to forecast. Snowiest place in the U.S.
U.S. News Dec 17, 1990 p. 34

Last of the red-hot waves. Highest and lowest temperatures ever recorded in each state.
U.S. News Jul 16, 1990 p. 66

Where disaster may strike. Hurricane paths. Landslides. Floods. Droughts. Tornadoes. Earthquake activity. Map of U.S.
Scholastic Update Dec 15, 1989 pp. 10–11

How a severe storm forms. What it does. Diagrams.
U.S. News Jul 24, 1989 pp. 50–51

Making the connection.
Colder-than-normal winter temperatures track closely with high solar activity. 1952-1980.
U.S. News Mar 6, 1989 p. 53

A new theory of the drought [cold water of La Niña]. Map and diagram.
Time Oct 31, 1988 p. 90

The changing face of planet Earth. 3 billion years ago. 100 million years ago. 18,000 years ago. 50 years from now. Chart.
U.S. News Oct 31, 1988 p. 60

CLOTHING AND DRESS

Dressing down. Millions of men's and boys' suits. Made in U.S. Imported. 1979-1991.
Business Week Aug 12, 1991 p. 42

The price of fun. Americans parted with more than $13 billion for sports clothing last year. Here is how much the average participant spent. [By activity].
U.S. News Oct 9, 1989 p. 81

Ties that bind. Company policy on ties and jackets for male executives.
U.S. News Apr 11, 1988 p. 68

CLOTHING INDUSTRY - IMPORTS AND EXPORTS

The payoff from U.S. production. Domestic $100 dress. Imported $100 dress. Retail profit margins. Wholesale costs.
Business Week Nov 7, 1988 p. 117

COACHING

Sideline salaries. Male college coaches earn more, on average, than female coaches. [Sport]. Average salary. Male. Female.
U.S. News Oct 28, 1991 p. 90

COCAINE

The flow goes on. Average monthly cocaine seizures in Panama. 1989-1991.
Time Aug 26, 1991 p. 18

Price of cocaine. Median wholesale price per kilogram. 1985-1990.
Time Dec 3, 1990 p. 47

Past the peak. Cocaine-related "medical crises" from emergency rooms in 21 cities. Emergency room visits. 1985-1990.
Newsweek Sep 24, 1990 p. 79

Good news in the drug war. City. Price of a kilo of cocaine. 6/89-6/90.
Newsweek Jul 2, 1990 p. 24

Percent of high school students who have used these drugs in the past 30 days. Marijuana. Cocaine. 1979-1989.
Time Feb 26, 1990 p. 38

The toll is heavy on minorities. Cocaine-related medical emergencies. Black. White. Hispanic. 1984, 1988.
Business Week Nov 27, 1989 p. 121

Where the cocaine trail begins. Coca-growing areas. Coca-processing areas. Map.
Scholastic Update Nov 17, 1989 p. 13

The growing toll. Cocaine deaths. Cocaine-related illnesses. 1984-1988.
Newsweek Apr 10, 1989 p. 21

Market value. Price range of cocaine. Cocaine seized. 1985-1988.
Newsweek Apr 10, 1989 p. 25

A consumer's guide to highs and lows. Alcohol. Cocaine. Nicotine. How it works. How it feels. How it hurts. How to get help.
Newsweek Feb 20, 1989 p. 56

Busting loose. Pounds of cocaine seized in [various European countries]. 1984-1988.
U.S. News Feb 20, 1989 p. 35

Cocaine related deaths have increased and so has federal funding for the war on drugs. 1983-1987.
Black Enterprise Aug 1988 p. 29

The price of failure. Range of wholesale prices per kilogram of cocaine. U.S. Miami. 1981-1987.
U.S. News Jul 11, 1988 p. 21

COFFEE

Classy beans: Heating up. Gourmet coffee sales in the U.S. Total coffee sales. 1987-1991.
Business Week Nov 18, 1991 p. 80

Coffee drinkers in different age groups. 1962, 1987.
FDA Consumer Jan 1988 p. 24

Consumption of coffee and other beverages. 1962, 1987.
FDA Consumer Jan 1988 p. 25

COLD (DISEASE)

Is it a cold or the flu? Symptoms.
FDA Consumer Nov 1988 p. 10

COLD FUSION

See FUSION, COLD

COLLEGE ATHLETES

College-athlete graduation rates. Football. Basketball. Percentage of Americans who bet on college sports. [Other selected statistics.]
U.S. News Jan 7, 1991 p. 12

College-athlete graduation rates. Average hours per week student athletes spend on sports in season. On preparing for and attending class. Colleges with the most players in the NFL.
U.S. News Dec 31, 1990 p. 12

COLLEGE EDUCATION, COST OF

Future shock. What you'll pay for college. Private. Public. 1992-2008.
Newsweek Oct 21, 1991 p. 53

If you're starting now. Rough guide to how much you should start putting away each month. Years to college. Monthly investment. Public. Private.
Newsweek Oct 21, 1991 p. 53

College. How much money you'll need to set aside for a child's college education. Years until student begins college. Costs. Monthly investment. Public college. Private college.
U.S. News Jul 30, 1990 p. 74

Slowing tuitions. Private college costs. Percent increase from previous academic year. 78-79. 81-82. 90-91.
Newsweek Mar 19, 1990 p. 8

The skyrocketing costs of college. Annual expenses. 1990 estimates. Percent increase from 1980.
Newsweek Feb 5, 1990 p. 6

A growing bite. Chart shows what colleges expect you to pay, out of your own pocket, for the school year 1990-91.
Newsweek Jan 29, 1990 p. 57

Revenge of the nerds. Costs at private four-year colleges will rise an average of 9 percent for the 1989-90 school year. Rates in public schools will increase by an average of 7 percent. 1988-1990.
Newsweek Aug 21, 1989 p. 65

Premium-price diplomas. Average college tuitions. Percentage change. Public schools. Private schools. Average guaranteed student loan. 1980-1989.
U.S. News Aug 29, 1988 p. 103

The price of knowledge. Average cost of a year at a college or university. Public. Private. 1980-1989.
U.S. News Apr 18, 1988 p. 75

Parental expectations. What a family of four at various levels of income would be expected to contribute to send one child to college for a year.
U.S. News Apr 18, 1988 p. 75

More donors. Total private contributions to four-year colleges. More competition. Percentage of private contributions going to public vs. private colleges. 1960-1986.
U.S. News Apr 11, 1988 pp. 54–55

COLLEGE GRADUATES

Diplomas aplenty. A record number of adults have high-school and college degrees. [Here are] the leading states. High-school graduates. College graduates.
U.S. News Dec 16, 1991 p. 95

The less educated. Median annual family income. By type of family head under 30. Dropout. High school grad. Some college. College grad. 1973, 1989.
Business Week Aug 19, 1991 p. 81

Estimated starting salaries for June 1990 graduates with bachelor degrees [by major].
Black Enterprise Feb 1991 p. 48

Choosy grads. Percentages of recent M.B.A. graduates who say they would not work for these employers.
U.S. News Oct 22, 1990 p. 82

Head of the class. Proportion of 22-year-old population receiving undergraduate degrees. Percentage of undergraduates receiving science and engineering degrees. U.S. Japan. West Germany.
U.S. News Jul 16, 1990 p. 26

The big ten. Number of black students entering medical school in 1989 from [ten large universities].
U.S. News May 28, 1990 p. 61

Master's degrees awarded to minorities. African American. Hispanics. Asian American. American Indian. 1978-1987.
Black Enterprise Apr 1990 p. 45

A matter of degrees. Ph.D.'s awarded to: Men. Women. 1960-2000.
U.S. News Dec 25, 1989 p. 66

Turning away from high tech. Science and engineering degrees awarded to women as a share of all such degrees in 1986 [by discipline and type of degree]. Baccalaureates. Masters. Doctorates.
Business Week Aug 28, 1989 p. 89

After-school money. School attended, job. Total cost of degree. Average first-year income.
Newsweek Jun 12, 1989 p. 6

[Percent] of blacks who are college graduates. 1967, 1987.
Time Mar 13, 1989 p. 68

Average starting salaries of college graduates, by discipline, 1987-1988.
Black Enterprise Feb 1989 p. 63

Paper-chase payoff. Estimated starting salaries for 1988-89 college graduates holding bachelor's degrees. Major and starting salary. Changes from 1987-88.
U.S. News Jan 30, 1989 p. 81

The payoff of education is huge. 1986 earnings of men 25-29 [by years of education].
Business Week Sep 19, 1988 p. 132

Doctorate degrees by race. Selected years. Black. White. 1975-1976. 1980-1981. 1984-1985.
Black Enterprise Sep 1988 p. 23

Paychecks for '88. Average starting salary, by degree. Bachelor's degrees. Master's degrees. Doctorates, other advanced degrees.
U.S. News Apr 25, 1988 p. 66

Starting salaries of college graduates by discipline, 1986-1987. Discipline. Salary. Change.
Black Enterprise Feb 1988 p. 77

Bachelor's degrees by race, 1975-76, 1984-85. White. Black. Hispanic. Asian. American Indian. Nonresident alien.
Black Enterprise Jan 1988 p. 39

Degrees conferred by historically black colleges and universities. 1981-82. 1984-85.
Black Enterprise Jan 1988 p. 39

COLLEGE STUDENTS

Campus cheats. 70 percent of all undergraduates say they've cheated at least once. Here's how.
U.S. News Jun 24, 1991 p. 71

Degrees of separation. College enrollment among Blacks. 1970, 1989. Master's degrees. Professional degrees. Doctorate degrees.
Newsweek May 6, 1991 p. 27

No deal. Freshmen choosing business careers. 1980-1990.
Newsweek Apr 1, 1991 p. 6

Against the odds. College enrollment (1988). Black males. Black females. White males. White females.
Black Enterprise Feb 1991 p. 41

An American dilemma. Who goes to college. Black males. Black females. White males. White females. As of 1988.
Newsweek Oct 15, 1990 p. 67

Students rate themselves above average in [various characteristics]. Freshmen at black colleges. Freshmen overall.
Black Enterprise Sep 1990 p. 45

Head of the class. Proportion of 22-year-old population receiving undergraduate degrees. Percentage of undergraduates receiving science and engineering degrees. U.S. Japan. West Germany.
U.S. News Jul 16, 1990 p. 26

Enrolled in college participation rates of dependent 18-to-24-year-old high school graduates. White. African American. Hispanic. 1973-1988.
Black Enterprise Apr 1990 p. 45

Reader's block. The percentage of [college] students who could name the authors of the following literary works.
U.S. News Nov 13, 1989 p. 89

Gone to college. Percentage of college students who are part-time students. 25 years or older. 1972-1986.
U.S. News Mar 20, 1989 p. 50

The rising drop-in rate. Percentage of 25-to-29-year-olds who completed: Less than 4 years of high school. 4 or more years of college. White. Black. 1950-1987.
U.S. News Mar 20, 1989 p. 50

The old college try. College students who: Dropped out or attended a two-year program. Put off going to college. Attended full time for four years. Took some time off in college. Mixed full and part-time schedules.
U.S. News Jan 9, 1989 p. 65

No room on campus. University campuses with most students. Fall, 1988 enrollment. Full-time students.
U.S. News Dec 19, 1988 p. 70

Greek comeback. Men initiated into college fraternities. Women initiated into college sororities. 1975-1987.
U.S. News Sep 19, 1988 p. 72

Percentage of 18-24 year old high school graduates enrolled in college, selected years. Black. White. Total. 1976, 1981, 1985.
Black Enterprise Sep 1988 p. 43

Lip service. U.S. college students enrolling in Japanese-language courses. 1977-1987.
U.S. News Mar 28, 1988 p. 41

Growing Asian enrollment. Percentage of American university students who are of Asian or Pacific-islander descent. 1976-1986.
U.S. News Mar 28, 1988 p. 53

COLLEGE STUDENTS - FOREIGN

Coming to America. More than 366,000 foreign students attended U.S. colleges and universities last year. The countries that sent the most students. 1988-89 academic year. 1979-80 academic year.
U.S. News Dec 18, 1989 p. 82

Foreign students studying in the United States. American students studying abroad. Engineering. Physical and life sciences.
U.S. News Dec 26, 1988 p. 112

COLLEGES AND UNIVERSITIES

U.S. News top 25 national universities. The biggest of the best offer a wide range of programs, give high priority to research and award many doctoral degrees.
U.S. News Sep 30, 1991 p. 93

U.S. News top 25 national liberal arts colleges.
U.S. News Sep 30, 1991 p. 97

U.S. News top regional universities. Test scores, acceptance rates and other vital data of the leaders in the North and South.
U.S. News Sep 30, 1991 p. 105

Integrating the ivory tower. African-American representation among students and full-time faculty at undergraduate and selected graduate institutions, 1989-90.
Black Enterprise May 1991 p. 74

The final ranking: The top 25 national universities.
U.S. News Oct 15, 1990 pp. 118–119

The final ranking: The top 25 national liberal-arts colleges.
U.S. News Oct 15, 1990 pp. 122–123

Best of the rest. National universities. The 295 other national universities and liberal-arts colleges that did not rank among the top 25 in their respective categories.
U.S. News Oct 15, 1990 p. 124

The final ranking: The top 60 regional colleges & universities. North. South. Midwest. West.
U.S. News Oct 15, 1990 pp. 128–129

The final ranking: The top 40 regional liberal-arts colleges. North. South. Midwest. West.
U.S. News Oct 15, 1990 pp. 132–133

Top specialty schools. The arts. Business. Engineering.
U.S. News Oct 15, 1990 p. 134

College without sweat. Percentage of four-year institutions that do not require taking courses in: Classics. Foreign languages. Philosophy. Mathematics. English or American literature. Natural and physical sciences. History. English composition. Social sciences.
U.S. News May 28, 1990 p. 75

America's best colleges. National universities. National liberal-arts colleges. Regional colleges and universities. Regional liberal-arts colleges. Specialty schools.
U.S. News Oct 16, 1989 pp. 66–82

Professorial shortfall. Faculty supply-and-demand projections in the social sciences and humanities. Supply. Demand. 1987-2012.
U.S. News Sep 25, 1989 p. 55

Cities of students. These cities have the most four-year colleges and universities within their boundaries.
U.S. News Aug 14, 1989 p. 70

A class of deadbeats. These four-year colleges and universities have the highest percentage of defaults on student loans made under the Guaranteed Student Loan program. 1986, 1987.
U.S. News Jul 3, 1989 p. 62

Largest individual cash gifts to selected United Negro College Fund schools. Contributor. Amount. Year.
Black Enterprise Feb 1989 p. 27

Universities are buying more. Average PC expenditures for students and faculty at 3,340 institutions.
Business Week Oct 24, 1988 p. 83

America's best colleges. The top 125 schools rated by student selectivity, faculty quality and academic resources. Survey of college presidents and deans. [Special section].
U.S. News Oct 10, 1988 pp. 2–32

Upping the ante. College and universities are setting—and achieving—record fund-raising goals. Total private contributions. 1980-1987.
Newsweek Sep 5, 1988 p. 67

Japanese-endowed chairs at MIT.
Business Week Jul 11, 1988 p. 70

Enrollment in institutions of higher education, by race/ethnicity of students, 1980 and 1984. White. Black. Hispanic. Asian. American Indian. Nonresident alien.
Black Enterprise Jan 1988 p. 39

COLLEGES AND UNIVERSITIES - ENROLLMENT

More and more youngsters seek a college education. Enrollment in institutions of higher education. 1975-1990.
Business Week Jun 3, 1991 p. 20

No room on campus. University campuses with most students. Fall, 1988 enrollment. Full-time students.
U.S. News Dec 19, 1988 p. 70

COLLEGES AND UNIVERSITIES - GRADUATE WORK

Top 25 business schools. 25 graduate schools of business with the highest scores in the *U.S. News* survey.
U.S. News Apr 29, 1991 p. 68

Top 25 law schools. 25 law schools with the highest scores in the *U.S. News* survey.
U.S. News Apr 29, 1991 p. 74

Top 25 in engineering. 25 graduate schools of engineering with the highest scores in the *U.S. News* survey.
U.S. News Apr 29, 1991 p. 82

Top 15 medical schools. 15 medical schools with the highest scores in the *U.S. News* survey.
U.S. News Apr 29, 1991 p. 88

COLLEGES AND UNIVERSITIES - RESEARCH

Corporate funding for R&D: The new lineup. Business contributions to research universities [ranked by university]. 1989, 1982.
Business Week May 20, 1991 p. 124

University research: The squeeze is on. Funding from federal government, state and local government, corporate. 1980-1991. Who foots the bill.
Business Week May 20, 1991 p. 125

Building empires of science. Major research universities [leading recipients of federal research funds] with their overhead charges as a percentage of funding.
U.S. News Mar 4, 1991 p. 52

From ivory tower. University spending on research and development for science and engineering. To big business. Patents issued to universities. 1980-1986
U.S. News May 2, 1988 p. 51

COMETS

Death from the sky. An asteroid or a comet perhaps 8 km (5 miles) in diameter may have hit the earth 65 million years ago. Diagram.
Time Jul 1, 1991 p. 60

COMIC BOOKS, STRIPS, ETC.

A time line of the funnies. 40 of the longest-running comic strips appearing in newspapers today. Year started.
U.S. News Jul 2, 1990 p. 62

Biff! Pow! Blowie! In a study of comic books, a national consumer group counted the following average number of violent acts per issue.
U.S. News Jul 24, 1989 p. 66

COMMODITY FUTURES

Declining prices. Commodity-futures prices, quarterly average. 1989, 1990.
U.S. News Dec 24, 1990 p. 51

COMMONWEALTH OF INDEPENDENT STATES

What in common? The new marriage of Slavic and Central Asian republics may be rife with tensions. Founders of the commonwealth. Agreed to join. Interested. Not interested. Map.
Newsweek Dec 23, 1991 p. 28

Re-forming itself. Founding members of the Commonwealth. Agreed to join. Considering the pact. Baltic states. Map.
Time Dec 23, 1991 p. 20

Chaos or commonwealth? Old animosities, economic chaos and the presence of some 27,000 nuclear weapons will make the transition from the Soviet Union to a Commonwealth of Independent States a difficult one. [Other selected statistics.]
U.S. News Dec 23, 1991 p. 38

State of the union. Republics. Population. Per capita GNP. [Date] established as a separate republic of the Soviet Union. [Description of recent political events]. Map.
Time Sep 9, 1991 pp. 30–31

See Also SOVIET UNION - BREAKUP OF THE UNION

COMMUNIST PARTY

Membership in all Communist parties. Card-carrying U.S. communists. U.S. circulation of the communist journal *People's Daily World.* [Other selected statistics.]
U.S. News Sep 9, 1991 p. 10

The party: Politburo. Central Committee. Party Congress. The government: Supreme Soviet. Council of Ministers. Congress of People's Deputies. Chart.
Time Feb 19, 1990 p. 34

The party's almost over. Communist Party membership in Eastern Europe [various countries].
U.S. News Feb 12, 1990 p. 33

COMPACT DISCS

Where the growth is. Cassettes. CDs. 1986-1990.
Business Week May 27, 1991 p. 38

COMPUTER CHIPS

Chip fabrication costs are rising out of sight. Plant and equipment costs to produce standard DRAM chip. 1971-2000.
Business Week Jul 29, 1991 p. 34

The U.S. loses more ground in chips. Share of worldwide semiconductor sales. 1985-1990.
Business Week Jan 8, 1990 p. 100

Fading memories. Market share of DRAM (dynamic random-access memory) produced by: U.S. Japan. Europe. Other. 1974-1988.
U.S. News Dec 18, 1989 p. 38

Japanese producers [of memory chips] are earning record profits. Sales. Increase. Net profits. Increase [by company].
Business Week Jul 4, 1988 p. 108

As the chip shortage keeps prices high, makers of small computers feel the pinch. Average selling prices for DRAM chips in the U.S. Estimated memory chip costs as percentage of total computer costs.
Business Week Jun 27, 1988 p. 29

Another pickup in chip sales. U.S. sales. 1983-1988.
Business Week Jan 11, 1988 p. 114

COMPUTER CRIMES

Computer crime: Only minor losses. Data or computer downtime lost because of malicious acts (hackers and viruses) over the past two years.
Business Week Aug 6, 1990 p. 71

COMPUTER DISPLAY TERMINALS - HEALTH ASPECTS

No problems here. The average magnetic fields around 30 typical monitors. Monitor type. Number tested. Measurement distance. Average readings in milligauss.
U.S. News Sep 10, 1990 pp. 84–85

Our 'average' monitor. Magnetic-field readings in milligauss. Top, back, right, left, front. Diagram.
U.S. News Sep 10, 1990 p. 85

COMPUTER INDUSTRY

Hardware jobs are disappearing. U.S. employment. Computer and office-equipment manufacturing. But employment in software is rising. U.S. employment. Computer software and data processing services. 1986-1991.
Business Week Aug 19, 1991 p. 106

Domestic PC sales flatten out. Units sold through computer stores, U.S. only. Sept. '89 - Sept. '90.
Business Week Nov 5, 1990 p. 140

Slower growth for computers. Annual growth in U.S. systems shipments. 1985-1990.
Business Week Jan 8, 1990 p. 97

A weak year for PC makers. Annual growth in dollar sales of personal computers.
Business Week Dec 25, 1989 p. 45

Signs of aging. Key indicators show that the business is graying. Computer system dollar sales. Increase in revenues by U.S.-based companies. 1983-1992.
Business Week Mar 6, 1989 p. 68

The soaring workstation market. Worldwide sales. 1983-1989.
Business Week Jan 16, 1989 p. 88

Slow growth in minicomputers. Dollar value of worldwide midrange shipments from U.S.-based manufacturers. 1984-1990.
Business Week Nov 21, 1988 p. 106

U.S. customers are paying higher prices. Average 1988 DRAM [dynamic random-access memory] prices. U.S. Japan. 256K. 1 megabit.
Business Week Jul 4, 1988 p. 108

Demand for midrange computers is growing. Value of shipments worldwide. 1987-1992.
Business Week Jun 27, 1988 p. 30

COMPUTER INDUSTRY - MARKET SHARE

The home-PC race: Mac by a mile. Apple Classic. IBM PS/1. U.S. unit sales. 1990, 1991.
Newsweek Aug 26, 1991 p. 65

Out of the lab, into the office. Total workstation sales. 1986-1993.
Business Week Apr 15, 1991 p. 74

Far out front [Hewlett-Packard Co.]. 1990 share of laser-printer market in U.S.
Business Week Apr 1, 1991 p. 79

The Japanese edge in laptops. 1990 U.S. market share. Japan. U.S. Other.
Business Week Mar 18, 1991 p. 121

Gross revenues for PC's and mainframes worldwide. 1981-1990.
U.S. News Dec 24, 1990 p. 47

Computer cash. U.S. market share for large multi-user computers.
Newsweek Sep 17, 1990 p. 50

The PC battle. U.S. PC market shares. IBM. Apple. Zenith. Compaq. Tandy.
Newsweek Mar 26, 1990 p. 35

David and Goliath. PCs and mainframes [market share]. 1984-1993.
Newsweek Dec 18, 1989 p. 48

U.S. computer makers' share of world markets. Geographical markets. Non-U.S. companies. U.S. companies. World market in computer systems. PCs. Small [computer systems]. Medium. Large.
Business Week Mar 6, 1989 p. 74

Slower gains for computer makers. Worldwide sales of U.S.-based manufacturers. 1984-1989.
Business Week Jan 9, 1989 p. 93

The top computer dealers. Company. U.S. outlets. U.S. sales. Growth. 1985-1989.
Business Week Oct 10, 1988 p. 114

How U.S. makers dominate Europe's computer markets. In mainframes. Midsize computers. Microcomputers. Market share. U.S. European. Japanese and other Asian.
Business Week Sep 19, 1988 p. 145

After taking over the memory chip business. Share of worldwide dynamic random-access memory (DRAM) market. Japan. U.S. 1981-1987.
Business Week Jul 4, 1988 p. 108

IBM is still the [computer] market leader. Market share.
Business Week Jun 27, 1988 p. 30

For growth, look to PCs again. Percent increase in U.S. dollar sales. PCs. Mainframes. Midrange. 1986-1988.
Business Week Jan 11, 1988 p. 111

COMPUTER NETWORKS

Leading in LANs [local area networks]. Share of U.S. shipments in 1990. Novell. Microsoft. Apple. Banyan. Others.
Business Week Sep 2, 1991 p. 78

The new architecture of computing. Personal-computing networks are fast replacing terminals linked to a mainframe.
U.S. News May 21, 1990 p. 49

The soaring data market. Global network information systems. U.S. Europe. Asia/Pacific Rim. 1988, 1992. Total market. 1988, 1992.
Business Week Oct 3, 1988 p. 34

COMPUTER OPERATING SYSTEMS

OS/2's battle from behind. OS/2. MS-DOS. Windows. Thousands of units. 1988-1990.
Business Week Apr 22, 1991 p. 33

More PCs are doing windows. Units shipped worldwide. DOS. Windows. 1988-1992.
Business Week May 21, 1990 p. 150

Software battlefield. PC operating systems. Worldwide unit shipments, in millions [market shares]. DOS. Windows. Macintosh. 1985-1990.
Newsweek May 21, 1990 p. 69

Worth the fight. UNIX market share. 1982-1992.
U.S. News May 30, 1988 p. 51

A bulging market share. Percent of all computers worldwide using UNIX. 1985-1991.
Business Week Mar 14, 1988 p. 96

COMPUTER PRINTERS

Far out front [Hewlett-Packard Co.]. 1990 share of laser-printer market in U.S.
Business Week Apr 1, 1991 p. 79

COMPUTER PROGRAMMERS

U.S. employment of programmers and software engineers/analysts. Bachelors' degrees in computer science granted by U.S. colleges and universities. 1977-1990.
Business Week Mar 11, 1991 p. 105

COMPUTER SOFTWARE INDUSTRY - MARKET SHARE

Software. Key acquisitions. 1991. Market share. 1985 ranking. 1990 ranking.
Business Week Oct 14, 1991 p. 87

Can the U.S. stay ahead in software? [Market share by country.]
Business Week Mar 11, 1991 pp. 98–99

COMPUTER VIRUSES

A computer virus: What it is and how it spreads. Creation. Distribution. Infection. Attack. Diagram.
Business Week Aug 1, 1988 p. 70

COMPUTERS

Number crunchers. Conventional computer. Parallel computer. Problem. Solution. Diagram.
Time Nov 11, 1991 p. 74

Productivity paradox. White-collar high-tech investment and output per worker in the service sector. 1980-1989.
U.S. News Dec 24, 1990 pp. 46–47

Inexpensive information. Average cost of processing 1 million instructions per second. Mainframes. PC's. 1986-1995.
U.S. News Dec 24, 1990 p. 48

Numbers crunch. Business purchases of computers. 1980-1991.
U.S. News Oct 23, 1989 p. 60

Outfitting the ergonomic office [products to ease VDT work].
U.S. News Jan 9, 1989 p. 61

COMPUTERS - EDUCATIONAL USE

The schools' buying spree slows. Number of computers in school. 1983-1992.
Business Week Jul 17, 1989 p. 112

The leaders in the education market. Share of personal computer shipments [by company].
Business Week Oct 24, 1988 p. 76

COMPUTERS - MAINTENANCE AND REPAIR

The shift away from hardware maintenance. Worldwide spending for maintenance on computer systems. 1988 worldwide revenues. 1992 projected worldwide revenues.
Business Week Mar 5, 1990 p. 82

COMPUTERS, PERSONAL

The home-PC race: Mac by a mile. Apple Classic. IBM PS/1. U.S. unit sales. 1990, 1991.
Business Week Aug 26, 1991 p. 65

Worldwide. U.S. Personal computer sales in billions of dollars. 1981-1991.
Business Week Aug 12, 1991 pp. 58–59

What it costs to log on. Compuserve. Genie. Prodigy. Start-up cost. Monthly/hourly cost. Extras.
Business Week Jun 17, 1991 p. 113

Inexpensive information. Average cost of processing 1 million instructions per second. Mainframes. PC's. 1986-1995.
U.S. News Dec 24, 1990 p. 48

Domestic PC sales flatten out. Units sold through computer stores, U.S. only. Sept. '89 - Sept. '90.
Business Week Nov 5, 1990 p. 140

A weak year for PC makers. Annual growth in dollar sales of personal computers.
Business Week Dec 25, 1989 p. 45

High-tech, low budget. Personal computers in the Soviet Union. Actual and planned production of personal computers in the Soviet Union. 1986-1990.
U.S. News Nov 20, 1989 p. 35

A small boom. Personal computers. Fax machines. Cellular phones. Home copiers.
Newsweek Apr 24, 1989 p. 59

More and more computing power for the buck. Cost of hardware to process one million instructions per second. Mainframes. Minicomputers. PCs. 1981-1993.
Business Week Mar 6, 1989 p. 70

PC profusion. Percentage of households with personal computers. 1980-1988. Time at the terminal. How many hours people spend using personal computers each week at home.
U.S. News Dec 26, 1988 p. 120

Electronic publishing: More and more a desktop game. Sales of electronic publishing systems by type of hardware. Apple. IBM PC & compatibles. Workstation. Mainframe.
Business Week Nov 28, 1988 p. 156

The laptop express. Laptop computers shipped to U.S. distribution channels. 1986-1990.
Business Week Nov 14, 1988 p. 121

Universities are buying more. Average PC expenditures for students and faculty at 3,340 institutions.
Business Week Oct 24, 1988 p. 83

For more workers, there's no place like home. White-collar workers operating out of home offices. Full-time. Part-time. PCs used in home offices.
Business Week Oct 10, 1988 p. 105

For growth, look to PCs again. Percent increase in U.S. dollar sales. PCs. Mainframes. Midrange. 1986-1988.
Business Week Jan 11, 1988 p. 111

CONGLOMERATE CORPORATIONS

Back to basics. Firms operating in one industry. Fewer conglomerates. Firms operating in 20+ industries. 1985, 1989.
U.S. News Sep 17, 1990 p. 47

CONGRESS

Americans who favor a 12-year congressional term limit. Members of Congress who have served 12 years or more. Registered lobbyists on Capitol Hill. [Other selected statistics.]
U.S. News Nov 4, 1991 p. 20

The golden rocking chair. Thanks to fat pensions, many retired officials make more than those still on the job. Ex-presidents. Ex-senators. Ex-congressmen.
Time Jun 10, 1991 p. 21

Out with the old. Proposals to limit terms in Congress to 12 years would put 213 legislators out of work, including these most senior members.
U.S. News Jan 14, 1991 p. 62

Who controls redistricting. Changes as a result of census. Current number of seats. Projected number in 1992.
Time Nov 19, 1990 p. 41

Capitol stuff. The number of legislative-branch employes. Legislative-branch costs. 1960-1990.
U.S. News Oct 22, 1990 p. 30

Cash and carry. Senate campaign costs. House campaign costs. [Number of] political action committees. Biggest PAC givers. 1974-1988.
U.S. News Oct 22, 1990 p. 31

Westward ho, Republicans! Population change in percent. Congressional seats. Gained. Lost. No change. 1980-1990.
Newsweek Sep 10, 1990 p. 33

More help at home. House staff based in district offices. 1974-1988.
U.S. News Aug 7, 1989 p. 54

Gabbing for dollars. The top 10 [congressional] earners in 1988. What they made. What they kept.
Newsweek Jul 3, 1989 p. 22

City within a city: An insider's guide to CongressWorld. The Capitol has its own peculiar places to eat, meet and carry out the business of the day. Map.
Newsweek Apr 24, 1989 p. 31

Capitol Hill. Map.
Scholastic Update Feb 24, 1989 p. 4

The 101st Congress at a glance. Representatives. Senators. Party affiliation. Average age. Profession before entering Congress. Sex. Race, ethnic group. Religious affiliation.
Scholastic Update Feb 24, 1989 p. 5

The perks of power. Pensions. Health benefits. Life insurance. Parking. Telephones. Health club. Haircuts.
Time Jan 23, 1989 p. 13

CONGRESS - ELECTIONS

Congress's self-protection racket. Percentage of incumbents seeking re-election who won. House. Senate. 1946-1986.
U.S. News Nov 7, 1988 p. 29

CONSTRUCTION INDUSTRY

See HOUSING - CONSTRUCTION

CONSULTANTS

The top five consulting firms. 1987 consulting revenues.
Business Week Jun 13, 1988 p. 82

CONSUMER PRICE INDEX

Feverish growth. Percent change since 1980. Total health expenditures. Consumer price index. Federal education expenditures. 1980-1991.
Time Nov 25, 1991 pp. 34–35

Did the Fed cause the recesssion? Consumer price index. Money supply. Federal funds rate. Commercial and industrial loans. Gross national product. 1987-1991.
U.S. News Jul 1, 1991 p. 50

A delicate balance. Percentage change in consumer price index. Percentage change in real GNP. 1989, 1990.
U.S. News Dec 17, 1990 p. 58

Office vacancies remain high. Values lag behind inflation. Home prices keep rising, although far from evenly. Median sale price of existing one-family homes. Northeast. Midwest. 1983-1988.
Business Week Dec 26, 1988 p. 172

CONSUMER PROTECTION

FDA enforcement actions 1987-1988. Legal actions. Import reviews.
FDA Consumer Apr 1989 p. 5

CONSUMER SPENDING

Consumers haven't retreated. Consumption expenditures as a share of GNP. 1981-1990.
Business Week Apr 30, 1990 p. 20

The 1980s' buying binge may be petering out. Discretionary consumer purchases. Percent of national income. 1976-1989.
Business Week Jul 31, 1989 p. 20

A leaner year for consumer spending. Growth rate in consumer spending. 1984-1989.
Business Week Jan 9, 1989 p. 84

CONSUMERS - CHINA

China's hunger for consumer products. Annual growth in per capita consumption. Total goods. 1981-1987.
Business Week Jan 25, 1988 p. 69

CONSUMERS - JAPAN

Shop till you drop. Consumer spending. Adjusted for inflation, in trillions of yen. 1980-1990.
Newsweek Aug 13, 1990 p. 47

CONSUMERS - SOVIET UNION

Attention, comrade shoppers. Raisa Gorbachev may shop for furs and fancy clothes, but ordinary Soviet citizens are keen on household products. Here are some purchases they hope to make in the next two years.
U.S. News May 7, 1990 p. 75

CONSUMERS - UNITED STATES

How bad will it get? Consumer confidence index. 1990.
Newsweek Jan 14, 1991 p. 32

Black vs. white consumers. Planned major purchases. Blacks. Whites.
Black Enterprise Oct 1990 p. 45

Recession's signs and portents. Consumer attitude. Index of consumer expectations. 1989, 1990.
U.S. News Sep 24, 1990 p. 80

In search of excellence. Of 1,005 consumers polled, most said they would pay more for higher quality. Extra amount consumers will pay for higher quality.
U.S. News Jul 3, 1989 p. 62

Smaller gains in consumer spending. Growth rate of U.S. consumer spending. 1984-1988.
Business Week Jan 11, 1988 p. 99

CONTRACEPTIVES

See BIRTH CONTROL

CONVENIENCE STORES

The nation's largest convenience store chains. Company. Number of stores.
Business Week Jun 13, 1988 p. 86

COPYRIGHT - INTERNATIONAL ASPECTS

Artful dodgers. Overseas violators of intellectual-property rights cost U.S. companies billions. [Listed are] the top 10 nemeses. Country. Disputed products. Losses.
U.S. News Nov 14, 1988 p. 50

CORN

See GRAIN

CORPORATE SOCIAL RESPONSIBILITY

Conscientious investors. Most popular [social policy] issues.
U.S. News Jul 8, 1991 p. 57

Stiff slaps. Fines for environmental offenses. 1983-1989.
Newsweek Sep 25, 1989 p. 35

Living up to high standards. Countries with high business ethics. Most ethical region of U.S. The most ethical professions.
U.S. News Mar 14, 1988 p. 76

CORPORATIONS

Less profit pressure. Corporate profit. 1973-1991. Shrinking bottom line. Strong overseas earnings.
U.S. News Jul 29, 1991 p. 46

Pay for performance: Who measures up and who doesn't.
Business Week May 6, 1991 p. 93

Plummeting profits. Corporate profits. 1985-1990.
U.S. News Nov 19, 1990 p. 58

Back to basics. Firms operating in one industry. Fewer conglomerates. Firms operating in 20+ industries. 1985, 1989.
U.S. News Sep 17, 1990 p. 47

The 20 highest-paid chief executives, 10 who aren't CEOs. [Executive]. Company. 1989 salary and bonus. Long-term compensation. Total pay. The 10 largest golden parachutes. Company. Reason. Total package.
Business Week May 7, 1990 p. 57

The decade's biggest CEO money-makers [by name]. Total pay 1980-89.
Business Week May 7, 1990 p. 60

The leaders in 1989. The top 25 in sales. The top 25 in earnings.
Business Week Mar 19, 1990 pp. 64–65

The leaders in 1988 sales and profits. Top 25 in sales. Top 25 in earnings. Profits. Percent change. Rank.
Business Week Mar 20, 1989 pp. 62–63

Pay for performance. Executives who gave shareholders the most for their pay. The least. Executives whose companies did the best relative to their pay. The worst. 1985-1987.
Business Week May 2, 1988 p. 53

The leaders in 1987 sales and profits. Top 25 in sales. Top 25 in earnings.
Business Week Mar 14, 1988 pp. 122–123

A spotlight on 1987 industry profits. Industries with the biggest dollar change in 1987. Winners and losers in the year's profit race.
Business Week Mar 14, 1988 p. 124

CORPORATIONS - ACQUISITIONS AND MERGERS

Stock and bond issues. Mergers and acquisitions. 1987-1991.
Time Sep 16, 1991 p. 46

The standout acquisitions of the 1980s. The best. The worst.
Business Week Jan 15, 1990 pp. 52–53

More junk bonds are being used. To fund more and larger mergers. Value of mergers and acquisitions. 1980-1987.
U.S. News Nov 7, 1988 p. 62

The 10 largest golden parachutes. [Executive]. Company. Reason for payment. Total package.
Business Week May 2, 1988 p. 51

1988's billion-dollar deals. Buyer. Target. Price.
Business Week Mar 21, 1988 p. 123

The year of the megadeal. The number of deals is down but the value is way up. 1986-1988.
Business Week Mar 21, 1988 p. 124

See Also LEVERAGED BUYOUTS

CORPORATIONS - DIRECTORS

Who's sitting on corporate boards? Women. Ethnic minorities. 1985-1989.
Black Enterprise Aug 1991 p. 35

CORPORATIONS - FINANCE

The number of deals financed by new debt are on the decline. 1987-1991.
Time Sep 16, 1991 p. 47

More companies are now raising funds by issuing new stock. Amount raised by the issue of securities, in billions of dollars. 1990, 1991.
Time Sep 16, 1991 p. 47

The road to recovery? Pharmaceuticals. Semiconductors. Home appliances. Net profits. 1989-1992.
U.S. News Jun 10, 1991 pp. 48–49

Intense interest. Corporations. Net interest payments as a percentage of cash flow. 1975-1990.
U.S. News Nov 19, 1990 p. 56

Stopgap solution. Monthly changes. Outstanding non-financial commercial paper. Commercial and industrial loans at large commercial banks. 1990.
U.S. News Nov 19, 1990 p. 58

The debt pile. Corporate debt. Net equity issued. 1970-1988.
U.S. News Feb 13, 1989 p. 62

The debt bill. Interest payments as a share of cash flow. 1960-1988.
U.S. News Feb 13, 1989 p. 63

The growing burden of corporate debt. Net interest payments as a percent of cash flow. 1978-1988.
Business Week Sep 12, 1988 p. 26

CORPORATIONS, INTERNATIONAL

A sampler of the largest foreign employers in the U.S. Foreign parent. Home office. U.S. subsidiary. U.S. employment.
Business Week Dec 17, 1990 p. 81

The foreign payroll. Percentage of U.S. employees at foreign-owned companies in the U.S. [by country].
Business Week Dec 17, 1990 p. 84

U.S. companies are bringing home more of the bacon. Repatriated earnings of nonfinancial corporations. Billions of dollars, annual rate. 1974-1988.
Business Week Aug 1, 1988 p. 17

How companies stack up globally [leading international companies]. Sales. Profits. Share-price gain.
Business Week Jul 18, 1988 p. 137

COSMETIC INDUSTRY - MARKET SHARE

A new factor in makeup. U.S. mass-market cosmetics, 1990 share estimates.
Time Apr 22, 1991 p. 61

Who's who in the beauty business. Manufacturer. Brands. Market share. 1988.
Business Week Jul 31, 1989 p. 33

COST OF LIVING - UNITED STATES

How groceries eat your cash. Here's the average amount spent on food by households in the largest metropolitan areas. Annual food bill. Percent of household income. 1988.
U.S. News Mar 26, 1990 p. 73

COST OF LIVING - WORLD

The new cost of Soviet living. [Selected] item. Old price. New price. Black market.
Newsweek Apr 15, 1991 p. 39

Costs of living. Net monthly income. Goods as percent of take-home pay. Germany. West. East.
U.S. News Nov 27, 1989 p. 43

Global groceries. Here's the cost of several foods per pound in some world capitals in May, 1989.
U.S. News Jul 24, 1989 p. 66

How living costs compare around the globe. Cost of a basket of 108 goods and services for a typical European family of three [by city].
Business Week Nov 7, 1988 p. 24

Cost of living. The Polish economy suffers from rampant inflation. Increase in consumer-price index since 1980. [By selected Eastern European country].
Newsweek Sep 5, 1988 p. 40

COUNTRY LIFE

The spread of city living. State residents who live outside metropolitan areas. [State]. Nonmetro-population. Percentage change since 1950.
U.S. News Mar 13, 1989 p. 73

CRACK (COCAINE)

Crack connection. Outposts of Jamaican Posses. Crips and Bloods movements. Jamaican Posses' overseas smuggling routes. Map.
U.S. News Aug 19, 1991 p. 46

Highs and lows: Counting costs in the drug market. Ice. Crank. Crack. Heroin. Cost. Duration. Effects. Side effects.
Newsweek Nov 27, 1989 p. 38

See Also DRUG ABUSE

CREATIVITY

Amateur artistry. Expressing one's artistic tendencies has become much more popular. 1975, 1987.
U.S. News Mar 5, 1990 p. 72

CREDIT

Borrowing has slowed sharply. Long-term borrowing by nonfinancial businesses. 1986-1991.
Business Week Aug 12, 1991 p. 20

Money growth stalls. M3 [A broad money supply measure]. 1990, 1991.
Business Week Aug 12, 1991 p. 20

Who's got the best credit. Top bond ratings [by state]. Worst bond ratings [by state].
U.S. News Feb 18, 1991 p. 48

See Also LOANS

CREDIT BUREAUS

Filing a beef. Complaints and inquiries about credit reports. 1989-1991.
Newsweek Oct 28, 1991 p. 42

CREDIT CARDS

Bank card volume grows. Visa and MasterCard charges. Delinquencies may have peaked. Outstanding debt overdue. 1988-1991.
Business Week Oct 21, 1991 p. 125

Defying gravity. Credit cards. New cars. Consumer finance rate. Fed discount rate. 1988-1991.
Time Oct 14, 1991 p. 44

Charge volume worldwide in billions of dollars. Cards issued worldwide in millions. 1985-1990.
Time Jul 1, 1991 p. 51

Green charge. Over 75 Green groups market an "affinity" Visa or Mastercard. A sampling of 1990 [dollar] totals.
U.S. News Apr 22, 1991 p. 71

Who's who among big card issuers and how banks are losing share. 1986-1990.
Business Week Apr 15, 1991 p. 29

[Credit cards]. Figuring your annual cost of credit. Interest rate. No annual fee. $20 annual fee. $50 annual fee. How a grace period affects your cost. Interest rate. No grace period. 25-day grace period.
U.S. News Jan 21, 1991 p. 79

Consumers are struggling to pay credit-card bills. Bank credit-card bills overdue more than 30 days as share of total amount owed. Percent. 1987-1990.
Business Week Dec 10, 1990 p. 205

Where credit is cheapest. The average interest rate on Visa and MasterCard purchases is 18.5 percent, but many banks offer better rates. These banks give the best deals.
U.S. News Oct 30, 1989 p. 82

Charge it. Popular credit cards. Cards. % using. How they are used. Purchases. % with card.
Newsweek Sep 18, 1989 p. 8

Bargain plastic. Banks with the best credit-card deals. Name. Phone number. Interest rate. Annual fee. Interest-free days.
U.S. News May 30, 1988 p. 72

CRIME

Comparing state stats. Violent crimes per 100,000 people. [By state].
Scholastic Update Jan 11, 1991 pp. 26–27

Cops and crime. Cops under fire. [Crime rates and number of police for] Washington D.C. Dallas. San Francisco. The 10 cities with the largest police forces.
U.S. News Dec 3, 1990 p. 34

Cities and crime. Cops under fire. The worst and best cities on crime. 1989.
U.S. News Dec 3, 1990 p. 35

Guns and crime. Cops under fire. Murders with guns. All firearms. Handguns. Knives or other cutting instruments were the next most common murder weapon. 1974-1989.
U.S. News Dec 3, 1990 p. 38

The portrait of a nightmare [rape statistics].
Newsweek Jul 23, 1990 p. 48

Social problems are getting worse. Percentage increases. High school dropouts. 1970-88. Violent crime. 1979-88. Poverty rate. 1975-85. Drug abuse cases. 1980-87.
Scholastic Update Feb 23, 1990 p. 3

Comparing state stats. Violent crimes per 100,000 people. [By state].
Scholastic Update Jan 12, 1990 pp. 30–31

The meanest streets. Midsized cities saw the worst of last year's surge in reported street crime. Crimes per 100,000 people. 1988-1989.
U.S. News Jan 8, 1990 p. 66

The crime rate is soaring. Number of crimes committed per 100,000 people. 1984-1988.
Scholastic Update Nov 17, 1989 p. 8

The states pay the price. While violent crime climbs, so do prison costs. Percent increase 1988 to 1989 in state spending on: Penal institutions. Medicaid. Education. Welfare. Overall spending.
Time Aug 21, 1989 p. 25

City crime: A matter of perception. Perceived safety. Violent crimes per 100,000 population [by city].
Newsweek Aug 14, 1989 p. 6

Inmates. Number of prisoners in state and federal prisons, in thousands. Crime rate. Number of crimes committed per 1,000 population. 1980-1988.
Time May 29, 1989 p. 29

The statistics: Separate and unequal [blacks and whites compared].
Newsweek Mar 13, 1989 p. 19

A tour of the urban killing fields. Cities with the highest per capita murder rates. The biggest increases. 1988.
Newsweek Jan 16, 1989 p. 45

Slaughter in the streets. Number of murders in each city. New York City. Washington. Houston. New Orleans. Miami. 1987 total. 1988 to date.
Time Dec 5, 1988 p. 32

U.S. Almanac: Comparing state stats. Violent crimes per 1,000 people. [By state].
Scholastic Update Dec 2, 1988 pp. 18–19

Uneven odds. Lifetime risk of being murdered. Black males. White males. Black females. White females.
U.S. News Aug 22, 1988 p. 54

Violence in the cities. Reported violent crimes per 100,000 residents in 1987. Change from 1986. [Various U.S. cities].
U.S. News Jul 25, 1988 p. 66

Costs to society. Alcohol. Drugs. Crime. Treatment. Lost productivity.
Time May 30, 1988 p. 14

Women and crime. Women as percentage of adult arrests for selected crimes in 1986. Violent crime. Property crime. Forgery, fraud and embezzlement. Vice.
Ms. Feb 1988 p. 26

How Miami's vices measure up. Crimes per 100,000 population. Greater Miami. New York. Chicago. Los Angeles.
Newsweek Jan 25, 1988 p. 27

See Also MURDER

CRIME AND YOUTH

Kids who kill. Firearm murders committed by offenders under age 18. 1984, 1987, 1989.
U.S. News Apr 8, 1991 p. 26

A tide of teen violence. Total arrests, under age 18, 1985-89. Murder/nonnegligent manslaughter. Aggravated assault. Total violent crimes.
Scholastic Update Apr 5, 1991 p. 4

Number of juveniles in custody facilities, 1985-89. Total U.S. Per 100,000 juveniles.
Scholastic Update Apr 5, 1991 p. 4

Demographic characteristics of 56,123 juveniles in custody, Feb. 15, 1989. Female. Male. Other. Hispanic. Black. White.
Scholastic Update Apr 5, 1991 p. 4

The juvenile justice system. Portrait of a crisis. Property and violent crime arrest rate by age group. Age 10 - 65.
Scholastic Update Nov 4, 1988 p. 6

Total number of offenders in juvenile facilities. 1975-1985.
Scholastic Update Nov 4, 1988 p. 7

The average yearly operating cost, per resident, of public and private juvenile detention facilities. 1975-1984.
Scholastic Update Nov 4, 1988 p. 7

The number of times that juveniles in detention have been arrested. Their number of prior admissions to juvenile detention facilities.
Scholastic Update Nov 4, 1988 p. 7

CRIME, CORPORATE

Tough laws, stiff penalties. Swelling fines for corporate criminals. Antitrust. Banking. Environment. Securities.
Business Week Apr 22, 1991 pp. 102–103

White collars, red faces. Arrests for embezzlement. [For] fraud. 1976-1986.
U.S. News Mar 14, 1988 p. 76

CRIME PREVENTION

To stop a thief. Responses of 589 convicted criminals who were asked how best to foil burglars.
U.S. News May 1, 1989 p. 76

CRIME, VICTIMS OF

Victims of crime. Cops under fire. Those at greatest risk of being crime victims.
U.S. News Dec 3, 1990 p. 41

Race and violence. When the assailant is black, the victim is: Black. White. When the assailant is white, the victim is: Black. White.
U.S. News Aug 22, 1988 p. 54

CRIMINAL JUSTICE, ADMINISTRATION OF

More cases. U.S. federal district court filings. 1960-1990.
Time Aug 26, 1991 p. 54

Crime and punishment. Homicide cases May 1-7, 1989. Potential prosecutions. Suspects identified. Suspects arrested. Convicted. Sentences imposed.
Time May 14, 1990 p. 31

Upstairs-downstairs justice. Average sentence for: Drugs. Fraud. 1979-1989.
U.S. News May 7, 1990 p. 24

Crime and punishment. Here is the average sentence given after a: Murder. Robbery. Rape. Aggravated assault. Burglary. Drug trafficking. Larceny. Trial by jury. Trial by judge. Guilty plea.
U.S. News Apr 23, 1990 p. 78

Crime and punishment. Criminal defendants per 100,000. Percent convicted. New York. Philadelphia. Detroit. Los Angeles. Chicago. Boston.
Newsweek Nov 27, 1989 p. 10

The jailhouse crunch. Corrections population. Out of custody. In custody. 1983-1989.
U.S. News Nov 20, 1989 p. 76

Courts and prisons are swamped. Number of prison inmates, state and federal. 1984-1988.
Scholastic Update Nov 17, 1989 p. 9

Most released criminals commit new crimes. Percent of prisoners released in 1983 who were rearrested, reconvicted, or returned to jail within 3 years. All prisoners. Males prisoners. Female prisoners. Prisoners age 17 or younger.
Scholastic Update Nov 17, 1989 p. 9

Average sentence length, in years. Murder. Rape. Robbery. Arson. Fraud.
Time Nov 6, 1989 p. 62

Crime and punishment. Average time served in state prisons (in months). Blacks. Whites.
U.S. News Aug 22, 1988 p. 54

CUBA AND THE SOVIET UNION

Comrades in arms. Aid from the Soviet bloc. Economic aid. Military aid. To Cuba. To Nicaragua.
Newsweek Dec 5, 1988 p. 30

CZECHOSLOVAKIA - ECONOMIC CONDITIONS

Adding up the damage. Bulgaria. Czechoslovakia. Hungary. Poland. Romania. Real economic growth. Hard-currency balance. Oil imports.
U.S. News Nov 5, 1990 pp. 55–56

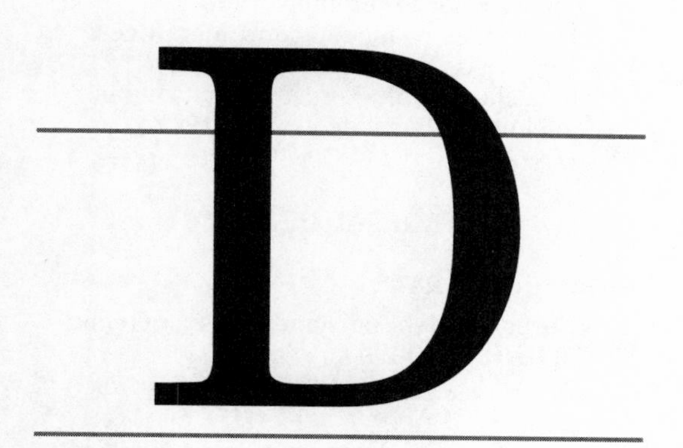

DADE COUNTY (FLORIDA)

Latin power: A demographic upheaval. Dade County's ethnic mix. Breakdown of the Latin population.
Newsweek Jan 25, 1988 p. 28

DAMAGES (LEGAL)

More money. Average punitive damages awarded by jury. 1965-69, 1980-84.
Time Aug 26, 1991 p. 54

Five of the largest punitive damage awards.
Business Week Mar 27, 1989 p. 54

DAY CARE

See CHILD CARE

DEATH AND DEATH RATES

Causes of death. Heart disease. Cancer. Stroke. Diabetes. Liver disease. Asthma. Tuberculosis. Whites. Blacks. 1988.
Time Sep 16, 1991 p. 51

Death. Deaths of children under age 5 (per 1,000 live births, 1988) [by country].
Scholastic Update Jan 25, 1991 pp. 4–5

Disease. Child deaths attributable to [various] diseases (1988).
Scholastic Update Jan 25, 1991 pp. 4–5

Children's crusade. Causes of death.
Newsweek Oct 8, 1990 p. 48

Death odds. Death by: Smoking (by age 35). Bicycle accident. Tornado. Lightning. Bee-sting. Falling airplane parts. Shark. Your odds.
Newsweek Sep 24, 1990 p. 10

Choosing death. Elderly suicides. People 65 years or older. National average. 1977-1987.
Newsweek Jun 18, 1990 p. 46

Fatal attraction. Lightning strikes the earth 100 times each second. Here is the number of lightning deaths in the U.S. 1940-1989.
U.S. News Jun 4, 1990 p. 78

Killer sports. Here are the death rates per 100,000 participants in deadlier sports.
U.S. News Jan 15, 1990 p. 67

Risky business. Deadly places to work [by selected occupation].
Newsweek Dec 11, 1989 p. 43

Fitness pays. Death rates. Fitness levels. Men. Women.
Newsweek Nov 13, 1989 p. 77

Fitness pays. Deaths per 10,000 people in a year. All causes. Cardiovascular. Cancer. Men. Women.
Time Nov 13, 1989 p. 13

Teen tragedies. Teenager's motor-vehicle deaths in 1988. The percentage of teenagers who own a car.
U.S. News Aug 14, 1989 p. 70

Deaths by firearms. Accidental deaths. Homicides. Suicides. 1950-1986.
Time Jul 17, 1989 p. 61

Rating threats. Accidental deaths. Causes in the U.S. 1985.
Time Mar 27, 1989 p. 27

On a collision course. Country [U.S. and various European countries]. Maximum speed limit. Traffic deaths per 100 million vehicle miles.
U.S. News Jan 9, 1989 p. 65

10 leading causes of death: United States, 1987. Rank. Cause of death. Number. Percent of total deaths.
FDA Consumer Nov 1988 p. 5

The heart-attack gap widens. Deaths from heart disease. Black males. Black females. White males. White females. 1970-1985.
Newsweek Oct 31, 1988 p. 66

The 65-mph question. Motor-vehicle traffic fatalities in 1987 [by state]. Deaths per 100,000 population. Change since 1986.
U.S. News Aug 15, 1988 p. 78

More, but safer, miles. Average miles per driver. 1970-1987. Death rate per 100 million vehicle miles. 1970-1987.
U.S. News Aug 15, 1988 p. 78

Survival of the fittest: How insurers see it. Annual deaths per 1,000 insured persons. 1958. Current. Smoker. Nonsmoker.
Business Week May 30, 1988 p. 105

The risks. AIDS [deaths]. Homicides. Suicides. Motor-vehicle deaths.
Newsweek Apr 11, 1988 p. 67

Which jobs are killers. Annual number of deaths per 1,000 workers. Blue-collar and service workers. White-collar and clerical workers.
U.S. News Jan 18, 1988 p. 75

Surviving the teen years. Leading causes of death. Ages 15-24. All ages.
Scholastic Update Jan 15, 1988 p. 2

DEATH PENALTY

See CAPITAL PUNISHMENT

DEBT

Heavy lifting. How America's debt burden threatens the economic recovery. Debt-to-GNP ratio. 1952-1989. Total U.S. debt. 1980-1989. U.S. net national savings as a percentage of GNP. 1953-1989.
U.S. News May 6, 1991 pp. 52–54

Personal consumption expenditures. Gross national product. Nonfinancial debt. Mortgage foreclosures started. Total bankruptcy filings. 1982-1990.
Business Week Aug 13, 1990 p. 23

Debt growth is slowing sharply. Total public and private debt. 1980-1989.
Business Week Feb 19, 1990 p. 22

The $5 trillion mountain of debt. Outstanding amounts of individual credit and insurance programs in trillions. 1965-1988.
Time Dec 18, 1989 p. 40

The case for lower interest rates. Percentage change in consumer prices. 1988-1990. Percentage change in household debt. Percentage change in total nonfinancial debt. 1980-1989.
U.S. News Oct 23, 1989 p. 56

See Also FEDERAL DEBT

DEBT, CORPORATE

Corporate debt outstanding as a percent of GNP. 1980-1990.
Time Nov 19, 1990 p. 79

The debt pile. Corporate debt. Net equity issued. 1970-1988.
U.S. News Feb 13, 1989 p. 62

The debt bill. Interest payments as a share of cash flow. 1960-1988.
U.S. News Feb 13, 1989 p. 63

Corporate debt. 1982-1988.
Time Dec 5, 1988 p. 70

Corporate America's big bet: More debt, less equity. Debt. Equity. 1983-1988.
Business Week Nov 7, 1988 pp. 138–139

Corporate debt rises to historic highs. Ratio of debt to net worth. 1972-1987.
Business Week Nov 7, 1988 p. 140

Corporate debt in trillions of dollars. 1983-1988.
Newsweek Nov 7, 1988 p. 80

Increased debt burden on the economy and U.S. corporations. Corporate debt as a share of gross national product. 1970-1988.
U.S. News Nov 7, 1988 p. 63

The growing burden of corporate debt. Net interest payments as a percent of cash flow. 1978-1988.
Business Week Sep 12, 1988 p. 26

An insatiable appetite for debt. Corporate debt as a percent of GNP. 1982-1988.
Business Week Apr 18, 1988 p. 22

Bankruptcy bonds. Public debt of bankrupt companies. 1984-1987.
U.S. News Jan 11, 1988 p. 58

DEBT, INTERNATIONAL

Mixed performance. Overseas IOU's. U.S. net foreign debt as a percentage of GNP. 1984-1989.
U.S. News Dec 25, 1989 p. 42

The poor get poorer. Per capita gross national income. Debt service as a percentage of exports of goods and services. [Selected world regions].
U.S. News Oct 2, 1989 p. 51

Caught in the debt trap. 1988 debt. 1982 debt. 1988 payments due. Argentina. Brazil. Mexico. Poland. Hungary. Ivory Coast. Nigeria. Philippines.
U.S. News Jul 24, 1989 pp. 20–21

Heap of trouble. Foreign debt [Latin American countries] in billions. 1987.
Time Jan 9, 1989 p. 33

The ever-mounting debt of developing nations. sub-Saharan Africa: External debt. 1981-1987.
Black Enterprise Jan 1989 p. 53

The poor must pay. Payments due the developed nations now exceed the flow of new loans. Sub-Saharan Africa. Asia. Latin America and the Caribbean.
Time Oct 10, 1988 pp. 86–87

How the Third World debt burden mounts. Countries are borrowing more. Developing nations' total foreign debt. 1983-1988.
Business Week Oct 3, 1988 p. 110

How the Third World debt burden mounts. Their capital keeps flowing out. Debt service and other foreign-exchange outflow from 15 leading debtors. 1983-1988.
Business Week Oct 3, 1988 p. 110

How the Third World debt burden mounts. The market is losing faith in them. Year-end open-market price for loans. Argentina. Brazil. Mexico. 1985-1988.
Business Week Oct 3, 1988 p. 110

The debt bomb. Total external debt of sub-Saharan Africa. Official development assistance to sub-Saharan Africa. 1981-1986.
U.S. News Jun 27, 1988 p. 31

Debt bombs. Total hard-currency debt. Poland. East Germany. Czechoslovakia. Hungary. Yugoslavia. Romania. Bulgaria.
U.S. News May 16, 1988 p. 31

The reddest ink. Per-capita net hard-currency debt. Debt-service ratio [by Eastern European country].
U.S. News Feb 22, 1988 p. 63

DEBT, PERSONAL

[Consumer] debt burden has grown. Total non-mortgage debt of U.S households. 1980-1991.
Business Week Nov 11, 1991 p. 134

Caught in the web of consumer debt. 1990 consumer debt. Mortgages. Auto. Credit cards. Other. Non-installment credit. Mobile homes.
Black Enterprise Sep 1991 p. 102

Heavy debt load. Household debt as a percentage of disposable income. 1957-1990.
U.S. News Jul 8, 1991 p. 51

Constrained consumers. Ratio of household debt to income. 1960-1989.
U.S. News May 6, 1991 p. 54

Personal bankruptcies are rocketing. Bankruptcies per thousand persons. 1980-1990.
Business Week Aug 20, 1990 p. 18

Debt. Amount and types of debt, excluding mortgage, per household. [By] age of head of household. Consumer credit. Margin debt. Personal loans.
U.S. News Jul 30, 1990 p. 72

DECISION MAKING

Decisions, decisions. In a survey of 1,000 Americans, the following percentage said they have trouble [making these specific decisions].
U.S. News Feb 5, 1990 p. 74

DEFENSE CONTRACTS

Military money. Company. Defense contracts, 1989 [in billions].
Newsweek Aug 6, 1990 p. 44

The top 10 defense contractors. Pentagon contracts received in fiscal 1989.
Business Week Jul 2, 1990 p. 69

The biggest guns. Top companies [defense contracts]. Top states [defense contracts].
Time Apr 30, 1990 p. 70

Where defense dollars go. Top 10 states by defense revenues. Value. Top contractor. 1988.
U.S. News Feb 12, 1990 p. 43

A thinner cushion for defense. Backlog of government orders for aircraft, engines, and parts. 1985-1990.
Business Week Jan 8, 1990 p. 70

Top guns. In 1988, these U.S. firms received $137 billion in Pentagon contracts. Leading contractors. Earnings from defense. Percent of total revenues.
U.S. News Dec 4, 1989 p. 52

Defense contracts go to all 50 states. State-by-state breakdown of prime Defense Dept. contracts awarded for 1988 [amount per state]. Map.
Scholastic Update Oct 6, 1989 p. 10

DEFENSE SPENDING

The shrinking U.S. military. Budget. 1990, 1995. Personnel. 1987, 1995.
U.S. News Apr 15, 1991 p. 42

The uneven impact of defense cuts. Defense-related job losses by state. 1991-1994. Map.
Business Week Jul 2, 1990 p. 64

Professions hit hardest by Pentagon cutbacks. Number of workers employed in key defense-plant occupations. Communications equipment. Guided missiles. Shipbuilding and repair. 1988-1994.
Business Week Jul 2, 1990 p. 67

How the federal government spends its R&D money. Federal R&D outlays. Basic. Applied. Development. Defense. Nondefense. 1980, 1985, 1989.
Business Week Jun 15, 1990 p. 47

U.S. military spending 1930-1990.
Bull Atomic Sci Mar 1990 p. 37

Defense spending. 1980-1990.
Newsweek Jan 22, 1990 p. 26

Defense budget, 1946-1990. Chart.
Scholastic Update Oct 6, 1989 p. 6

Bigger bucks for big bombers. B-2. B-1B. B-52. B-29. Last purchased. Number built. Cost per plane. Range.
U.S. News Jul 31, 1989 p. 10

Stable or on a spree? Defense spending. Defense as a percent of total federal spending. 1970-1989.
Newsweek Jan 23, 1989 p. 17

Build up, build down. U.S. defense spending, adjusted for inflation. 1980-1991.
U.S. News Mar 7, 1988 p. 34

Build up, build down. Funding for [defense] support equipment. Munitions. Spare parts. 1980-1989 (est.).
U.S. News Mar 7, 1988 p. 34

Conventional arms. Defense spending as percentage of gross domestic product. Military personnel as a percentage of total population [by country].
U.S. News Mar 7, 1988 p. 42

The pentagon tightens its purse strings. Defense dept. procurement outlays. 1983-1988.
Business Week Jan 11, 1988 p. 83

See Also DEFENSE CONTRACTS

DEFENSE SPENDING - BUSH ADMINISTRATION

U.S. nonnuclear defense missions. The defense budget.
U.S. News Oct 14, 1991 p. 29

DEFENSE SPENDING - EUROPE

Europe reins in defense spending. Defense expenditures by NATO Europe. 1985-1990.
Business Week Feb 19, 1990 p. 35

DEFENSE SPENDING - REAGAN ADMINISTRATION

What did we get for $2 trillion? Pentagon spending increased 65 percent during the Reagan years. Army. Air Force. Navy. Strategic nuclear launchers. Total nuclear warheads.
U.S. News Jan 16, 1989 p. 21

Ronald Reagan's military budget dreams. Defense Department five-year budget requests. Energy Department five-year budget requests. Nuclear weapons. Fiscal years 1982-1989.
Bull Atomic Sci Dec 1988 p. 52

DELIVERY INDUSTRY

Instant delivery. Cost to have mail delivered overnight [by carrier/company].
U.S. News May 23, 1988 p. 76

DEMOCRACY

The spread of democracy. Central & South America. Eastern Europe. U.S.S.R. People's Republic of China. Africa. [Color-coded map indicating] nations that are free. Nations that are partly free. Nations that are not free.
Scholastic Update Mar 9, 1990 pp. 6–7

Freedom House's latest analysis of political liberty around the world. Free. Partly Free. Not Free. Map
U.S. News Jan 8, 1990 p. 16

DEMOCRATIC PARTY

Which party is better for the economy? How the economy fared under recent presidents [1960-1988]. Democrats. Republicans. Real GNP growth. Unemployment rate. Inflation rate. Stock prices. Prime rate.
U.S. News Oct 17, 1988 p. 82

Presidential politics in black and white. Who votes for the Democrats? Percent of total black vote. Percent of total white vote. 1968-1984.
Business Week Jul 4, 1988 p. 97

DENMARK - MILITARY STRENGTH

Dwindling force. Danish defenses. 1967, 1977, 1987.
U.S. News May 2, 1988 p. 46

DENTISTRY

Drilling for danger? A debate over the safety of 'silver' fillings. Mercury vapors escape from standard amalgam fillings. Contents of a 'silver' amalgam filling. Diagram.
Newsweek Oct 15, 1990 p. 80

As many as 100,000 Americans will undergo surgery to be fitted with dental implants by 1992. Blade-shaped implant. Screw-shaped implant. Tripodal pin implant.
FDA Consumer Jan 1989 p. 14

Look, ma, no cavities! Percentage of children [by age] who have never had a cavity. 1979-80. 1986-87.
U.S. News Dec 26, 1988 p. 120

A toothy smile. Percentage of children free of cavities and decay. 1986-1987.
U.S. News Jul 4, 1988 p. 12

DEPARTMENT OF ENERGY

The DOE's [Department of Energy] expensive studies. Costs of RIFS work at 12 DOE sites. DOE site. Original estimate. Cost to date.
U.S. News Feb 19, 1990 p. 28

DEPRESSION (MENTAL)

Weapons in the war against depression. Brand [of drugs]. Side effects. Market share.
Newsweek Mar 26, 1990 p. 40

DESKTOP PUBLISHING

Electronic publishing: More and more a desktop game. Sales of electronic publishing systems by type of hardware. Apple. IBM PC & compatibles. Workstation. Mainframe.
Business Week Nov 28, 1988 p. 156

DEVELOPING COUNTRIES

The poor get poorer. Per capita gross national income. Debt service as a percentage of exports of goods and services. [Selected world regions].
U.S. News Oct 2, 1989 p. 51

The poor must pay. Payments due the developed nations now exceed the flow of new loans. Sub-Saharan Africa. Asia. Latin America and the Caribbean.
Time Oct 10, 1988 pp. 86–87

DEVELOPING COUNTRIES - DEFENSES

Weapons: How and what the Third World produced, 1982-87. On the ground. In the air. At sea. [By country].
Bull Atomic Sci May 1990 p. 21

Third World ballistic missiles. Country. Designation. Range. First flight. Comments.
Bull Atomic Sci Jun 1988 p. 19

DIABETES

Disabling disease. Possible long-term complications. In a healthy person. In Type I diabetes. In Type II diabetes. Diagram.
Time Nov 26, 1990 p. 52

DIET

See NUTRITION

DIETING

Broader figures. Number of dieters in the U.S. 1986, 1991.
Time Jul 8, 1991 p. 51

An all-consuming passion. Percent of the female population [who] dieted before reaching 18. Percent of all high-school seniors [with] an eating disorder. Bulimia nervosa. Anorexia nervosa.
Newsweek May 13, 1991 p. 58

Eating out, slimming down. Most [customers] have changed what they order when dining out.
U.S. News Jun 27, 1988 p. 64

Pounding away. Rating popular diets.
U.S. News Jun 20, 1988 p. 78

DINNERS AND DINING

Piecemeal dinners. The family meal is a vanishing species. Percent of families who dine together.
U.S. News Apr 23, 1990 p. 78

New-age cooks. Restaurants are responding to their customers' interest in healthier eating [various cooking methods]. 1984, 1989.
U.S. News Feb 26, 1990 p. 63

A taste for real meals. Twelve hundred adults were asked how many of their dinners during a typical week were eaten at home and prepared mostly from scratch using fresh ingredients. Here is the percentage.
U.S. News Jan 15, 1990 p. 67

Pigging out at the mall. Food. Sales share. Annual sales.
Newsweek Sep 25, 1989 p. 6

Eating out, slimming down. Most [customers] have changed what they order when dining out.
U.S. News Jun 27, 1988 p. 64

When do people eat fresh fruit? Why are people eating more fresh fruit?
FDA Consumer May 1988 p. 13

Nibbling away at the grocer's pie. Food eaten away from home [at restaurants, fast-food outlets, cafeterias, bars and taverns] as a share of total U.S. food sales. 1983-1988.
Business Week Jan 11, 1988 p. 86

DINOSAURS

Dinos on parade [types of dinosaurs]. Triassic [period]. Jurassic [period]. Cretaceous [period]. Timeline.
Newsweek Oct 28, 1991 pp. 56–57

DISARMAMENT

See ARMS CONTROL

DISASTERS

Lives lost in U.S. tornadoes. Lives lost in floods. Lives lost in hurricanes. [Other selected statistics.]
U.S. News Jul 8, 1991 p. 6

Where disaster may strike. Hurricane paths. Landslides. Floods. Droughts. Tornadoes. Earthquake activity. Map of U.S.
Scholastic Update Dec 15, 1989 pp. 10–11

DISCRIMINATION

Growing tolerance? Reported discriminatory acts against: Asians. Hispanics. Blacks. Jews. 1980-1987.
U.S. News Aug 29, 1988 p. 103

DISCRIMINATION IN EMPLOYMENT

How companies are hiring. The following percentages of surveyed companies have written affirmative action plans or goals and timetables for hiring protected groups. Size of business. Type of business.
Black Enterprise Sep 1989 p. 46

DISCRIMINATION IN HOUSING

Black and white disparities. 10 metro areas with the *highest* ratio of black-to-white loan rejections. Black rejection rate. White rejection rate. Black-white-rejection ratio.
U.S. News Feb 27, 1989 p. 26

10 metro areas with the *closest* ratio of black-to-white loan rejections. Black rejection rate. White rejection rate. Black-white-rejection ratio.
U.S. News Feb 27, 1989 p. 27

DISEASES

Reported cases of tuberculosis in the U.S. 1960-1990.
Time Dec 2, 1991 p. 85

Reported U.S. Lyme disease cases. Map.
FDA Consumer Oct 1991 p. 5

Number of cholera cases in the continental United States and Latin America. Map.
FDA Consumer Sep 1991 p. 15

Ailments. Disease. Where it strikes. Number of cases in the U.S. Rheumatoid arthritis. Diabetes (Type 1). Graves' disease. Systemic lupus erythematosus. Multiple sclerosis. Myasthenia gravis.
Time Jun 10, 1991 p. 55

The epidemic [of cholera]. Epicenter, widespread infection. Fewer cases. At risk. Map.
Newsweek May 6, 1991 p. 44

Tuberculosis increases. Reported U.S. TB cases. Earlier CDC projections. 1970-1989.
FDA Consumer Mar 1991 p. 23

Disease. Child deaths attributable to [various] diseases (1988).
Scholastic Update Jan 25, 1991 pp. 4–5

Disabling disease. Possible long-term complications. In a healthy person. In Type I diabetes. In Type II diabetes. Diagram.
Time Nov 26, 1990 p. 52

America's hepatitis chart. New infections per year. Chronic carriers.
U.S. News Aug 13, 1990 p. 14

A year's cases: Estimated outcomes. Hepatitis B infections. Chart.
FDA Consumer May 1990 p. 16

The proliferation of Lyme disease. Reported cases. Highest incidence. Light to moderate. None. Map
Newsweek May 22, 1989 p. 69

Matters of life and death. Chronic conditions that afflict [women and men] between the ages of 45 and 64. Women. Men. Nonfatal. Fatal.
U.S. News Aug 8, 1988 p. 53

Old ills reborn. Cases of measles, mumps and whooping cough. 1980-1987.
U.S. News Jul 4, 1988 p. 61

[Map of U.S. showing spread of Lyme disease].
FDA Consumer Jul 1988 p. 23

DISEASES, HEREDITARY

How diseases are inherited. Mode of inheritance. What it means. Examples.
FDA Consumer Dec 1990 pp. 19–21

DISK DRIVES (COMPUTERS)

As small drives take hold, the big ones lose out. Worldwide sales. 3.5-inch hard disk drives. 5.25-inch hard disk drives. 1983-1992.
Business Week Oct 17, 1988 p. 106

The leaders in PC disk drives. Independent producers of disk drives. 1987 sales.
Business Week Oct 17, 1988 p. 106

DIVORCE

Calling it quits. The following number of states consider these acts legal grounds for divorce.
U.S. News Nov 6, 1989 p. 108

Matrimonial misses. How long marriages last. Years of marriage. Percentage of couples who will divorce in this or later years. Average expected length of marriage.
U.S. News Feb 27, 1989 p. 75

Where love lasts best. Marriage and divorce rates [by state] per 1,000 people for 1987.
U.S. News May 30, 1988 p. 72

U.S. in focus: Comparing the states. Percent population change. Metropolitan areas. Divorces. Marriages. [By state].
Scholastic Update Jan 15, 1988 pp. 16–17

See Also CHILDREN OF DIVORCED PARENTS

DNA

See GENETIC ENGINEERING

DOCTORS

See PHYSICIANS

DOGS

Putting on the dog. Here's the average charge in 1988 for [grooming] the following dogs.
U.S. News Aug 7, 1989 p. 62

Top dogs. Of the 1.2 million new dog registrations in 1987, the top 10 breeds.
U.S. News Jul 4, 1988 p. 58

A look at the favorites. Vital statistics on the three most popular breeds. Cocker spaniel. Miniature poodle. Labrador retriever. Profiles of 13 behavioral traits.
U.S. News Jul 4, 1988 p. 60

DRAFT, MILITARY

See MILITARY SERVICE, COMPULSORY

DREXEL BURNHAM LAMBERT, INC.

[Junk bond chronology]. 1969-1989.
Time Feb 26, 1990 p. 48

DRINKING WATER

The chlorination quandary. Solutions. Diagram.
U.S. News Jul 29, 1991 p. 50

The nitrate threat. Solution. Diagram.
U.S. News Jul 29, 1991 p. 51

Lead poisoning. Solutions. Diagram.
U.S. News Jul 29, 1991 p. 52

Microbial contaminants. Solutions. Diagram.
U.S. News Jul 29, 1991 p. 53

Chemical menaces. Solution. Diagram.
U.S. News Jul 29, 1991 p. 54

DROPOUTS

Diplomas for dropouts. Here are the state-by-state results for high-school equivalency tests in 1989. Percentage who passed. Graduates over age 40.
U.S. News Jun 25, 1990 p. 66

Status dropout rate, ages 16-24, by race/ethnicity, by sex: October 1968 to 1988.
Black Enterprise Dec 1989 p. 53

The rising drop-in rate. Percentage of 25-to-29-year-olds who completed: Less than 4 years of high school. 4 or more years of college. White. Black. 1950-1987.
U.S. News Mar 20, 1989 p. 50

Putting the poor and minorities at a disadvantage. Reading scores. Dropout rate. White. Black. Hispanics. Poor.
Black Enterprise Feb 1989 p. 70

Leaving early. High-school dropouts. 1976-1986.
Newsweek May 2, 1988 p. 60

See Also HIGH SCHOOL STUDENTS

DROUGHT

Rain, rain, come again [California drought]. Little rainfall. Reservoir levels down. Snowpack down. Map.
Newsweek Feb 18, 1991 p. 56

Arid farmlands. Rainfall from Oct. 1, 1990 to Feb. 1, 1991. Hardest hit [crops].
Time Feb 18, 1991 p. 55

The long drought is draining California's reservoirs. Reservoir storage. Precipitation. 1986-1991.
Business Week Feb 11, 1991 p. 30

What's stopping the rain? A lingering high-pressure area and split air currents prevent precipitation. June, 1988.
U.S. News Jul 4, 1988 p. 46

The browning of America. Extremely dry. Abnormally dry. Normal. Wet. [Color-coded] map.
Time Jun 27, 1988 p. 21

DRUG ABUSE

U.S. babies born in 1989 with cocaine in their blood. Mothers arrested to date for using cocaine while pregnant. Amount the U.S. government will spend fighting drugs in 1991. [Other selected statistics.]
U.S. News Sep 23, 1991 p. 14

U.S. antidrug spending. Estimated number of U.S. cocaine addicts in 1989. Drug-related arrests. Cocaine-related deaths. High-school seniors reporting drug use in past 30 days. [Other selected statistics.]
U.S. News Dec 24, 1990 p. 12

Past the peak. Cocaine-related "medical crises" from emergency rooms in 21 cities. Emergency room visits. 1985-1990.
Newsweek Sep 24, 1990 p. 79

Substance abuse: The toll on blacks. Smoking-attributable deaths. Cirrhosis of the liver caused by alcohol abuse. Deaths caused by drug abuse.
Black Enterprise Jul 1990 p. 39

DEA's mixed success. Total arrests. Drug seizures. Retail purity. Wholesale price. Retail price. 1985, 1989.
U.S. News Apr 30, 1990 p. 25

Social problems are getting worse. Percentage increases. High school dropouts. 1970-88. Violent crime. 1979-88. Poverty rate. 1975-85. Drug abuse cases. 1980-87.
Scholastic Update Feb 23, 1990 p. 3

Drug abuse is declining. Americans living in households currently using illegal drugs. 1985, 1988.
Business Week Nov 27, 1989 p. 121

The toll is heavy on minorities. Cocaine-related medical emergencies. Black. White. Hispanic. 1984, 1988.
Business Week Nov 27, 1989 p. 121

Highs and lows: Counting costs in the drug market. Ice. Crank. Crack. Heroin. Cost. Duration. Effects. Side effects.
Newsweek Nov 27, 1989 p. 38

Infant deaths. Drug addiction among pregnant women is driving up the U.S. infant mortality rate. Deaths in first year of life, per 1,000 live births [by city].
Newsweek Oct 16, 1989 p. 10

The growing toll. Cocaine deaths. Cocaine-related illnesses. 1984-1988.
Newsweek Apr 10, 1989 p. 21

Cocaine related deaths have increased and so has federal funding for the war on drugs. 1983-1987.
Black Enterprise Aug 1988 p. 29

Trouble in Peoria. Drug-related incidents reported to police. 1981-1988. Price of illicit drugs [by U.S. city].
U.S. News Jun 27, 1988 p. 15

Costs to society. Alcohol. Drugs. Crime. Treatment. Lost productivity.
Time May 30, 1988 p. 14

Antidrug money. Pentagon spending on drug interdiction. 1982-1988.
U.S. News May 30, 1988 p. 18

America's drug problem at a glance. U.S. drug programs. The federal government budget for the war on drugs, in billions of dollars. Prevention, education, and treatment. Law enforcement. Total federal spending. 1981-1988.
Scholastic Update May 20, 1988 p. 4

People who didn't say no. Drug-related emergency-room visits in 1986 and change since 1983. Emergency-room visits involving particular drugs in 1986 and change since 1983.
U.S. News Mar 21, 1988 p. 73

The drug pipeline. Sources of drugs to U.S. Drugs of choice. Current users [by sex and race]. Drug deaths. Map.
U.S. News Mar 14, 1988 p. 23

DRUG ENFORCEMENT ADMINISTRATION

DEA's mixed success. Total arrests. Drug seizures. Retail purity. Wholesale price. Retail price. 1985, 1989.
U.S. News Apr 30, 1990 p. 25

DRUG INDUSTRY

See DRUGS (PHARMACEUTICAL)

DRUG TRAFFICKING

The flow goes on. Average monthly cocaine seizures in Panama. 1989-1991.
Time Aug 26, 1991 p. 18

Crack connection. Outposts of Jamaican Posses. Crips and Bloods movements. Jamaican Posses' overseas smuggling routes. Map.
U.S. News Aug 19, 1991 p. 46

New routes for the drug trade. Major processing and refining areas. Shipping routes. Established routes. Developing routes. Map of South America.
Newsweek Jul 1, 1991 p. 33

Price of cocaine. Median wholesale price per kilogram. 1985-1990.
Time Dec 3, 1990 p. 47

The new drug front. Mexico's Northern Border Response Force. Helicopter bases. Alternate bases. Map.
U.S. News Dec 3, 1990 p. 54

Looking over the horizon. The OTH [over-the-horizon] radar system in Maine can scrutinize 11.7 million sq. km (4.5 million sq. mi.), up to 3,300 km (2,000 mi.) away. Diagram.
Time May 7, 1990 p. 83

The drug pipeline. Major routes. Cocaine. Marijuana. Heroin. 1989 production in metric tons. Coca leaves. Marijuana. Map.
Time Feb 19, 1990 p. 62

Where the cocaine trail begins. Coca-growing areas. Coca-processing areas. Map.
Scholastic Update Nov 17, 1989 p. 13

The cartels in America. Federal authorities have identified Colombian drug organizations in 16 states. Map.
Newsweek Nov 13, 1989 p. 38

Dangerously cheap. Median wholesale price per kg. Cocaine seized by DEA. 1981-1989.
Time Nov 13, 1989 p. 81

Cocaine connections. This is the way Andean drug lords get the drug to market. [Map].
U.S. News Sep 11, 1989 p. 20

Targeting the traffickers. Map.
Newsweek Sep 4, 1989 p. 21

The big suppliers. 1989 production in metric tons. Coca leaves. Opium. Marijuana.
Time Aug 28, 1989 p. 11

The road to America: The global smuggling maze. Suspected opium production. Map.
U.S. News Aug 14, 1989 p. 32

Bennett's world: The front line. The new drug czar is responsible for coordinating the activities of 58 different offices. A look at where some of the key forces are being deployed. Map.
Newsweek Apr 10, 1989 pp. 22–23

Busting loose. Pounds of cocaine seized in [various European countries]. 1984-1988.
U.S. News Feb 20, 1989 p. 35

The price of failure. Range of wholesale prices per kilogram of cocaine. U.S. Miami. 1981-1987.
U.S. News Jul 11, 1988 p. 21

Where drugs come from. Sources of drugs to the U.S. Opium (heroin). Cocaine. Marijuana and hashish. Map.
Scholastic Update May 20, 1988 pp. 12–13

Where U.S. drugs come from [by country]. Marijuana. Cocaine. Heroin.
Scholastic Update May 20, 1988 pp. 12–13

How drugs are smuggled into the U.S. [by type of transportation]. Marijuana. Cocaine.
Scholastic Update May 20, 1988 pp. 12–13

The drug pipeline. Sources of drugs to U.S. Drugs of choice. Current users [by sex and race]. Drug deaths. Map.
U.S. News Mar 14, 1988 p. 23

DRUGS AND AUTOMOBILE DRIVERS

In high drive. Percentage of injured drivers treated in emergency rooms who showed signs of: Alcohol and drugs. Drugs only. Alcohol only. No drugs or alcohol.
U.S. News Dec 25, 1989 p. 8

DRUGS AND CRIME

Arrests. Number arrested for drug-abuse violations. 1985-1989.
Time Dec 3, 1990 p. 46

One measure of the war on drugs. How homicide totals changed during 1989 in 10 cities with the highest 1988 per capita murder rates. 1989 murders.
U.S. News Apr 16, 1990 p. 14

Drugs and crime go hand in hand. Percent of people who tested positive for hard drugs at time of arrest in 11 cities. Male. Female.
Scholastic Update Nov 17, 1989 p. 8

Costs to society. Alcohol. Drugs. Crime. Treatment. Lost productivity.
Time May 30, 1988 p. 14

DRUGS AND EMPLOYMENT

Wanted: Drug-free applicants. Of the job seekers tested for drug use in these industries in 1988, the following percentages tested positive.
U.S. News Oct 30, 1989 p. 82

DRUGS AND GANGS

The spread of drugs and violence. Loosely organized gangs are spreading drugs and violence from large urban areas to smaller cities and even to remote areas of the country. Most powerful drug gangs. Map.
Newsweek Mar 28, 1988 p. 22

DRUGS AND PREGNANCY

U.S. babies born in 1989 with cocaine in their blood. Mothers arrested to date for using cocaine while pregnant. Amount the U.S. government will spend fighting drugs in 1991. [Other selected statistics.]
U.S. News Sep 23, 1991 p. 14

DRUGS AND SPORTS

Teenagers blasé about steroid use. Chart 1: Most frequent steroid-associated health problems cited by current and former users. Chart 2: Some reasons cited by current users as to why they disagree with experts about risks.
FDA Consumer Dec 1990 p. 2

An apothecary chest of substances banned at the games. Effects. Sport in which use is common. Testing. Minor complications. More severe reactions. Chart.
Time Oct 10, 1988 p. 76

DRUGS AND YOUTH

Teenagers blasé about steroid use. Chart 1: Most frequent steroid-associated health problems cited by current and former users. Chart 2: Some reasons cited by current users as to why they disagree with experts about risks.
FDA Consumer Dec 1990 p. 2

Percent of high school students who have used these drugs in the past 30 days. Marijuana. Cocaine. 1979-1989.
Time Feb 26, 1990 p. 38

Drug use among U.S. teenagers. Use of 10 major drugs by high school seniors, 1986. Drug. Ever used. Past month. Past year.
Scholastic Update May 20, 1988 p. 4

Easy to get. High school seniors who say it would be "fairly easy" or "very easy" for them to get the following drugs.
Scholastic Update May 20, 1988 p. 4

Drug-related juvenile arrests. Detroit. New York City. Washington. Los Angeles. 1980-1987.
Time May 9, 1988 p. 23

DRUGS (PHARMACEUTICAL)

Common drugs that bend the mind. 100 most prescribed brand-name prescription drugs. Use. Side effects. Frequency/comments.
U.S. News Oct 28, 1991 p. 84

The top 10 drug books.
U.S. News May 20, 1991 p. 81

Storming the wall. Scientists are trying to trick the cells of the blood-brain barrier, which bar therapeutic drugs from entering the brain. Diagram.
U.S. News Oct 15, 1990 p. 90

The downside of relief. Pain relievers. Possible side effects. Cold, allergy medications. Possible side effects.
Newsweek Mar 12, 1990 p. 82

Pill bills. % change in price index since 1980. Prescription drugs. Consumer price index. 1980-1989.
Time Jan 8, 1990 p. 58

Generic-drug sales in billions. FDA inspections in thousands. 1980-1988.
Time Aug 28, 1989 p. 56

The generic drug difference. The six most prescribed generic drugs on the market and their brand-name counterparts. Generic drug. Brand name [drug]. Price. Use.
U.S. News Aug 28, 1989 p. 26

The protein connection. Conventional drug. Rationally designed drug. Diagram.
U.S. News May 8, 1989 p. 58

What the doctors ordered. Drugs most often prescribed. 1982, 1988.
U.S. News May 1, 1989 p. 76

The drug testing gauntlet. Phase I. Phase II. Size. Duration. Aim. Percentage of all new drugs that pass.
U.S. News Jan 23, 1989 p. 50

The pharmaceutical harvest. Sales of drugs, prescription and nonprescription. Research and development financed by pharmaceutical companies. 1977-1987.
U.S. News Jun 6, 1988 p. 41

From beakers to best sellers. Top-selling prescription drugs in the U.S. and their patent-expiration dates.
U.S. News Jun 6, 1988 p. 41

Prescription for profits. Merck's net income in millions. 1983-1987.
Time Feb 22, 1988 p. 45

Another brisk year for drug companies. Worldwide sales of U.S.-made pharmaceuticals. 1983-1988.
Business Week Jan 11, 1988 p. 119

DRUNK DRIVING

Where you could lose your license on the spot. Map.
Time Oct 7, 1991 p. 27

Deadly consequences. Traffic deaths, 1989. Total traffic fatalities. Alcohol-related traffic fatalities, age 15 to 19.
Scholastic Update Nov 16, 1990 p. 4

In high drive. Percentage of injured drivers treated in emergency rooms who showed signs of: Alcohol and drugs. Drugs only. Alcohol only. No drugs or alcohol.
U.S. News Dec 25, 1989 p. 8

Stearing clear of trouble. Mandatory minimum jail term for drunk-driving conviction [by state]. Jail term. Number of offenses before imprisonment.
U.S. News Dec 12, 1988 p. 96

DUCKS - NORTH AMERICA

Ducks are down [numbers by kinds of duck]. 1957-87 average. 1988. Duck-breeding ponds. 1974-87. 1988. North Central U.S. Prairie Canada.
U.S. News Oct 24, 1988 p. 72

EARTH - INTERNAL STRUCTURE

The earth's turbulent interior. This schematic diagram outlines the basic features of scientists' new view of the earth's internal engine. Diagram.
U.S. News Oct 30, 1989 pp. 38–39

A patchwork of dancing continents. The reigning model of how the earth moves is called plate tectonics. Recent research suggests, however, that this model is too simple. Diagram.
U.S. News Oct 30, 1989 p. 40

EARTHQUAKES

Rules of the road. An elevated highway consists of many parts. It must be designed so that a quake does not undermine one segment, bringing the entire structure down. Diagram.
Newsweek Oct 30, 1989 p. 35

Slip-sliding away. The earth is covered by a dozen tectonic plates, which meet at faults. There are at least two types. Diagrams.
Newsweek Oct 30, 1989 p. 41

Clusters of quakes. Major earthquakes. Year 500-2000.
U.S. News Jul 25, 1988 p. 57

EARTHQUAKES - CALIFORNIA

The rescuers' battleground. Map.
Newsweek Oct 30, 1989 p. 31

Hazardous odds. Probability of damaging earthquakes within the next 60 years.
Newsweek Oct 30, 1989 p. 38

Percent probability of a major earthquake along fault segments from 1988 to 2018. Map.
Time Oct 30, 1989 pp. 44–45

A rocky relationship's hidden stresses. California. Forecasting the big ones. Diagram.
U.S. News Oct 30, 1989 p. 45

Newly identified fault lines. Los Angeles. Newly identified faults. Known faults. Map.
U.S. News Dec 19, 1988 p. 11

EAST GERMANY

See GERMANY (EAST)

EASTERN EUROPE

The new Eastern Europe [by country]. Map.
Scholastic Update Sep 7, 1990 p. 10

The party's almost over. Communist Party membership in Eastern Europe [various countries].
U.S. News Feb 12, 1990 p. 33

Five troubled regions in an ethnic patchwork. Poland. Rumania. Yugoslavia. Bulgaria. Moldavia.
Time Jan 29, 1990 p. 50

Eastern Europe at a glance. U.S.S.R. before World War II. Territory gained by the U.S.S.R. Soviet-dominated East Europe. Other Communist nations. Map.
Scholastic Update Oct 20, 1989 pp. 6–7

Eastern Europe at a glance [vital statistics]. Bulgaria. Yugoslavia. Hungary. Romania. German Democratic Republic. Poland. Czechoslovakia. Albania.
Scholastic Update Oct 20, 1989 pp. 6–7

EASTERN EUROPE - ECONOMIC CONDITIONS

Eastern Europe's economies are shrinking. Poland. Czechoslovakia. Hungary. Industrial production. Private economy as percent of total GNP. New private business. 1989-1991.
Business Week Apr 15, 1991 p. 48

Adding up the damage. Bulgaria. Czechoslovakia. Hungary. Poland. Romania. Real economic growth. Hard-currency balance. Oil imports.
U.S. News Nov 5, 1990 p. 55

Worker paradise. Hourly industrial wage 1988 [by country].
Newsweek Dec 18, 1989 p. 50

An economic portrait of Eastern Europe. East Germany. Poland. Czechoslovakia. Hungary. Spain [used as a comparison]. Per capita GNP. Hourly wages. Education level. Population. Industrialization. Debt. Hard currency assets. 1987, 1988.
Business Week Nov 27, 1989 pp. 62–63

Cost of living. The Polish economy suffers from rampant inflation. Increase in consumer-price index since 1980. [By selected Eastern European country].
Newsweek Sep 5, 1988 p. 40

Debt bombs. Total hard-currency debt. Poland. East Germany. Czechoslovakia. Hungary. Yugoslavia. Romania. Bulgaria.
U.S. News May 16, 1988 p. 31

EASTERN EUROPE - NATIONALISM

Five troubled regions in an ethnic patchwork. Poland. Rumania. Yugoslavia. Bulgaria. Moldavia.
Time Jan 29, 1990 p. 50

EATING DISORDERS

An all-consuming passion. Percent of the female population [who] dieted before reaching 18. Percent of all high-school seniors [with] an eating disorder. Bulimia nervosa. Anorexia nervosa.
Newsweek May 13, 1991 p. 58

ECLIPSES, SOLAR

The blackness of day. Hawaii's total eclipse. Longest blackout. Graying of the continent. Percentage of sun eclipsed. Map.
U.S. News Jul 15, 1991 pp. 44–45

ECONOMIC ASSISTANCE

Who's sending care packages to Moscow. Government. Type of aid. Value (billions).
Business Week Oct 28, 1991 p. 42

The donor derby. Official aid to developing countries. Official aid as a share of gross national product [by country].
U.S. News Mar 14, 1988 p. 29

Less money to play with. Third World security aid not committed by Congress. 1955-1987.
U.S. News Jan 25, 1988 p. 46

ECONOMIC ASSISTANCE, AMERICAN

Getting a helping hand: The top 10. Top recipients of direct U.S. economic and military aid, fiscal 1990 [by country].
Newsweek Apr 16, 1990 p. 23

What the U.S. has spent on Nicaragua. 1979-1989.
Time Mar 12, 1990 p. 12

U.S. foreign aid. [To] Israel. Egypt. Poland-Hungary. Pakistan. Turkey. Philippines. Panama. Map. Total. % of aid that is military.
Time Mar 12, 1990 p. 14

ECONOMIC ASSISTANCE, DOMESTIC

Bank runs and bailouts: A short and costly history lesson. Crisis. Solution. FDR's bank holiday. Penn Central. New York City. Chrysler Corporation.
U.S. News Feb 20, 1989 pp. 50–51

ECONOMIC ASSISTANCE, JAPANESE

The strong yen gives a boost to foreign aid. Japan's official development assistance. Billions of dollars. Percentage of GNP. 1982-1987.
Business Week Jan 18, 1988 p. 41

ECONOMIC CONDITIONS - REAGAN ADMINISTRATION

Reaganomics. The end of 'tax and tax,' but not of 'spend and spend.' Effective personal income-tax rates. Federal budget outlays. Defense. Other. 1977-1987.
Business Week Feb 1, 1988 pp. 60–61

Reaganomics. The resulting stimulus paid off. Gross national product. Real disposable personal income per capita. Nonagricultural employment. 1977-1987.
Business Week Feb 1, 1988 pp. 60–61

Reaganomics. The cost was high. Percent of after tax income. Investment as a percentage of GNP. Federal budget deficit. 1977-1988.
Business Week Feb 1, 1988 pp. 60–61

Reaganomics. The successful fight against inflation. Inflation-adjusted yield on treasury securities. Consumer price index annual change. Change in the value of foreign currency per dollar. 1977-1987.
Business Week Feb 1, 1988 pp. 60–61

Reaganomics. [It] also brought its share of bad news. Merchandise trade balance. Net foreign investment position of the U.S. Manufacturing employment as a percent of total. 1977-1987.
Business Week Feb 1, 1988 pp. 60–61

ECONOMIC CONDITIONS - UNITED STATES

Federal slippage. Domestic per capita spending, 1989 dollars. Federal. State and local. 1979, 1989.
U.S. News Dec 23, 1991 p. 52

Bush's nightmare. Civilian unemployment rate. Real disposable income. Consumer confidence index. Dow Jones Industrial average.
U.S. News Dec 2, 1991 p. 20

Driving down inflation. Economic growth. Commodities Futures Price index. Money supply.
U.S. News Oct 7, 1991 p. 58

Turning back the clock? Inflation rate. 1914-1990.
U.S. News Oct 7, 1991 p. 60

How bad was it? Real gross national product. Civilian unemployment rate. Industrial production. Change during recession. 1990-91. Average.
Newsweek Aug 5, 1991 p. 64

Did the Fed cause the recesssion? Consumer price index. Money supply. Federal funds rate. Commercial and industrial loans. Gross national product. 1987-1991.
U.S. News Jul 1, 1991 p. 50

The road to recovery? Pharmaceuticals. Semiconductors. Home appliances. Net profits. 1989-1992.
U.S. News Jun 10, 1991 pp. 48–49

Closing the gap. U.S. merchandise trade deficit, percentage of GNP. 1980-1991.
U.S. News Jun 10, 1991 p. 57

Heavy lifting. How America's debt burden threatens the economic recovery. Debt-to-GNP ratio. 1952-1989. Total U.S. debt. 1980-1989. U.S. net national savings as a percentage of GNP. 1953-1989.
U.S. News May 6, 1991 pp. 52–54

A sick economy. Economic growth 1989-1990 over three-month periods.
Scholastic Update Mar 8, 1991 p. 3

Awesome assets. U.S. assets in 1989. Financial. Tangible. GNP.
U.S. News Jan 21, 1991 p. 68

The forces shaping a disappointing year. Civilian unemployment. Annual real growth rate of U.S. trading partners. Industrial production index. U.S. corporate aftertax profits. 1986-1991.
Business Week Jan 14, 1991 p. 62

Economic slump to rival '81-82 debacle. Dollar drops. Value of U.S. dollar. Deficit deepens. U.S. budget deficit. 1980-1991.
U.S. News Jan 7, 1991 p. 46

Loans 90 days overdue at banks. Corporate junk-bond defaults. Personal saving rate in. Credit-card debt per U.S. household.
U.S. News Oct 22, 1990 p. 16

Where the big economies stand. United States. Japan. West Germany. Economic growth. Inflation. Cost of credit. 1987-1990.
Business Week Aug 20, 1990 p. 29

Sluggish growth or a worsening downturn? Slow-growth scenario. Recession scenario. Oil Prices. Inflation. Unemployment. Real GNP. Federal budget deficit. 1990-1993.
U.S. News Aug 20, 1990 pp. 32–33

Personal consumption expenditures. Gross national product. Nonfinancial debt. Mortgage foreclosures started. Total bankruptcy filings. 1982-1990.
Business Week Aug 13, 1990 p. 23

Stuck in the mire. % change in real GNP, at annual rates. 1988-1991 [forecast].
Time Jul 23, 1990 p. 58

Paying down debt is getting harder. Net cash flow as a percent of debt outstanding. 1980-1989.
Business Week Jun 4, 1990 p. 30

Who says inflation is rising? Year-to-year change in consumer price index excluding food and energy. 1979-1990.
Business Week May 21, 1990 p. 31

The pressures under inflation. Excluding food and energy. Producer prices. Consumer prices. 1988-1990.
Business Week Apr 2, 1990 p. 27

Nonfinancial debt. Personal savings. Debt and equity-financed costs. Nonresidential investment. 1985-1990.
Business Week Mar 12, 1990 p. 112

The '80s: A great decade for financial assets. U.S. treasury bills. Long-term U.S. bonds. Stocks. Average annual rate of return. 1960s. 1970s. 1980s.
Business Week Dec 25, 1989 pp. 74–75

An investor's time line. Stocks, paintings and criminal acts soared in a decade seemingly defined by excess. 1980-1989.
U.S. News Dec 4, 1989 pp. 70–71

West South Central [Texas, Oklahoma, Arkansas, Louisiana]. Employment. Population. Median home price. Annual rate of change from previous year. 1989-1990.
U.S. News Nov 13, 1989 p. 55

Up, up, up. Current recovery since 1982. Post 1954. Post 1970 [a comparison of three economic recovery periods].
Newsweek Sep 11, 1989 p. 44

This is already the longest peacetime expansion. 1945-1982.
Business Week Mar 6, 1989 p. 50

Four keys to success as the expansion matures. Monthly ratio of inventories to sales in manufacturing. Exports. Index of nonfarm productivity. Employment cost index for private industry workers. 1984-1989.
Business Week Jan 9, 1989 pp. 64–65

Office vacancies remain high. Values lag behind inflation. Home prices keep rising, although far from evenly. Median sale price of existing one-family homes. Northeast. Midwest. 1983-1988.
Business Week Dec 26, 1988 p. 172

Which party is better for the economy? How the economy fared under recent presidents [1960-1988]. Democrats. Republicans. Real GNP growth. Unemployment rate. Inflation rate. Stock prices. Prime rate.
U.S. News Oct 17, 1988 p. 82

Hit-or-miss economy. See-sawing stock markets. Rising bank failures. The diving dollar.
U.S. News Apr 11, 1988 p. 47

Race gap then and now. Living in poverty. Unemployment. Median income. Blacks. Whites. 1986. Latest.
U.S. News Mar 14, 1988 p. 11

Rise & fall of theories. Unemployment rate. Money supply. Personal-savings rate. 1970-1987.
U.S. News Feb 1, 1988 p. 44

See Also RECESSION (FINANCIAL)

ECONOMIC CONDITIONS - WORLD

Economic map of the world. Per capita GNP in U.S. dollars. Map.
Scholastic Update Oct 18, 1991 pp. 28–29

Economic outlook. Why the world won't run out of capital. Current account surpluses as percentage of GNP. Long-term bond yields. U.S. Germany.
U.S. News Apr 15, 1991 p. 57

New demands for capital won't be a burden. German reunification. Japanese public investment. Kuwait. U.S. infrastructure. Eastern Europe.
Business Week Apr 1, 1991 p. 64

Economic map of the world. Per capita GNP in U.S. dollars. Map.
Scholastic Update Nov 2, 1990 p. 18

Carving up the global pie. U.S. Japan. Europe. United Germany. Standard of living. Trade balance. Percent of the world economy 1990, 2000.
U.S. News Jul 16, 1990 pp. 22–23

Economic map of the world. Per capita G.N.P. in U.S. dollars.
Scholastic Update Sep 22, 1989 pp. 16–17

Economic map of the world. Per capita G.N.P. in U.S. dollars. Map.
Scholastic Update Sep 23, 1988 pp. 12–13

Shifting economic power. U.S. U.S.S.R. Japan. China. 1950, 1990, 2010 (est.).
U.S. News Jan 25, 1988 p. 46

EDUCATION

Putting in the school days. [Required number of school days in selected countries].
Time Sep 2, 1991 p. 64

Comparative report card. New York City. Public schools. Catholic schools. Students. Student-to-teacher ratio. Percentage graduated on time. Special-education classes. Spending. Average teacher salary. Administrators.
Time May 27, 1991 p. 49

Comparing state stats. Public school spending per student. Percent graduated from high school. [By state].
Scholastic Update Jan 11, 1991 pp. 26–27

Children's crusade. Education. Illiteracy. Developed countries. Nondeveloped countries. Global total.
Newsweek Oct 8, 1990 p. 48

Bookworms. Days spent in grade school per year. U.S. Japan. West Germany.
U.S. News Jul 16, 1990 p. 26

A dearth of role models. Minority students earning education degrees. Minority share of public school children. 1977-1987.
Business Week May 7, 1990 p. 120

A mixed picture. High school graduation rate (for those 25-34 years old). Black. White. 1970-1988.
U.S. News Jan 22, 1990 p. 28

Comparing state stats. Public school spending per student. Percent graduated from high school. [By state].
Scholastic Update Jan 12, 1990 pp. 30–31

Percentage of Hispanic population with a high-school education, by year of immigration.
U.S. News Sep 25, 1989 p. 31

More teacher per pupil. Students in at least 47 states may be getting more individual attention from teachers than they used to. Here is the ratio of public-school pupils to teachers by state. 1982 ratio. 1988 ratio.
U.S. News Sep 11, 1989 p. 87

Study shortcuts. The percentage of Cliffs Notes users by grade. Of the 210 titles published by Cliffs Notes, here are their 10 bestsellers.
U.S. News Aug 21, 1989 p. 66

Progress report. Average teacher's salary. Expenditure per pupil. Standardized-test scores. Arkansas. South Carolina. West Virginia. 1982, 1988.
Time Aug 14, 1989 p. 46

Putting the poor and minorities at a disadvantage. Reading scores. Dropout rate. White. Black. Hispanics. Poor.
Black Enterprise Feb 1989 p. 70

Fighting for what's been left behind. In Chicago, Dallas, Los Angeles and Miami white students continue to abandon public schools, while blacks and Hispanics compete for scarce resources. Black. White. Hispanic.
Newsweek Dec 12, 1988 p. 29

U.S. Almanac: Comparing state stats. Public school spending per student. Percent graduated from high school. [By state].
Scholastic Update Dec 2, 1988 pp. 18–19

The leaders in the education market. Share of personal computer shipments [by company].
Business Week Oct 24, 1988 p. 76

School enrollment. 1965, 1975, 1985.
Black Enterprise Oct 1988 p. 137

The payoff of education is huge. 1986 earnings of men 25-29 [by years of education].
Business Week Sep 19, 1988 p. 132

How education lowers the risk of unemployment. Jobless rates. High school dropouts. High school graduates. College graduates. 1967-1987.
Business Week Feb 29, 1988 p. 20

The good old days. Leading school discipline problems: 1940s. 1980s.
Time Feb 1, 1988 p. 54

School profiles. City. Minorities. Dropouts. Assaults. Counselors. [By selected cities].
Time Feb 1, 1988 p. 55

See Also PUBLIC SCHOOLS

EDUCATION - FINANCE

Feverish growth. Percent change since 1980. Total health expenditures. Consumer price index. Federal education expenditures. 1980-1991.
Time Nov 25, 1991 pp. 34–35

Paying to learn. Average salary of public school teachers. Spending per pupil. 1979-1991. [Other selected statistics.]
U.S. News Sep 16, 1991 p. 13

Comparing state stats. Public school spending per student. Percent graduated from high school. [By state].
Scholastic Update Jan 11, 1991 pp. 26–27

Back of the class. Per pupil education spending as a percent of per capita income [by country].
U.S. News Aug 6, 1990 p. 46

The funding gap. Per-pupil expenditures for school districts with over 100 pupils, 1986-87. California. Connecticut. Illinois. Kentucky. Montana. New Jersey. New York. Texas.
Business Week Jun 4, 1990 p. 99

The U.S. lags in basic education spending. Outlays for kindergarten through 12th grade. U.S. Britain. France. West Germany. Canada. Japan. Norway.
Business Week Jan 29, 1990 p. 22

Comparing state stats. Public school spending per student. Percent graduated from high school. [By state].
Scholastic Update Jan 12, 1990 pp. 30–31

Schools and transportation get less. Elementary and secondary education. Transportation. 1987-1989.
Business Week Jun 5, 1989 p. 88

More dollars, smaller classes. Average annual amount spent per student in public schools. Students per teacher in public schools. 1955-1985.
U.S. News Mar 20, 1989 p. 51

Educational spending by state. Map.
Scholastic Update Dec 2, 1988 p. 9

U.S. Almanac: Comparing state stats. Public school spending per student. Percent graduated from high school. [By state].
Scholastic Update Dec 2, 1988 pp. 18–19

What we spend on education and who pays. State government. Local government. Private sources. Federal government.
Black Enterprise Oct 1988 p. 136

EDUCATION - TESTS AND MEASUREMENTS

A formula for failure. Math-achievement levels of high-school seniors. Below basic. Basic. Proficient. Advanced.
Newsweek Oct 14, 1991 p. 54

How the states rank. Rankings on the eighth-grade math-assessment test. [By state].
Newsweek Jun 17, 1991 p. 65

Diplomas for dropouts. Here are the state-by-state results for high-school equivalency tests in 1989. Percentage who passed. Graduates over age 40.
U.S. News Jun 25, 1990 p. 66

Back to basics. Proficiency levels of students, age 13, in selected countries. Math. Science.
Time Sep 11, 1989 p. 69

How students score. Average scores of 17-year-olds on proficiency tests. Reading. Math. Science. White. Hispanic. Black. 1970-1986.
U.S. News Mar 20, 1989 p. 51

An 'F' for the U.S. Among 18-to 24-year-olds, Americans scored dead last [out of nine countries] in identifying 16 geographic locations. Map.
Newsweek Aug 8, 1988 p. 31

EDUCATION, ELEMENTARY

Bookworms. Days spent in grade school per year. U.S. Japan. West Germany.
U.S. News Jul 16, 1990 p. 26

It's arithmetic, in a landslide. Favorite subjects of children in school. Girls. Boys. Math. English. Art. Reading. Science.
U.S. News Sep 12, 1988 p. 71

EDUCATION, HIGHER

Professorial shortfall. Faculty supply-and-demand projections in the social sciences and humanities. Supply. Demand. 1987-2012.
U.S. News Sep 25, 1989 p. 55

EDUCATION, PRESCHOOL

Preschool boom. Percentage of children enrolled in school. Ages 5 and 6. Ages 3 and 4. 1965-1985.
U.S. News Mar 20, 1989 p. 50

EDUCATION, SECONDARY

No pain, no gain. Time [spent] on homework [by] seventeen-year-old students. 1988. Chart.
Newsweek May 27, 1991 p. 62

EGYPT - MILITARY STRENGTH

Quality versus quantity. Country. Population in millions. Armed forces: Regular, reserves. Tanks. Combat aircraft.
Newsweek May 1, 1989 p. 44

ELDERLY

See OLDER AMERICANS

ELECTIONS

Midterm-election voters. Percentage of registered adults. Number of women running. Cost of 30-second political ad.
U.S. News Oct 29, 1990 p. 16

ELECTIONS - UNITED STATES

A bigger foothold for the G.O.P. Texas voters in primaries. Florida registered voters. California registered voters. 1980-1990.
Time Apr 23, 1990 p. 22

Congress's self-protection racket. Percentage of incumbents seeking re-election who won. House. Senate. 1946-1986.
U.S. News Nov 7, 1988 p. 29

Opening positions. States are shown in proportionate size to their electoral vote. Map.
Time Aug 29, 1988 p. 28

The Democrats' electoral game plan [electoral votes by state]. Map.
Newsweek Jul 25, 1988 p. 18

See Also PRESIDENTIAL ELECTIONS

ELECTIONS - UNITED STATES, 1988

Election results. State. Governor's name. U.S. Senators [political party]. U.S. Reps [number and political party].
Scholastic Update Dec 2, 1988 pp. 16–17

ELECTIONS - UNITED STATES, 1990

The scorecard. House seats. Senate seats. Governorships. Democrats. Republicans. Old [before election]. New [after election].
Business Week Nov 19, 1990 p. 43

Failing midterms for the party in power. Change in seats for President's party in midterm elections. House. Senate. 1962-1990.
Business Week Nov 19, 1990 p. 45

Bush's box score. Bush campaigned for 63 candidates since May. Here's how they fared.
Time Nov 19, 1990 p. 31

Unbeatable house of incumbents. Who ran. Who won. Who ran unopposed. Who ran financially unopposed.
Time Nov 19, 1990 p. 32

ELECTRIC POWER

Power puzzle. Net generation of electricity by U.S. utilities. Nuclear. Coal. Natural gas. Petroleum. Renewal sources.
Time Apr 29, 1991 pp. 56–57

Nuclear power. Fossil fuel power. Hydroelectric power. Solar power. Wind power. Pros. Cons. Comments.
Time Apr 29, 1991 pp. 56–57

Who's gone nuclear. Percent of electricity derived from nuclear power. [By country].
Time Apr 29, 1991 p. 61

Small is beautiful. Energy generated for electric utilities by independent producers. 1980-1987.
U.S. News Aug 22, 1988 p. 39

Tough choices. Electricity generated from: Nuclear power. Coal. Natural gas. Hydropower. Petroleum. Geothermal and other. Sweden. U.S. 1987.
U.S. News Jul 18, 1988 p. 43

See Also NUCLEAR POWER PLANTS

ELECTRIC POWER - CONSUMPTION

Watt's up. Electricity sales (in kilowatt-hours). 1973-1987.
U.S. News Aug 22, 1988 p. 39

ELECTRIC POWER - SWEDEN

Tough choices. Electricity generated from: Nuclear power. Hydropower. Coal. Petroleum. Sweden. 1987.
U.S. News Jul 18, 1988 p. 43

ELECTRIC UTILITIES

A challenge to public utilities. Share of U.S. electric power capacity owned by independent producers. 1985-1990.
Business Week Jan 8, 1990 p. 92

ELECTROMAGNETIC WAVES

A journey across the spectrum. Equipment to view energy emitted by heavenly bodies. Gamma rays. Visible light. Near infrared. Far infrared.
Newsweek Jun 3, 1991 p. 48

Wavelengths (in meters). X-ray. Ultraviolet. Microwaves. Television.
Newsweek Jun 3, 1991 pp. 48–49

ELECTRONIC MAIL SYSTEMS

E-mail is growing. Here's why. Billions of messages. 1988-1991. Cost of sending a three-page letter. Federal Express. Telex. Fax. Electronic mail.
Business Week Feb 20, 1989 p. 36

ELEPHANTS

Home on the savanna. Elephant populations. Endangered. Stable. Flourishing. Map.
Newsweek Nov 18, 1991 p. 87

EMPLOYEE BENEFITS

Compensation, anyone? A breakdown of employee benefits.
Black Enterprise Oct 1991 p. 108

The trend toward flex. How companies offer benefits as part of fixed or flex packages.
Black Enterprise Oct 1991 p. 110

Benefits outpace pay. Annual percentage change. Benefits. Wages. 1981-1990.
U.S. News Nov 5, 1990 p. 58

Pack up the pooch. An employe asked to relocate can usually count on reimbursement for normal moving expenses, but what about extras? Percentage of companies that will pay [various extras].
U.S. News Jul 16, 1990 p. 66

Pay and benefits in private industry. Percentage change in worker compensation. 1984-1989.
U.S. News Apr 9, 1990 p. 42

The relentless growth of fringe benefits. Pensions and profit-sharing. Health insurance. Other. Percent of total compensation. 1950-1987.
Business Week Nov 13, 1989 p. 34

Say goodbye to the age of free rides. Types of group health plans. 1984, 1988.
Newsweek Jan 30, 1989 p. 47

Work hard and play hard. Average number of working hours and vacation days in major (international) cities. [City]. Hours worked per week. Vacation days per year.
U.S. News Nov 14, 1988 p. 82

The growing benefits burden. Medium and large U.S. firms with major medical plans. Out-of-pocket payments by employes on benefits. Employers' spending on employe benefits.
U.S. News Mar 28, 1988 pp. 58–59

Still more coverage. A sampling of state laws passed in 1987 requiring new or expanded health-care coverage.
U.S. News Mar 28, 1988 p. 61

How do my benefits stack up? Large company. Medium company. Small company. [By specific benefit].
U.S. News Mar 28, 1988 pp. 71–74

EMPLOYEE DISMISSAL

Where parting is sweetest. Percentage of annual salary typically awarded employes who are fired or laid off [by country]. Manufacturing workers. Chief executives.
U.S. News Dec 19, 1988 p. 70

The canning business. Days of the week employes get fired. Time of day employes get fired.
U.S. News Apr 18, 1988 p. 83

EMPLOYEE TRAINING

American workers are not only undertrained. Percent of workers with qualifying training for their jobs, by job category.
Black Enterprise Feb 1990 p. 70

They don't get enough retraining. Percent of workers who have received educational upgrading for their jobs, by job category.
Black Enterprise Feb 1990 p. 72

Formal training by sex and race of trainee. Share of training. Share of the work force. Males. Females. Whites. Blacks. Hispanics. [By] age.
Black Enterprise Aug 1989 p. 41

The cost of training. Elementary and secondary. Post secondary. Employee informal training. Employee formal training. Government training.
Black Enterprise May 1989 p. 70

The looming mismatch between workers and jobs. Actual skill levels of new workers. Percent of 21-to 25-year-olds entering the labor market from 1985 to 2000. Skill levels needed for new jobs.
Business Week Sep 19, 1988 pp. 104–105

EMPLOYEES

Time is money. Per-employe cost of various breaks in the workday. Employe's annual income. 10-minute meeting delay. 1-hour meeting. 2-hour lunch.
U.S. News Aug 8, 1988 p. 65

EMPLOYMENT

See JOBS

EMPLOYMENT REFERENCES

Loose lips bring lawsuits. Percentage of employers who are: Reluctant to give bad reference. Reluctant even when they have proof of employe wrongdoing.
U.S. News Oct 16, 1989 p. 125

ENDANGERED SPECIES

Home on the savanna. Elephant populations. Endangered. Stable. Flourishing. Map.
Newsweek Nov 18, 1991 p. 87

A vanishing menagerie. Casualties of a shrinking planet. Winners and losers in a fight to survive.
U.S. News Oct 2, 1989 pp. 54–57

ENERGY CONSERVATION

The big bang from energy efficiency. Innovation [in] energy use in six areas that account for nearly half of America's energy demand. Airplanes. Electric motors. Cars & light trucks. Lighting. Refrigeration. Heating, cooling, ventilation.
Business Week Sep 16, 1991 pp. 86–87

The big gains in efficiency have leveled off. Relative change in: Gross national product. Energy use. 1973-1990.
Business Week Sep 16, 1991 p. 90

The world's economic engine gets better mileage. B.t.u.'s used to generate each dollar of real gross domestic product. Canada. United States. Britain. West Germany. France. Japan. Percentage change (1978-88).
U.S. News Aug 27, 1990 p. 37

ENERGY CONSUMPTION

U.S. consumption of world's energy. Percentage of world's energy consumed by U.S. Percentage of world's population in U.S. 1960-1988.
Scholastic Update Apr 19, 1991 p. 4

Heavy users. U.S. total energy consumption. 1975-1989.
Newsweek Jan 7, 1991 p. 29

Gas pains. Total energy consumption. Amount coming from petroleum. Residential. Transportation. Industrial. Electric utilities.
U.S. News Sep 10, 1990 p. 68

Energy: Just say no. Item. Extra cost at purchase. Time to pay back. Carbon saved.
Newsweek Jun 18, 1990 p. 51

Energy eaters. Per capita energy consumption (in oil equivalent, barrels per year): U.S. Sweden. Soviet Union. West Germany. Japan. China. India. Kenya. 1970,1989.
U.S. News Apr 23, 1990 p. 70

Uncle Sam against the world. The U.S. has 5% of the earth's population, but [has the following % of oil use, nitrogen oxide and carbon dioxide emissions, toxic waste generated].
Time Dec 18, 1989 p. 63

Fuel guzzlers. Average annual growth rate in energy consumption. U.S. Japan. South Korea. China. Taiwan. 1978-83. 1984-88.
U.S. News Nov 6, 1989 p. 70

Energy *deja vu*. Net U.S. oil imports. 1977-1988.
U.S. News Mar 27, 1989 p. 55

The energy-GNP ratio tilts up again. Total energy consumption in U.S. divided by GNP. 1980-1988.
Business Week Sep 5, 1988 p. 18

ENERGY, COST OF

Smaller shock. Energy costs. 1973-1991.
U.S. News Oct 22, 1990 p. 59

ENERGY SUPPLY

U.S. energy supply, by type of fuel. Oil and other petroleum products. Natural gas. Coal. Nuclear power. Hydropower. Geothermal and other. 1960-1988.
Scholastic Update Apr 19, 1991 p. 5

Sources of U.S. oil, by country and region. 1960-1988.
Scholastic Update Apr 19, 1991 p. 5

ENGINEERING SCHOOLS

Top 25 in engineering. 25 graduate schools of engineering with the highest scores in the *U.S. News* survey.
U.S. News Apr 29, 1991 p. 82

ENGINEERS

Percentage of U.S. scientists and engineers employed in defense industries. Percentage of 1989 engineering Ph.D.'s from U.S. colleges granted to foreigners. Percentage of college freshmen who plan to be scientists.
U.S. News Feb 4, 1991 p. 14

Engineering women. Scientists. Engineers. Number of women in field. Percentage in field. 1978, 1988.
U.S. News Jul 16, 1990 p. 66

ENTERTAINMENT INDUSTRY

Year the first nickelodean opened. Weekly movie attendance. 1950-1990. Number of movie screens. Average movie ticket price. Money spent by Americans at movies. [Other selected statistics.]
U.S. News Aug 19, 1991 p. 10

Entertainment: An export machine. U.S. entertainment sales to foreign countries. 1986-1991.
Business Week Jan 14, 1991 p. 97

Score card of hits. Books. Best-selling hardbacks. Fiction. Nonfiction. Television. Most watched shows. Movies. Biggest box-office successes. Videocassettes. Top rentals. Pop-music compact disks. Bestsellers.
U.S. News Dec 26, 1988 p. 120

Take me out to the ball game. Price of a ticket [to various events].
U.S. News Apr 4, 1988 p. 73

ENVIRONMENT

Firing lines. The West. Environmentalists and others challenge business interests for control of the land. Key battlegrounds. Map.
Newsweek Sep 30, 1991 p. 29

Not so green. Here is the percentage of households that [practice various conservation methods].
U.S. News Feb 4, 1991 p. 71

The Geritol solution. Can tiny sea creatures get us out of the greenhouse fix? Diagram.
Newsweek Nov 5, 1990 p. 67

The President's environmental report card. *U.S. News* asked leaders of 10 conservation groups and 10 business and trade associations to grade the Bush administration's environmental efforts in 12 key areas. Issue. Grade.
U.S. News Mar 19, 1990 p. 22

Uncle Sam against the world. The U.S. has 5% of the earth's population, but [has the following % of oil use, nitrogen oxide and carbon dioxide emissions, toxic waste generated].
Time Dec 18, 1989 p. 63

Stiff slaps. Fines for environmental offenses. 1983-1989.
Newsweek Sep 25, 1989 p. 35

Legacy of a spill: The damage done...and not done. First 48 hours. By 14 days. After 14 days.
U.S. News Sep 18, 1989 p. 64

Threats to the environment. Name. Description. Causes. Effects. Chart.
Scholastic Update Apr 21, 1989 p. 3

A guide to some of the scariest things on earth. What environmentalists consider some of the worst problems in the past year [by geographic region]. Map.
Scholastic Update Apr 21, 1989 pp. 4–5

The Earth's thermostat. Carbon. A complex cycle regulates its distribution around the planet. Diagram.
U.S. News Oct 31, 1988 p. 58

The global carbon budget. Carbon added to atmosphere (in metric tons per year). Carbon removed from atmosphere (in metric tons per year).
U.S. News Oct 31, 1988 p. 58

America's troubled coasts. Toxic chemicals. PCB or pesticide contamination. Areas of oxygen depletion. Areas closed to commercial shellfish harvest. Map.
Time Aug 1, 1988 p. 50

ENVIRONMENTAL ORGANIZATIONS

Green charge. Over 75 Green groups market an "affinity" Visa or Mastercard. A sampling of 1990 [dollar] totals.
U.S. News Apr 22, 1991 p. 71

World membership in environmental groups. 1988, 1989.
Time Dec 18, 1989 p. 61

EQUAL PAY FOR EQUAL WORK

Unequal pay. Female-to-male. Black-to-white. 1970-1990.
U.S. News Nov 4, 1991 p. 63

The biggest gaps. Women's earnings. [As a] percent of men's earnings. [By selected occupations].
Business Week Oct 28, 1991 p. 35

Sideline salaries. Male college coaches earn more, on average, than female coaches. [Sport]. Average salary. Male. Female.
U.S. News Oct 28, 1991 p. 90

Who earns what. Median weekly earnings for executive, administrative and managerial positions. Men. Women. 1983-1990.
Black Enterprise Aug 1991 p. 42

The pay gap between men and women. Annual salary after graduation. Discrepancy in pay scales. By school. By industry.
Business Week Oct 29, 1990 p. 57

The gender gap shrinks. Ratio of women's earnings to men's. Ratio of bachelor's degrees earned by women vs. men. 1960-1988.
U.S. News Apr 2, 1990 p. 45

Salary daze. State-government action on pay equity.
Ms. Sep 1989 p. 88

Pink-collar wage gap. Women as percent of occupation. Women's wages as percent of men's.
Ms. Mar 1988 p. 69

See Also WAGES AND SALARIES - WOMEN

ERGONOMICS

Outfitting the ergonomic office [products to ease VDT work].
U.S. News Jan 9, 1989 p. 61

ETHIOPIA

Dry season. Ethiopia. Famine area. Food routes. Map.
Newsweek May 14, 1990 p. 33

EUROPE

Cold shoulders. Immigrants accepted this year under regular immigration procedures. Refugees applying for asylum this year. Percent of refugees accepted. [By European country].
Newsweek Dec 9, 1991 p. 36

[European security]. NATO members. EC members. Applied to EC. Other potential EC members. Claiming independence. Territorial disputes. Map.
U.S. News Sep 9, 1991 p. 47

Europe's wealth curtain. 1988 per capita gross domestic product. [European countries].
U.S. News Jul 17, 1989 p. 38

EUROPE - DEFENSES

Before and after [CFE - Treaty on Conventional Armed Forces in Europe]. Weapon. Alliance/country. Current holdings. Holdings after CFE.
Bull Atomic Sci Jan 1991 pp. 33–34

Winding down the cold war. Selected armed forces in Europe [by country]. 1990. Projected.
U.S. News Nov 26, 1990 p. 39

NATO nuclear weapons in Western Europe, 1990. U.S. nuclear warheads deployed by country. Nuclear delivery systems deployed by country. NATO nuclear forces in Germany. U.S. and NATO nuclear airbases in Europe.
Bull Atomic Sci Oct 1990 pp. 48–49

The last nuclear weapons in Europe. NATO. Number. Range. Warsaw Pact. Number. Range. 1987.
Newsweek May 8, 1989 p. 16

A mean bean count. Mean of estimates [military power]. Warsaw Pact. NATO. Warsaw Pact advantage.
Bull Atomic Sci Mar 1989 p. 31

In-place and immediate reinforcing divisions in central Europe. NATO country. Warsaw Pact country. Total divisions. Total armor division equivalents.
Bull Atomic Sci Mar 1989 p. 32

So who's counting [conventional forces in Europe]. Troops. Tanks. Combat aircraft. Warsaw Pact. NATO.
Bull Atomic Sci Sep 1988 p. 17

EUROPE - ECONOMIC CONDITIONS

The carnage on Europe's Bourses. Percent change since Jan. 1. through Sept. 28 [1990]. In local currency. In dollars. [By city].
Business Week Oct 15, 1990 p. 46

What's bothering Europe. Inflation is heating up. Consumer prices. Percent change. 1986-1991.
Business Week Oct 15, 1990 p. 47

What's bothering Europe. Capital is cooling. Capital investment. Percent change. 1986-1991.
Business Week Oct 15, 1990 p. 47

European profits are on a roll. Profit change from prior year [by country]. 1987, 1988.
Business Week Apr 24, 1989 p. 42

The consumer starts to feel the heat. Consumer price index. Britain. France. Italy. West Germany. 1988, 1989.
Business Week Mar 20, 1989 p. 54

Who has the edge in financial services [Europe, by various countries]. Commercial loans. Consumer loans. Home mortgages. Auto insurance.
Business Week Dec 12, 1988 p. 73

Growth race [annual percentage change in GNP or gross domestic product]. Britain. France. Spain. Portugal. West Germany. Italy. Greece. 1982-1987.
U.S. News Jul 18, 1988 p. 27

EUROPEAN ECONOMIC COMMUNITY

Redrawing Europe's borders. European economic community. European free trade agreement. Eastern European countries. Map.
Business Week Nov 13, 1989 p. 43

What's at stake for Europe and its major trading partners [Japan, United States]. Direct investments. Exports.
Business Week Dec 12, 1988 p. 49

U.S. investment in the EC is soaring again. Billions of U.S. dollars. 1982-1988.
Business Week Dec 12, 1988 p. 55

The twelve as traders equal EEC [by country]. Year of entry. Per capita GDP. Trade within EEC. EEC trade as share of total trade. Trade with U.S.
U.S. News Feb 29, 1988 p. 40

Doing well by doing good. Eastern Europe's two-way trade with the European Economic Community, 1986. Six Warsaw Pact countries.
U.S. News Feb 15, 1988 p. 41

EVERGLADES - ECOLOGY

The ever endangered Everglades. The original Everglades covered 4 million acres of southeast Florida. Map.
U.S. News Apr 2, 1990 p. 26

EVOLUTION

The birth of modern humans. The "Multiregional model" and the "Out of Africa model." Map. Diagram.
U.S. News Sep 16, 1991 pp. 56–57

Human evolution: A new insight. 100,000 [years ago]. Present. Chart.
U.S. News Feb 29, 1988 p. 9

Evolution's long march. Scientists now believe that the split between the ancestral hominid and chimpanzee was relatively recent—no more than 8 million years ago. [Evolution dateline]. Ancestral chimpanzee [to] modern homo sapiens. Chronological chart.
Newsweek Jan 11, 1988 p. 48

Two views of evolution. Punctuated model. Gradual model. Chart.
U.S. News Jan 11, 1988 p. 51

EXCELLENCE

In search of excellence. Of 1,005 consumers polled, most said they would pay more for higher quality. Extra amount consumers will pay for higher quality.
U.S. News Jul 3, 1989 p. 62

EXECUTIVES

Measuring the glass ceiling. Why white males monopolize the top of the corporate ladder. Men. Women. Minorities. Employees. Managers. Top execs.
U.S. News Aug 19, 1991 p. 14

Breakdown of U.S. employees in executive, administrative and managerial positions, 1990. [Men, women, white, black].
Black Enterprise Aug 1991 p. 42

Representation of women in corporate management. Professionals. Managers. Senior managers.
Black Enterprise Aug 1991 p. 42

EXECUTIVES - SALARIES, PENSIONS, ETC.

Who earns what. Median weekly earnings for executive, administrative and managerial positions. Men. Women. 1983-1990.
Black Enterprise Aug 1991 p. 42

The 20 highest-paid chief executives and 10 who aren't CEOs. Company. 1990 salary and bonus. Long-term compensation. Total pay.
Business Week May 6, 1991 p. 91

Pay for performance: Who measures up and who doesn't.
Business Week May 6, 1991 p. 93

The widening pay gap: CEOs vs. the others. Average total compensation. Chief executive. Engineer. School teacher. Factory worker. 1960-1990.
Business Week May 6, 1991 p. 96

Winners and losers. Sampling of corporate heads who got: Pay raise. Pay cut. 1989, 1990.
Newsweek May 6, 1991 p. 50

Hottest executive perks [chauffeurs, country club memberships, etc.].
Black Enterprise Feb 1991 p. 120

How paychecks compare. Median annual salary and bonus for middle managers. Germany. Japan. France. U.S. Britain.
Business Week Nov 26, 1990 p. 71

The 20 highest-paid chief executives, 10 who aren't CEOs. [Executive]. Company. 1989 salary and bonus. Long-term compensation. Total pay. The 10 largest golden parachutes. Company. Reason. Total package.
Business Week May 7, 1990 p. 57

The decade's biggest CEO money-makers [by name]. Total pay 1980-89.
Business Week May 7, 1990 p. 60

Incentive pay takes off. Bonuses. Options and other long-term incentives. Percent of senior executives' base salary.
Business Week Jan 8, 1990 p. 26

Paycheck punch. Purchasing power of chief executives. [By country].
U.S. News Jan 16, 1989 p. 70

The 25 highest-paid executives. Company. 1987 salary and bonus. Long-term compensation. Total pay.
Business Week May 2, 1988 p. 51

The 10 largest golden parachutes. [Executive]. Company. Reason for payment. Total package.
Business Week May 2, 1988 p. 51

Pay for performance. Executives who gave shareholders the most for their pay. The least. Executives whose companies did the best relative to their pay. The worst. 1985-1987.
Business Week May 2, 1988 p. 53

EXERCISE

Fitness test. These short tests for strength, aerobic fitness and flexibility can give parents a general idea of a child's fitness level.
U.S. News May 20, 1991 pp. 90–91

Fit for life. Physically fit person. Sedentary person. Activity level. Age.
Newsweek May 13, 1991 p. 61

What's your zone? [Aerobic heart rate formulas]. Heart rate (beats per minute). [Recommended lows and highs from] American Heart Association. American College of Sports Medicine. Institute for Aerobic Research.
U.S. News Oct 1, 1990 p. 86

Dumpling decade. Body mass index ages 10 and 11. Boys. Girls. 1980-1988
Newsweek Aug 27, 1990 p. 63

Exercises in fantasy. To relieve the drudgery of swimming laps or using a stair climber, you can dream about how far your efforts might take you. Equivalent. Distance [and] time.
U.S. News May 7, 1990 p. 75

Stretching the right way. Different activities put stress on different muscles. Fitness activities. Sports. Chart.
U.S. News Mar 5, 1990 pp. 66–67

Fitness pays. Death rates. Fitness levels. Men. Women.
Newsweek Nov 13, 1989 p. 77

Fitness pays. Deaths per 10,000 people in a year. All causes. Cardiovascular. Cancer. Men. Women.
Time Nov 13, 1989 p. 13

Walk for your life. Exercising more than a moderate amount pays diminishing dividends. Deaths per 10,000 person years by level of fitness. Men. Women.
U.S. News Nov 13, 1989 p. 18

Popular ways to work out. [Type of] exercise. 1987,1988.
U.S. News May 29, 1989 p. 72

Family fitness. Percentage of parents who exercise with their children for 20 minutes or more in a day during a typical week. Mothers. Fathers.
U.S. News Dec 19, 1988 p. 70

Paying the piper. Time needed to burn the calories consumed in [various foods and by various exercises].
U.S. News Jul 18, 1988 p. 54

An exercise in truth seeking [exercise myths and facts].
U.S. News Jul 18, 1988 p. 54

Fitness dropouts. Number of people who participated in these activities in 1987. Fitness walking. Running or jogging. Bicycling. Swimming.
U.S. News Jul 18, 1988 p. 54

EXERCISE EQUIPMENT

Tracking fitness gadgets. Percentage changes in retail sales of exercise equipment since 1984.
U.S. News Apr 3, 1989 p. 76

EXXON VALDEZ (SHIP) OIL SPILL, 1989

The spill. Where the oil went. The spread. Maps.
Newsweek Sep 18, 1989 p. 53

Drop by drop: A box score [damage and costs].
Newsweek Sep 18, 1989 p. 55

Disturbing numbers. Estimates vary on what has become of the 260,000 barrels of oil. Center for Marine Conservation. Exxon.
U.S. News May 15, 1989 p. 14

'A disaster of enormous potential.' In one week, the oil had spread over 900 square miles, coating thousands of marine mammals, birds, fish, and threatening hatcheries and parklands. Map.
Newsweek Apr 10, 1989 p. 55

See Also OIL SPILLS

EYE

Anatomy of tears. The tear film that bathes and protects the human eye is made up of three layers. Diagram.
FDA Consumer Feb 1989 p. 29

EYE - DISEASES AND DEFECTS

Effects of cataract formation. Diagram.
FDA Consumer Jan 1990 p. 28

EYE - SURGERY

Uses of laser surgery. Diagram.
FDA Consumer Jul 1990 p. 17

EYEGLASSES

Blind and visually impaired people. Number of guide-dog schools. Percentage of U.S. population that wears prescription glasses. [Other selected statistics.]
U.S. News Nov 18, 1991 p. 22

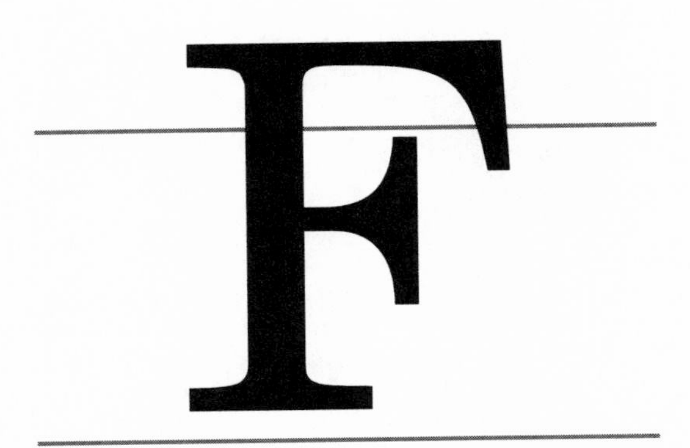

FACSIMILE MACHINES

See FAX MACHINES

FACTORIES

The manufacturing revolution. Many U.S. companies have overhauled their factories in recent years and are now producing goods in dramatically different ways. New-age production. Old style production. Diagram.
U.S. News Oct 22, 1990 pp. 52–53

Recession's signs and portents. Factory orders. Monthly increase in unfilled orders. 1989, 1990.
U.S. News Sep 24, 1990 p. 81

FAIRS

The fairest fairs of all. Late summer and early fall is high season for state, county and other large fairs. Those that drew the biggest crowds in 1988.
U.S. News Aug 21, 1989 p. 66

FAMILY

A portrait of change. The U.S. Census Bureau defines seven different kinds of families and households.
Scholastic Update Sep 6, 1991 p. 4

Broken families. 1990, 1970. [% of] married couple families. Nonfamily households. Female householders. Male householders.
Newsweek Mar 18, 1991 p. 6

Piecemeal dinners. The family meal is a vanishing species. Percent of families who dine together.
U.S. News Apr 23, 1990 p. 78

Changing households. Married couples. Singles. Female head. Percent of households. 1950, 1988.
Business Week Sep 25, 1989 p. 102

'I need more money!' Here's the salary needed to match the lifestyle of a family of four with a $55,000 income living in an average suburban town.
Newsweek Jul 3, 1989 p. 40

Sibling smarts. When the Mensa society, a group open to anyone who scores in the top 2 percent on a standardized IQ test, surveyed its members, here is how they fell in their family's lineup.
U.S. News May 22, 1989 p. 81

The continued breakup of the black family. Distribution of black and white families by type, 1940-1985. Other male head. Female head. Husband-wife. Black families. White families.
Black Enterprise Jan 1989 p. 52

Father facts. [Balancing job and home life.]
Newsweek Dec 19, 1988 p. 64

Family fitness. Percentage of parents who exercise with their children for 20 minutes or more in a day during a typical week. Mothers. Fathers.
U.S. News Dec 19, 1988 p. 70

The company we keep. Number of paid weeks [for family leave] and percent of normal pay. Canada. France. W. Germany. Italy. Japan. Sweden.
Black Enterprise Oct 1988 p. 33

Fewer moms at home. Percentage of families with two or more wage earners. Percentage of all preschool children with working moms. 1980-1987.
U.S. News Aug 29, 1988 p. 102

FAMILY - INCOME

See INCOME - FAMILY

FAMINES

Dry season. Ethiopia. Famine area. Food routes. Map.
Newsweek May 14, 1990 p. 33

FARMS AND FARMING

Leaner days down on the farm. Net cash farm income. 1986-1991.
Business Week Jan 14, 1991 p. 110

America abandons the farm. Counties where farming accounts for 20% or more of income. 1950, 1986.
Business Week Sep 25, 1989 p. 93

The declining number of black farmers. 1900-1982.
Black Enterprise Jul 1988 p. 21

The farm economy. Total farm income. Total farm debt. 1981-1988.
Business Week May 23, 1988 p. 53

FAST FOOD RESTAURANTS

McDonald's: Slowdown for a fast-food champ. Share of fast-food market. McDonald's. Pepsico. Domino's Pizza. Little Caesar's Pizza. 1985 and 1990.
Business Week Oct 21, 1991 p. 116

Fast-food fast-trackers. Fast-growing chains, with estimated systemwide 1991 sales (in millions).
Business Week Oct 21, 1991 p. 122

Make mine McLean. [Calorie and fat content of] McDonald's. Burger King. Wendy's. KFC. Best. Worst.
Newsweek May 27, 1991 p. 51

Midprice eateries still set the pace. Average same-store sales growth. Fast-food hamburger chains. Midprice restaurant chains. 1986-1991.
Business Week Jan 14, 1991 p. 92

While gaining ground in the burger battle. Sales per store. McDonald's. Burger King. 1986-1990. Burger King is losing the fast-food war. Market share of U.S. fast-food market. 1988-1990.
Business Week Oct 22, 1990 p. 61

An abundance of restaurants. U.S. restaurants, excluding those in hotels and convenience stores. 1985-1990.
Business Week Jan 8, 1990 p. 90

Fast food's steady climb. Fast-food revenues. 1984-1989.
Business Week Jan 9, 1989 p. 86

Family restaurants lag in growth. Family-style. Fast food. Sales. 1985-1988.
Business Week Sep 5, 1988 p. 63

McDonald's: Staying ahead. Systemwide restaurant revenues. McDonald's. Burger King. Wendy's. 1983-1987.
Business Week May 9, 1988 p. 92

FATHERS

U.S. families headed by single males. Percentage [of U.S. private companies] that offer paternity leave. Average number of days allowed for paternity leave.
U.S. News Jun 17, 1991 p. 10

Bringing up baby. Here's how much the women surveyed report their husbands help with the baby.
U.S. News Oct 2, 1989 p. 70

Father facts. [Balancing job and home life.]
Newsweek Dec 19, 1988 p. 64

FAX MACHINES

Who's faxing whom? Of all fax machines used in the U.S., the greatest percentages are found in these businesses.
U.S. News Oct 2, 1989 p. 70

E-mail is growing. Here's why. Billions of messages. 1988-1991. Cost of sending a three-page letter. Federal Express. Telex. Fax. Electronic mail.
Business Week Feb 20, 1989 p. 36

Fax machines used in home offices. 1987-1989 est.
Business Week Oct 10, 1988 p. 105

Fax crazy. Facsimile machines sold in U.S. 1983-1988.
Business Week Mar 21, 1988 p. 136

FEDERAL BUDGET

[The young] get less of the overall federal pie. Total federal spending per age group. Percent of total federal spending. Percent of U.S. population. Age 65 and older. Under age 18. Fiscal 1990.
Business Week Aug 19, 1991 p. 85

Discretionary spending. Mandatory spending. Defense. Domestic. Foreign. 1991, 1992.
Time Feb 18, 1991 p. 42

Where federal spending is on a double-digit tier. Interest payments. Medicare. Federal salaries. Health and human services. Unemployment insurance.
Business Week Aug 13, 1990 p. 23

Uncle Sam's handouts. Fiscal year 1991 cost. Medicare. Medicaid. Welfare and nutrition programs. Farm aid. Food stamps. Social Security. Veterans' benefits programs.
Business Week Jul 16, 1990 pp. 26–27

FEDERAL DEBT

Skyrocketing debt. Gross federal debt, percentage of GNP. 1980-1991.
U.S. News Dec 9, 1991 p. 65

Borrowing binge. U.S. government debt as a percentage of GNP.
U.S. News May 6, 1991 p. 58

The nation's debt will keep soaring. Total federal debt. 1991-1995.
Business Week Nov 5, 1990 p. 51

Big money. Amount of the national debt owed by every American man, woman and child. 1980-1990.
Time Oct 15, 1990 p. 46

FEDERAL DEFICIT

Budget ties that bind. Budget deficit projections. Total marketable public debt. Federal net interest payments.
U.S. News Sep 9, 1991 pp. 52–53

Stark reality. Uncertain future. Federal budget percentage of GNP. Spending. Revenues.
U.S. News Aug 19, 1991 p. 60

The S&L debacle haunts U.S. Total budget deficit with S&L cleanup. Deficit excluding S&L cleanup. 1989, 1990, 1991.
U.S. News Jul 8, 1991 p. 8

Deficit deepens. U.S. budget deficit. 1980-1991.
U.S. News Dec 31, 1990 p. 46

The deficit that isn't disappearing. Federal outlays including Social Security. Federal revenues including payroll tax. 1990-1995.
Business Week Nov 12, 1990 p. 43

The deficit won't shrink soon. Deficit with agreement [deficit-reduction agreement]. 1991-1995.
Business Week Nov 5, 1990 p. 51

Time to bite the bullet. The deficit. The deficit as a percent of GNP. 1987-1991.
Newsweek Oct 15, 1990 p. 27

Does anyone care about deficits? Federal budget deficits. Actual. 1985 Gramm-Rudman-Hollings target. 1987 Gramm-Rudman-Hollings target. 1985-1990.
U.S. News Oct 8, 1990 p. 68

Mounting debt. 1991 budget deficit projections in billions. Administration projections. Congressional projections. Jan. 1990. Current.
Time Sep 10, 1990 p. 42

The deficit's bite. Federal deficit as a percentage of GNP. 1979-1991.
U.S. News Jul 30, 1990 p. 16

The deficit is rising again. Baseline deficit including Social Security trust fund surplus. 1981-1992.
Business Week Jul 9, 1990 p. 24

The real deficit. Official deficit. RTC spending (S&L bailout). Social-Security surplus. Real deficit (total).
Newsweek Jul 9, 1990 p. 18

Masking the monster. Reported deficit. Reduced by Social Security trust funds. Reduced by other trust funds. 1985-1995.
Time Feb 19, 1990 p. 49

How Social Security masks the deficit. Federal budget deficit. Excluding Social Security surplus. 1985-1994.
Business Week Feb 5, 1990 p. 26

The Social Security take is gaining on income taxes. Individual income taxes. Social Security taxes. Deficit without Social Security. Social Security surplus. Unified budget deficit. 1980-1989.
Business Week Jan 29, 1990 p. 67

Masking the deficit. CBO base-line estimate for federal budget deficit + Social-Security Trust-Funds surplus.
Newsweek Jan 22, 1990 p. 45

The budget gap is narrowing, but the trade deficit is due to widen. Unified budget deficit. Merchandise trade deficit. 1985-1989.
Business Week Dec 25, 1989 p. 85

The budget gap. Deficit projections (by fiscal years). Congressional Budget Office deficit estimates. Gramm-Rudman deficit targets. 1991, 1992, 1993.
U.S. News Oct 16, 1989 p. 32

How interest outlays unbalance the budget. Surplus. Deficit excluding net interest payments. Total federal deficit. 1980-1988.
Business Week May 29, 1989 p. 20

With the budget growing more slowly Bush is relying on heftier tax receipts to lighten the deficit's impact. Outlays. Receipts. Deficit. Deficit as percentage of Gross National Product. 1982-1991.
Business Week Mar 13, 1989 p. 101

Two points of view. Federal deficit in billions of dollars. Deficit as a % of GNP. 1970-1989.
Time Jan 30, 1989 p. 47

The surplus scenario. How the surplus masks the deficit. Reported budget deficit. Projected Social Security surplus. Real projected budget deficit. 1989-1994.
U.S. News Dec 26, 1988 p. 30

The budget deficit: Is it falling fast enough? 1977-1989.
Business Week Nov 28, 1988 p. 31

Budget deficit. 1988, 1989, 1990.
Newsweek Nov 28, 1988 p. 31

While the deficit should decline only slightly. Fiscal year. Billions of dollars. 1986-1994.
Business Week Oct 10, 1988 p. 128

It [federal deficit] may well fall faster as a share of GNP. Deficit as percent of nominal GNP. Fiscal year. Percent. 1986-1994.
Business Week Oct 10, 1988 p. 129

The Social Security surplus is growing and helping to finance the deficit. Trust fund. Deficit without Social Security surplus. Unified budget deficit. 1986-1993.
Business Week Jul 18, 1988 p. 93

Two views of the federal deficit. Office of management & budget. Consensus of private economists. Billions of dollars. 1988, 1989.
Business Week Jun 6, 1988 p. 28

The deficit stays stubbornly high. Federal budget deficit. 1987-1990.
Business Week Feb 22, 1988 p. 43

FEDERAL DEPOSIT INSURANCE CORPORATION

The number of bank failures is declining, but the assets involved are growing, depleting the bank insurance fund. Total assets of failed banks. Bank insurance fund balance. 1987-1991.
Business Week Dec 9, 1991 p. 30

The evaporating bank-rescue fund. Yearend balance of FDIC's insurance fund. Current trend. Pessimistic scenario. 1987-1992.
Business Week Jul 15, 1991 p. 123

As bad bank loans explode, the bank insurance fund shrinks. Net charge-offs by insured commercial banks. Yearend balance. 1980-1991.
Business Week Jan 21, 1991 p. 25

A weakening safety net. Amount in FDIC fund for every $100 in insured deposits, year-end. 1934-1991.
Time Jan 21, 1991 p. 54

Estimated cost of FDIC rescues [various banks].
Time Sep 24, 1990 pp. 66–67

Can't get out of the red. Net income or deficit of the Federal Deposit Insurance Corporation in billions of dollars. 1983-1990.
Time Sep 24, 1990 p. 67

Busting the bank. Red-ink blues. Net spending by the FDIC. 1980-1988.
U.S. News Aug 15, 1988 p. 42

FEDERAL DRUG ADMINISTRATION

FDA enforcement actions 1987-1988. Legal actions. Import reviews.
FDA Consumer Apr 1989 p. 5

FEDERALLY INSURED PROGRAMS

The next crisis. The government is already suffering losses on many other guarantees. What it backs in all. Program. Taxpayer liability (in billions).
Newsweek May 21, 1990 p. 24

FEMINISM

See WOMEN'S MOVEMENT

FERTILIZATION IN VITRO, HUMAN

Couples who have difficulty conceiving a child. In vitro fertilizations. Average paid for human sperm. Fertilized human embryos in frozen storage.
U.S. News Dec 30, 1991 p. 21

A test-tube-baby census. Success rates at infertility clinics. 1987.
U.S. News Apr 3, 1989 p. 75

FETUS - SURGERY

In the operating womb. First major operation on a fetus. Diagram.
Newsweek Jun 11, 1990 p. 56

A delicate patch job [prenatal surgery]. Diagram.
Time Jun 11, 1990 p. 55

FIBER IN DIET

Fiber facts. [Foods] richer in water-soluble fiber and more likely to reduce heart disease. [Foods] richer in water-insoluble fiber and more likely to reduce colon cancer.
U.S. News Nov 27, 1989 p. 91

FIREWORKS

The rockets' dread scare. Fireworks the states allow. All Class C. Some Class C. None.
U.S. News Jul 4, 1988 p. 63

Handle with care. Fireworks-related injuries. 1987-1976.
U.S. News Jul 4, 1988 p. 63

FITNESS

See EXERCISE

FLAGS - UNITED STATES

Old Glory flying high. Number of flags hoisted above the U.S. Capitol on behalf of congressional constituents. 1937-1987.
U.S. News Jul 4, 1988 p. 63

FLU (DISEASE)

Predominant influenza strains and reported cases of Reye syndrome—United States, 1974 and 1977-1989.
FDA Consumer Nov 1990 p. 23

Contagion from coast to coast. Occurrence of influenza reported to the Centers for Disease Control at the end of January [1990]. Map.
Time Feb 12, 1990 p. 61

Is it a cold or the flu? Symptoms.
FDA Consumer Nov 1988 p. 10

FOOD

The big four: Redefining the basic food groups. The old order. The new controversy. Chart.
Newsweek May 27, 1991 p. 48

Make mine McLean. [Calorie and fat content of] McDonald's. Burger King. Wendy's. KFC. Best. Worst.
Newsweek May 27, 1991 p. 51

U.S. per capita red meat consumption. Poultry consumption. Fish consumption.
U.S. News Apr 22, 1991 p. 8

Falling foods. Supermarket sales . Items [which] showed the biggest percentage drops.
U.S. News Dec 10, 1990 p. 79

U.S. food irradiation rules. Product. Purpose of irradiation. Dose permitted. Date of rule.
FDA Consumer Nov 1990 p. 14

Where consumers looked for food facts in 1989.
FDA Consumer May 1990 p. 3

Rising U.S. food imports. 1973-1986.
FDA Consumer Sep 1988 p. 14

Comparing grocery bills. Weekly food costs at home [by size of family].
U.S. News May 2, 1988 p. 77

Favorites of the lunch bunch. Share of children age 6-12 citing these foods as their favorites for lunch.
U.S. News May 2, 1988 p. 77

How to get your protein. Amount needed [of selected foods] for 20 grams of protein. Cost of 20 grams of protein. 1987.
U.S. News Mar 21, 1988 p. 73

Nibbling away at the grocer's pie. Food eaten away from home [at restaurants, fast-food outlets, cafeterias, bars and taverns] as a share of total U.S. food sales. 1983-1988.
Business Week Jan 11, 1988 p. 86

FOOD - CONTENT

Where's the fat? TV ads claim that pork, "the other white meat," is about as low in fat as chicken and turkey. Here's the breakdown for a 3-ounce serving of: Pork. Veal. Turkey. Chicken. Fat. Cholesterol. Calories.
U.S. News Jun 4, 1990 p. 78

Debunking bananas. Bananas are commonly considered a rich source of potassium. Other foods actually offer more [by various foods].
U.S. News Apr 16, 1990 p. 62

Fiber facts. [Foods] richer in water-soluble fiber and more likely to reduce heart disease. [Foods] richer in water-insoluble fiber and more likely to reduce colon cancer.
U.S. News Nov 27, 1989 p. 91

A case for cornflakes. Most fast-food breakfasts are loaded with fat, calories and salt. Here's how a few favorites measure up. Fat. Calories. Sodium.
U.S. News Jun 26, 1989 p. 74

What's in a nut. Fat type and content of some popular nuts.
U.S. News Jun 5, 1989 p. 69

Preventive medicine [various foods]. Fiber (in grams). Soluble. Insoluble.
U.S. News May 22, 1989 p. 73

Garden of eating. Fruits and vegetables with the most calories.
U.S. News May 15, 1989 p. 73

Cholesterol count. Food. Saturated fat. Cholesterol.
U.S. News Apr 24, 1989 p. 74

Trading up at the table. How to add fiber to your diet [various foods].
U.S. News Mar 6, 1989 p. 66

Facts on fat. 3.5 oz. of [various foods]. Saturated fat. Cholesterol. Calories.
Time Dec 12, 1988 p. 68

Battling fat and cholesterol. Be careful and enjoy. Grams fat. Milligrams cholesterol. [Selected foods].
Newsweek Feb 8, 1988 p. 58

FOOD - LABELING

Diet-health trends. Shoppers who "pay attention" to ingredient lists. Shoppers who use lists to "avoid" or limit.
FDA Consumer May 1988 pp. 7–8

The name's the thing. Government standards for labeling of meat, dairy and juice products.
U.S. News Mar 7, 1988 p. 77

FOOD - PRESERVATION

How long will it keep? Storage guidelines for some of the foods that are regulars on America's dinner tables. Product. Storage period. In refrigerator. In freezer.
FDA Consumer Jan 1991 p. 21

When good snacks go bad. Packaging and preservatives determine how long snack goods stay on the supermarket shelves. Here's how long different snacks can be sold before they're pulled.
U.S. News Jun 11, 1990 p. 74

FOOD ADDITIVES

Sulfite consumer complaints by types of foods. Food type. No. of complaints. No. of serious reactions.
FDA Consumer Mar 1988 p. 11

FOOD ALLERGIES

Adverse reactions to food ingredients. Aspartame. Sulfites. All other.
FDA Consumer Oct 1988 p. 17

Sulfite consumer complaints by types of foods. Food type. No. of complaints. No. of serious reactions.
FDA Consumer Mar 1988 p. 11

FOOD CONTAMINATION

Organisms that can bug you. Disease and organism that causes it. Source of illness. Symptoms.
FDA Consumer Jan 1991 pp. 22–23

How to recognize can defects. Guidelines have been adapted for recognizing defects in cans made of plastic and other materials, as well.
FDA Consumer Sep 1990 p. 18

Must be something I ate. Major food-borne disease. Frequent source. Symptoms. Cases reported to the Centers for Disease Control. 1978, 1987.
Time Mar 27, 1989 p. 35

FDA's monitoring for pesticide residues. Number of food samples tested. Import. Domestic. 1982-1987.
FDA Consumer Oct 1988 p. 11

FOOD, FROZEN

Upscale frozen food cools down. Sales of premium frozen entrees. 1983-1987.
Business Week Apr 25, 1988 p. 89

FOOD INDUSTRY

Tepid times for food processors. Average growth in earnings per share for the 12 largest U.S. packaged food makers. 1985-1990.
Business Week Jan 8, 1990 p. 80

Food grows faster than markets. Worldwide production and volume traded (in metric tons). Wheat. Corn. Red meat. Cheese. Sugar. 1982-1988.
U.S. News May 23, 1988 pp. 56–57

A taste of money. Net profits for U.S. food industry. 1977-1988.
U.S. News May 23, 1988 p. 70

McDonald's: Staying ahead. Systemwide restaurant revenues. McDonald's. Burger King. Wendy's. 1983-1987.
Business Week May 9, 1988 p. 92

Whetting appetites. [Number of] new food products introduced each year. 1981-1987.
U.S. News Feb 29, 1988 p. 75

FOOD POISONING

See FOOD CONTAMINATION

FOOD PREFERENCES

Factors rated by shoppers as "very important" in selecting foods.
FDA Consumer Sep 1988 p. 2

When do people eat fresh fruit? Why are people eating more fresh fruit?
FDA Consumer May 1988 p. 13

FOOD SUPPLY - AFRICA

Food production down. Index of African food production, per capita, from 1965-1986.
Scholastic Update Jan 27, 1989 p. 10

FOOD SUPPLY - SOVIET UNION

Bitter harvest. Soviet farmers still cannot produce enough food to feed the nation. Value of agricultural output. Fixed-capital stock. 1970-1988.
U.S. News Nov 20, 1989 p. 32

Half a loaf, on a good day. Grain spilled by railroad cars. Percent of each Soviet wheat harvest used as seed for the next crop. Grain lost before it gets to market. Diagram.
U.S. News Nov 20, 1989 p. 36

FOOTBALL, COLLEGE

Holiday bonanza: Big games for big bucks. Bowl. Teams. Projected purse value (in millions).
Newsweek Jan 2, 1989 p. 63

FOOTBALL, PROFESSIONAL

Percentage of overall blacks vs. blacks in front-office management. NBA. NFL. MLB.
Black Enterprise Dec 1991 p. 47

Who holds the ball. (White and black head coaches and managers). NBA. NFL. MLB.
Black Enterprise Dec 1991 p. 47

Football salaries gain yardage. Average player salary. 1987-1991.
Business Week Oct 7, 1991 p. 40

Pigskin prices. Tickets, parking, food and mementos at a pro football game. The price range [by city/team].
U.S. News Sep 16, 1991 p. 69

The money is bigger than ever but it doesn't last long. Average salary. Average length of professional-sports career. Baseball. Basketball. Football. Hockey. 1981, 1991.
Business Week Jun 3, 1991 p. 55

[Retired] National Football League players [and] major league baseball players. [Other selected statistics.]
Business Week Jun 3, 1991 p. 58

Superbucks. Average expenditure per person (during Super Bowl week in New Orleans).
U.S. News Jan 21, 1991 p. 84

Player pay: Gaining yardage. Average annual salary. Increase per year.
Business Week Jun 25, 1990 p. 56

Football's road to riches. Average signing bonus. Average salary of first-round picks. 1985-1989.
Business Week Apr 9, 1990 p. 44

N.F.L. [television] contract. Increase from 1987.
Time Mar 26, 1990 p. 66

FORD FOUNDATION

Where Ford's money goes. Ford Foundation commitments. U.S. & international affairs programs. Developing countries programs.
Black Enterprise Jul 1990 p. 56

FOREIGN AID

See ECONOMIC ASSISTANCE

FOREIGN INVESTMENTS

See INVESTMENTS, FOREIGN

FOREST PRODUCTS INDUSTRY

Cutting up the woodpile. One cord of wood produces any of the following [products]. Paper consumed per person [by country].
U.S. News Feb 29, 1988 p. 75

Boom times for forest products. Net income of 26 U.S. forest products companies. 1983-1988.
Business Week Jan 11, 1988 p. 125

FORESTS AND FORESTRY

Trees that turn first. Approximate number of trees in the nation's parks, yards and streets. Trees used annually for Sunday editions of the New York Times. [Other selected statistics.]
U.S. News Oct 14, 1991 p. 15

Number of forested acres in the United States. Publicly owned. Old-growth forests still remaining. Percentage of U.S. timber supply coming from public lands. [Other selected statistics.]
U.S. News Jul 22, 1991 p. 8

What old-growth trees do for the ecosystem and for the economy. Diagram.
Time Jun 25, 1990 p. 62

FRANCE - ECONOMIC CONDITIONS

What's ailing France. Change in consumer spending. Housing starts. Change in capital spending. Change in gross national product. 1985-1991.
Business Week Mar 11, 1991 p. 45

Italy's growth is tailing off. Italy. France. Germany. Real GDP growth rates. Annual consumer price increases, 1990. Budget deficit as percent of GDP, 1990.
Business Week Jun 11, 1990 p. 34

FRANCHISES

A snapshot of the franchise buyer. Annual income. Net worth. Age. Total initial investment. Franchise fee.
Business Week May 6, 1991 p. 70

The top 10 Franchise-50 companies by percentage of black franchisees. Company. Type of franchise. Black franchisees.
Black Enterprise Sep 1990 p. 50

The Franchise-50. Franchise companies by industry.
Black Enterprise Sep 1990 p. 50

Types of franchised businesses owned by blacks, 1986.
Black Enterprise Sep 1989 p. 39

Break down of franchised businesses by ethnic background of owner. Black. Hispanic. American Indian. Oriental.
Black Enterprise Sep 1989 p. 39

Types of franchised businesses owned by all franchisees, 1986.
Black Enterprise Sep 1989 p. 39

FRATERNITIES AND SORORITIES

Greek comeback. Men initiated into college fraternities. Women initiated into college sororities. 1975-1987.
U.S. News Sep 19, 1988 p. 72

FREE TRADE AND PROTECTION

The fallout from protectionist measures. Annual cost to consumers. Total cost. Cost per job saved.
U.S. News Mar 21, 1988 p. 51

FRUIT

Garden of eating. Fruits and vegetables with the most calories.
U.S. News May 15, 1989 p. 73

When do people eat fresh fruit? Why are people eating more fresh fruit?
FDA Consumer May 1988 p. 13

FRUIT JUICES - MARKET SHARE

Tropicana pours it on. Market share, by volume, of total U.S. orange juice market. 1985, 1990.
Business Week May 13, 1991 p. 48

FURNITURE

Furnishings. Average annual expenditure for home furnishings and appliances. Income. Expenditure. 1987, 1988.
U.S. News Jul 30, 1990 p. 73

FURS

Pelt pile-up. Number of animals killed to make an average full-length fur coat [by animal].
U.S. News Dec 5, 1988 p. 93

FUSION

Brookhaven's new path to fusion. A small electrostatic accelerator fires billions of clusters of deuterium-containing water molecules at a target that also contains deuterium.
U.S. News Oct 2, 1989 p. 62

FUSION, COLD

What it all means [nuclear fusion and cold fusion]. Fusion in a bottle. The Muon theory. The promises. The problems. Diagram
Newsweek May 8, 1989 pp. 50–51

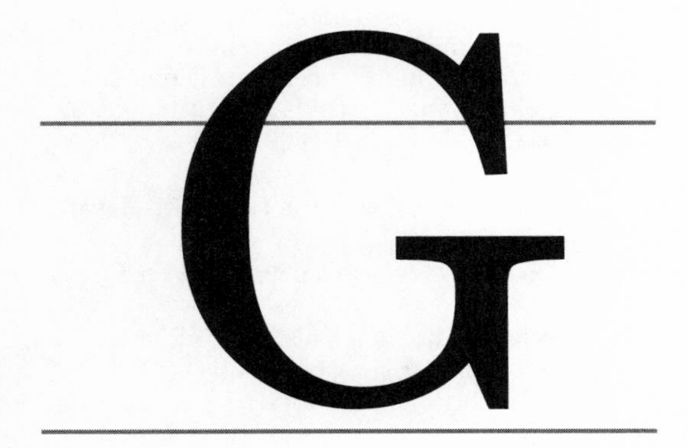

GALILEO PROJECT

The long road to Jupiter [Galileo mission timetable]. Diagram.
U.S. News May 15, 1989 p. 58

GAMBLING

Percentage of adults who said they had gambled. Lost more than they won. Bought a lottery ticket. Amount state lotteries collected in 1980.
U.S. News Mar 25, 1991 p. 12

Atlantic City [various statistics].
Newsweek Oct 29, 1990 p. 82

Las Vegas [various statistics].
Newsweek Oct 29, 1990 p. 83

Serious money. Amount Americans bet legally in 1988, in billions.
Time Jul 10, 1989 p. 17

GAMES

Top-selling children's board game. Languages that Dungeons & Dragons has been translated into. Top mind games judged by Mensa. [Other selected statistics.]
U.S. News Feb 25, 1991 p. 12

GARBAGE

An environmental scorecard. Land. Material recovered—and recycled—from municipal solid waste each year. 1960-1984. Total garbage generated by cities, counties and towns each year.
Scholastic Update Apr 21, 1989 p. 16

GARBAGE INDUSTRY

The largest trash haulers. Total 1987 solid waste revenues of publicly held companies.
Business Week Sep 12, 1988 p. 112

GARDENING

Mapping the perfect garden. The Department of Agriculture's new plant-hardiness map divides the U.S. into 20 minimum-temperature zones. Map.
U.S. News Mar 26, 1990 p. 63

Green-thumb choices. Most popular flowers grown in home gardens. Seeds. Bedding plants. Bulbs.
U.S. News Mar 20, 1989 p. 90

GAS, NATURAL

It's [natural gas] cheaper. Residential heating costs (per million B.t.u.). 1980-1990.
U.S. News Jul 30, 1990 p. 39

U.S. gas demand is rising. U.S. consumption, including imports. U.S. capacity is declining. Domestic production capacity. Trillions of cubic feet. 1983-1988.
Business Week Oct 24, 1988 p. 24

GASOLINE - CONSUMPTION

Paying through the hose. Gasoline consumption (gallons per day). 1982-1988.
U.S. News Nov 28, 1988 p. 57

GASOLINE PRICES

Gas: Still cheap. Average retail price of gasoline. Per gallon, in constant (1990) dollars. 1950-1990.
Newsweek Feb 11, 1991 p. 44

[Percent] increase in the price of crude oil. % increase in the price of gasoline. Week before the invasion of Kuwait. Week ending Oct. 26.
Time Nov 5, 1990 p. 52

In the U.S., 1990 gas at 1950 prices [adjusted for inflation]. Average retail price per gallon of gas. 1950-1990.
U.S. News Oct 8, 1990 p. 62

Driving is cheaper the American way. Retail prices for gasoline. Los Angeles. London. Tokyo. Paris. Stockholm. Milan.
U.S. News Oct 8, 1990 p. 63

Think you pay a lot at the pump? Price per gallon gasoline as of Aug. 25 [various countries].
Time Sep 10, 1990 p. 35

The pump it yourself payoff. Here's what self-service would have saved you over the years. Average fuel consumed. Average price per gallon. Full serve. Self-serve. Savings. 1985-1989.
U.S. News Mar 5, 1990 p. 72

Price spike. Retail gasoline. Retail home-heating oil (cents per gallon). 1989.
U.S. News Jan 22, 1990 p. 49

The real cost of a fill-up. Inflation-adjusted pump price. Average annual pump price. 1970-1989.
U.S. News Jun 12, 1989 p. 69

U.S. gasoline taxes are far behind the pack. Dollars per gallon. Tax portion [by country]. 1988.
Business Week Jan 30, 1989 p. 20

Pumps compared. Country. Total price per gal. Part that is tax.
Time Jan 23, 1989 p. 43

Paying through the hose. Average gasoline price per gallon. 1981-1988.
U.S. News Nov 28, 1988 p. 57

GEMS

Gem prices: A rocky road. Wholesale price. Diamond. Emerald. Sapphire. 1979-1988.
Business Week Aug 22, 1988 p. 104

GENERAL MOTORS CORPORATION

GM's total profits and losses in billions. GM's share of the U.S. car market. 1980-1991.
Time Dec 31, 1991 p. 56

GM has halted its market share slide. GM's share of U.S. auto and light truck market. Total sales of autos and light trucks in the U.S. Total net earnings [GM]. 1986-1991.
Business Week Feb 4, 1991 p. 95

Detroit's biggest family. Where GM's brand names came from, and how they're running. Chevrolet. Oldsmobile. Buick. Pontiac. Cadillac. Saturn.
Time Oct 29, 1990 p. 76

Share of car sales in the U.S. GM. Japan. 1980-1990.
Time Oct 29, 1990 p. 76

Battling to hold on as No. 1. U.S. retail sales of passenger cars. Imports. GM. Ford. Chrysler. 1980-1989.
Newsweek Oct 22, 1990 p. 51

Battle for position. GM's market share. 1980-1985.
Newsweek Feb 20, 1989 p. 39

GENETIC ENGINEERING

Gene farming. Giving animals human DNA is becoming routine. [Three-step process.] Diagram.
Newsweek Sep 9, 1991 p. 55

The plan of attack. In one of the experiments that received approval last week, researchers will use genetically engineered blood cells to fight skin cancer. Diagram.
Time Aug 13, 1990 p. 61

The DNA assembly line. Once a time–consuming process, making millions of copies of a fragment of DNA can now be accomplished in matter of hours with a new technique called polymerase chain reaction (PCR). Diagram.
U.S. News Jul 16, 1990 p. 53

GENETIC MAPPING

Mapping chromosomes. Sequencing genes. Chart.
Time Mar 20, 1989 pp. 64–65

GENETICS

The anatomy of a gene. Packed into nearly every cell in the body is a blueprint of sorts. Diagram.
U.S. News Nov 4, 1991 p. 66

From gene to protein. The machinery of the cell deciphers the genetic code to manufacture proteins. Diagram.
U.S. News Nov 4, 1991 p. 69

All in the family. A family tree for the human species. Genetic groups. Major language families.
U.S. News Nov 5, 1990 p. 65

Genetic liabilities. Some of the most common hereditary disorders that strike Americans, usually incurring heavy medical costs. Lifetime rates.
U.S. News Jul 23, 1990 p. 58

GEOGRAPHY

So where is Twin Peaks? Here are some of the questions asked last month at the National Geographic Society's second annual geography bee.
U.S. News Jun 11, 1990 p. 74

An 'F' for the U.S. Among 18-to 24-year-olds, Americans scored dead last [out of nine countries] in identifying 16 geographic locations. Map.
Newsweek Aug 8, 1988 p. 31

Percentage of Americans age 18 to 24 who could correctly identify these spots [various countries] on an unlabeled world map.
U.S. News Aug 8, 1988 p. 11

GEOLOGY

570 million years ago. Map.
Time Apr 8, 1991 p. 66

GERMAN REUNIFICATION

Estimated cost of unification over 10 years. [In various areas.]
Time Oct 8, 1990 p. 58

East German debt that will be absorbed by new government. Population in millions. Divorces per 10,000 people. Work force in millions. Expected East German unemployment.
U.S. News Oct 8, 1990 p. 14

One country, two armies. Active military. Reserves. Tanks. Combat aircraft. Artillery pieces. West Germany. East Germany.
U.S. News Oct 1, 1990 p. 48

Costs of division. A sampler of 1989 prices. White bread. Coffee. Subway ticket. Best opera ticket. Automatic washer. West. East [Germany].
Newsweek Jul 9, 1990 p. 32

Combined clout. Population. Percent employed. Agriculture. Industry. Service. Average monthly earnings in industry. West. East [Germany].
Newsweek Jul 9, 1990 p. 36

Rebuilding East Germany. Estimated annual investment through the 1990s. Annual GNP growth. Yield on 10-year West German government bonds.
Business Week Jun 4, 1990 p. 75

A brand-new superpower. What will each Germany bring to the union? How will united Germany rank next to other major industrial countries? [Other various statistics.]
Scholastic Update May 4, 1990 pp. 6–7

One Germany. Exports. Gross investment. Compensation.
Business Week Apr 2, 1990 pp. 46–47

How to pay for reunification. Annual price tag. Possible sources of funds.
Business Week Apr 2, 1990 p. 50

Putting together a powerhouse. West [Germany]. East [Germany]. Autos. Chemicals/pharmaceuticals. Electronics. Financial services. Steel. Map.
Business Week Apr 2, 1990 p. 51

How the parties are seated [percentages]. Christian Democrats. Social Democrats. Democratic Awakening. German Social Union. Alliance of Free Democrats. Party of Democratic Socialism. Chart.
Time Apr 2, 1990 p. 30

Putting on weight. Area. Production. Trade. Economy. Military. Cities. Population.
Time Mar 26, 1990 p. 36

What unification will cost West Germany. Direct subsidies and indirect costs.
Business Week Feb 26, 1990 p. 54

'There are no better-fed refugees.' Federal Republic of Germany (West). German Democratic Republic (East) [comparative statistics].
Newsweek Nov 20, 1989 p. 34

GERMANY - ECONOMIC CONDITIONS

Germany's dilemma. Rising inflation. Change in consumer prices. Slowing trade. Trade surplus. Slumping growth. GNP increase. 1989-1991.
Business Week Oct 28, 1991 p. 50

German budget blues. German budget balance, percentage of GDP. 1987-1991.
U.S. News Jul 22, 1991 p. 47

Where the big economies stand. United States. Japan. West Germany. Economic growth. Inflation. Cost of credit. 1987-1990.
Business Week Aug 20, 1990 p. 29

Italy's growth is tailing off. Italy. France. Germany. Real GDP growth rates. Annual consumer price increases, 1990. Budget deficit as percent of GDP, 1990.
Business Week Jun 11, 1990 p. 34

GERMANY - HISTORY

Germany [and its changing borders]. 1648. 1815. 1871. Maps.
Time Jul 9, 1990 p. 70

GERMANY (EAST) - ECONOMIC CONDITIONS

Costs of living. Net monthly income. Goods as percent of take-home pay. Germany. West. East.
U.S. News Nov 27, 1989 p. 43

East Germany's economic slide. Growth is slowing and exports are falling. 1984-1989.
Business Week Sep 4, 1989 p. 29

GERMANY (WEST) AND THE UNITED STATES

Doing well, doing good. How West Germany compares with the U.S. on Soviet trade—and aid. Imports. Exports. Aid.
Time Jul 16, 1990 p. 17

How the competition stacks up. Labor. Japan. West Germany. U.S. Productivity index. 1980-1990. Unit labor costs. 1980-1990. Scientists and engineers engaged in R&D. 1980-1987.
U.S. News Jul 16, 1990 pp. 24–25

Biggest banks. Home base of the world's top 100 banks. U.S. Japan. West Germany.
U.S. News Jul 16, 1990 p. 24

Bright ideas. Percentage of total U.S. patents granted. U.S. Japan. West Germany. 1975, 1989.
U.S. News Jul 16, 1990 p. 24

How the competition stacks up. Capital. Japan. West Germany. U.S. Cost of capital. 1980-1988. Real long-term interest rates. 1975-1990.
U.S. News Jul 16, 1990 p. 24

How the competition stacks up. Investment. Japan. West Germany. U.S. Investment in plant and equipment. 1980-1990. Nondefense R&D expenditures. 1980-1988. Public investment. 1967-1985.
U.S. News Jul 16, 1990 p. 25

Gray power. Percent of population now over 65. U.S. Japan. West Germany. World average.
U.S. News Jul 16, 1990 p. 25

Time clock. Average number of working hours per year. U.S. Japan. West Germany.
U.S. News Jul 16, 1990 p. 25

Head of the class. Proportion of 22-year-old population receiving undergraduate degrees. Percentage of undergraduates receiving science and engineering degrees. U.S. Japan. West Germany.
U.S. News Jul 16, 1990 p. 26

Bookworms. Days spent in grade school per year. U.S. Japan. West Germany.
U.S. News Jul 16, 1990 p. 26

Striking out. Number of workdays lost to labor disputes (in thousands). U.S. Japan. West Germany.
U.S. News Jul 16, 1990 p. 26

GERMANY (WEST) - ECONOMIC CONDITIONS

Costs of living. Net monthly income. Goods as percent of take-home pay. Germany. West. East.
U.S. News Nov 27, 1989 p. 43

A flood of emigres from the East bloc. Total East bloc emigres. Soviet Union, Poland, others. East Germany. 1986-1989.
Business Week Oct 23, 1989 p. 74

Will [emigres] fuel growth in West Germany. With immigration. Without immigration. Index of estimated gross national product. 1988-1995.
Business Week Oct 23, 1989 p. 74

West Germany. Surplus. Exports. Imports. 1988.
Newsweek Jul 17, 1989 p. 39

GIFTS

Untithed. Religious denomination. % of income members give to charity.
Newsweek Jan 22, 1990 p. 8

GLASS CEILING

See CAREER PLATEAUS

GLOBAL WARMING

See GREENHOUSE EFFECT

GOLD MINES AND MINING

Foreigners strike gold. Non-U.S. corporations own or control most of the biggest gold mines in the United States. The 15 top foreign-owned mines.
U.S. News Oct 28, 1991 p. 46

All that glitters. Gold production. 1980-1990.
Newsweek Jan 11, 1988 p. 39

GOLF AND GOLFERS

Fore! U.S. players. 1980, 1988.
Time May 8, 1989 p. 66

The $20 billion golf industry is growing, while more active sports are fading. Golf. Tennis. Jogging. 1984-1987.
Business Week Mar 27, 1989 p. 77

The changing golfer. Income. Age. Sex. College grads. Rounds per year. 1985, 1987.
Business Week Mar 27, 1989 p. 82

GOLF, PROFESSIONAL

Golf's precocious seniors. Senior PGA tour revenues. 1985-1989.
Business Week Jun 26, 1989 p. 122

GOVERNMENT CONTRACTS

Dollar value of federal contracts awarded to 8(a) [socially and economically disadvantaged business persons] firms, 1983-87. Number of firms. Contracts awarded.
Black Enterprise Mar 1988 p. 19

GOVERNMENT EMPLOYEES

State and local workers still get fatter pay hikes. State and local government workers. Private-sector workers. Change in total compensation. 1982-1991.
Business Week Sep 23, 1991 p. 26

War of the raises. Federal employes pay hike. Here's how past increases compare with average private-sector raises. 1980-1990.
U.S. News Dec 24, 1990 p. 66

Public servants gain. Changes in compensation. Public workers. Private workers. 1982-1990.
U.S. News Nov 5, 1990 p. 58

GOVERNMENT SERVICES

How America compares [to seven industrial countries]. Health care costs. Public spending on health. Infant mortality. National maternity-leave policy. Public spending on education. Labor force participation.
Business Week Dec 17, 1990 pp. 88–89

GRAIN

Love those traditional crops. Acres planted in: Corn. Soybeans. 1987-1991.
Business Week May 20, 1991 p. 46

Worldwide grain production. International crop production leaders. Number of farms in the U.S. Total U.S. farm debt. U.S. farmers' net cash income.
U.S. News Mar 25, 1991 p. 17

Global grain supplies in days. 1970-1989.
Newsweek Aug 15, 1988 p. 36

Global harvest. Leading grain producers' shares of world production [various countries]. Wheat. Corn. Soybeans.
U.S. News Jul 4, 1988 p. 45

U.S. stocks (in bushels). Wheat. Corn. Soybeans. 1977-1989.
U.S. News Jul 4, 1988 p. 45

Grain sales are growing faster. Consumption of American grain. Corn. Wheat. Soybeans. U.S. Export. 1985-1988.
Business Week Mar 28, 1988 p. 81

Less surplus in the farm belt. U.S. inventories of major grain crops. 1983-1988.
Business Week Jan 11, 1988 p. 123

GREAT BRITAIN - ECONOMIC CONDITIONS

Major's challenge. Change in consumer prices. Jobless rate. Current-account balance. GNP growth. Percent. 1979-1991.
Business Week Dec 10, 1990 p. 25

Maggie's economic report card. Unemployment. Inflation. Economic growth. Prime lending rate. 1987-1990.
Time Nov 26, 1990 p. 43

GREAT BRITAIN - ETHNIC GROUPS

Annual contribution to household income by ethnic group. Ethnic group. Total contributions. Employment. Self-employment.
Black Enterprise Dec 1989 p. 98

GREAT BRITAIN - POLITICS AND GOVERNMENT

Thatcher's report card. Accomplishments. Failures. [1979-1990]. Chart.
Business Week Dec 10, 1990 p. 26

Maggie's long run [highlights of Thatcher's career as Prime Minister]. 1975-1990. Chart.
Time Dec 3, 1990 p. 62

GREENHOUSE EFFECT

An atmospheric Catch-22. There are two crises in the skies. [Results of] more CFCs. [Results of] less CFCs. Diagram.
Newsweek Nov 4, 1991 p. 49

The nights get warmer. Average annual night temperature for U.S. 1901-1987.
Business Week Feb 4, 1991 p. 83

The Geritol solution. Can tiny sea creatures get us out of the greenhouse fix? Diagram.
Newsweek Nov 5, 1990 p. 67

Biggest greenhouse contributors. According to revised calculations, three of the five largest contributors of greenhouse gases are nonindustrialized countries.
U.S. News Jun 18, 1990 p. 47

The theory [of how global warming occurs]. Some uncertainties. Diagram.
Time Apr 30, 1990 p. 84

Your contribution to global warming. Pounds of carbon dioxide added to the atmosphere by these household appliances.
U.S. News Apr 23, 1990 p. 69

Rising temperatures. Table below shows the past and the predicted number of days with temperatures above 90 degrees. Recent norm. 2030.
Newsweek Jul 11, 1988 p. 17

Tomorrow's forecast: Unseasonably hot and dry. The greenhouse effect is expected to warm the earth between 3 and 8 degrees Fahrenheit over the next 50 years. 1987, 2000. Map.
Newsweek Jul 11, 1988 p. 18

See Also OZONE

GRENADA-AMERICAN INVASION, 1983 - CASUALITIES

U.S. battle deaths. Killed in action. Vietnam. Lebanon. Grenada. Panama. Persian Gulf.
Newsweek Mar 11, 1991 p. 28

GROSS DOMESTIC PRODUCT

Major world economies. Yearly change in gross domestic product. U.S.A. Britain. France. Germany. Japan. 1988, 1991.
U.S. News Dec 16, 1991 p. 34

World prospects. Growth rates of real GDP. Developing countries. Industrialized countries. 1980-2000.
U.S. News Dec 10, 1990 p. 60

The new superpower. Gross domestic product. Japan. United Germany.
Newsweek Feb 26, 1990 p. 30

GDP in trillions of dollars. U.S. E.C. Japan.
Time Sep 18, 1989 p. 40

Europe's wealth curtain. 1988 per capita gross domestic product. [European countries].
U.S. News Jul 17, 1989 p. 38

Leading the world. Gross domestic product per employed person [by country]. Gaining new ground. Changes in manufacturing productivity [by country].
U.S. News Feb 29, 1988 p. 52

GROSS DOMESTIC PRODUCT - AFRICA

Africa fights back. Growth rates of real GDP per capita. 1965-2000.
U.S. News Dec 10, 1990 p. 60

GROSS DOMESTIC PRODUCT - FRANCE

Growing economy. Percentage change in real GDP. Germany. France. 1986-1990.
U.S. News Dec 17, 1990 p. 73

GROSS DOMESTIC PRODUCT - GERMANY

Growing economy. Percentage change in real GDP. Germany. France. 1986-1990.
U.S. News Dec 17, 1990 p. 73

GROSS DOMESTIC PRODUCT - LATIN AMERICA

Latin America rebounds. Growth rates of real GDP per capita. 1965-2000.
U.S. News Dec 10, 1990 p. 60

GROSS NATIONAL PRODUCT - ASIA

Gross national product. Per capita. Average annual growth (1975-1987) [various Asian countries].
U.S. News Jun 5, 1989 p. 23

GROSS NATIONAL PRODUCT - EUROPE

What's bothering Europe. Growth is falling. Gross national product. 1986-1991.
Business Week Oct 15, 1990 p. 47

GROSS NATIONAL PRODUCT - SOVIET UNION

Shattered economy. Percent change in Soviet GNP. 1983-1991.
U.S. News Sep 9, 1991 p. 36

Slow growth. Despite *perestroika's* promises, Soviet production is stagnant. Average annual rate of GNP growth. 1961-1989.
U.S. News Nov 20, 1989 p. 32

GROSS NATIONAL PRODUCT - UNITED STATES

Average annual percentage increases in real GNP during these presidencies. [Truman to Bush].
U.S. News Dec 30, 1991 p. 37

Growth slows down. % change in real GNP, annually. 1988-1991.
Time Dec 9, 1991 p. 22

[Percent] change in real GNP. 1990, 1991.
Time Oct 7, 1991 p. 45

The tax bite just gets bigger. Taxes as a share of GNP. Federal. State & local. 1970-1990.
Business Week Sep 2, 1991 p. 32

How bad was it? Real gross national product. Civilian unemployment rate. Industrial production. Change during recession. 1990-91. Average.
Newsweek Aug 5, 1991 p. 64

Percentage change in real GNP. 1987-1991. GNP decline. Business investment. Residential construction. Government spending.
U.S. News May 6, 1991 p. 19

A delicate balance. Percentage change in consumer price index. Percentage change in real GNP. 1989, 1990.
U.S. News Dec 17, 1990 p. 58

Vital exports. Exports as a percentage of real GNP (in 1982 dollars). 1980-1990.
U.S. News Dec 3, 1990 p. 65

Stuck in the mire. % change in real GNP, at annual rates. 1988-1991 [forecast].
Time Jul 23, 1990 p. 58

It [federal deficit] may well fall faster as a share of GNP. Deficit as percent of nominal GNP. Fiscal year. Percent. 1986-1994.
Business Week Oct 10, 1988 p. 129

The energy-GNP ratio tilts up again. Total energy consumption in U.S. divided by GNP. 1980-1988.
Business Week Sep 5, 1988 p. 18

GROSS NATIONAL PRODUCT - WORLD

Asia ascending. Japan. U.S.A. China. Per capita GNP. 1991, 2041.
Newsweek Dec 9, 1991 p. 6

Global growth. Can the U.S. take up the slack? U.S. Japan. Germany. World. Percent change in real GNP. 1991, 1992.
Business Week Jul 15, 1991 p. 24

The dollar is taking off. GNP growth. U.S. Germany. Japan. 1990-1992.
Business Week Apr 8, 1991 p. 23

Growth slows overseas. Percent change in GNP. Japan. Germany. France. Britain. Canada. 1989-1990.
Business Week Dec 24, 1990 p. 62

Growth rate of GNP per capita (annual average, 1965-87). Singapore. South Korea. Japan. West Germany. U.S.
Scholastic Update Sep 7, 1990 p. 20

The new superpower. GNP per capita. United States. Japan. United Germany. Soviet Union.
Newsweek Feb 26, 1990 p. 21

Lagging output. Per capita GNP (1988). U.S. Japan. West Germany. East Germany. Singapore. U.S.S.R. Hungary. Bulgaria.
U.S. News Nov 20, 1989 p. 32

Economic map of the world. Per capita G.N.P. in U.S. dollars.
Scholastic Update Sep 22, 1989 pp. 16–17

Economic map of the world. Per capita G.N.P. in U.S. dollars. Map.
Scholastic Update Sep 23, 1988 pp. 12–13

America's slice of the pie. Percentage of world G.N.P. by country.
Scholastic Update Sep 9, 1988 p. 2

Growth. Increase in per capita GNP. Average annual percent change, 1973-86. U.S. Japan. China. Hong Kong. Singapore. S. Korea. Per capita GNP, 1986.
Newsweek Feb 22, 1988 p. 44

Shifting economic power. U.S. U.S.S.R. Japan. China. 1950, 1990, 2010 (est.).
U.S. News Jan 25, 1988 p. 46

GUIDED MISSILES

See MISSILES

GUN CONTROL

On the firing line. Ruger .38 revolver. Beretta 9-mm pistol. Uzi submachine gun. Colt submachine gun. Weight. Length. Capacity. Features.
Newsweek Nov 19, 1990 pp. 58–59

GUNS AND YOUTH

Kids who kill. Firearm murders committed by offenders under age 18. 1984, 1987, 1989.
U.S. News Apr 8, 1991 p. 26

GUNS IN THE UNITED STATES

U.S. murders by handguns. 1985-1989.
Time May 20, 1991 p. 26

America's arsenal. Handguns involved in murders, rapes, robberies, assaults. Children killed with handguns. Law-enforcement officers killed with handguns.
Newsweek Apr 8, 1991 p. 31

Guns and crime. Cops under fire. Murders with guns. All firearms. Handguns. Knives or other cutting instruments were the next most common murder weapon. 1974-1989.
U.S. News Dec 3, 1990 p. 38

On the firing line. Ruger .38 revolver. Beretta 9-mm pistol. Uzi submachine gun. Colt submachine gun. Weight. Length. Capacity. Features.
Newsweek Nov 19, 1990 pp. 58–59

Americans and their guns. How does the gun-owning population differ from the U.S. as as whole? [Selected statistics].
Time Jan 29, 1990 p. 20

Deaths by firearms. Accidental deaths. Homicides. Suicides. 1950-1986.
Time Jul 17, 1989 p. 61

Playing with firepower. Semiautomatic rifles range from military weapons to guns used for hunting. Assault rifle (banned). Assault rifle (legal). Hunting rifle (legal). Diagrams.
Newsweek Mar 27, 1989 p. 28

Deadly numbers. People killed by handguns in 1985. Canada (population). Britain. Japan. U.S.
Time Feb 6, 1989 p. 20

Street favorites. Assault weapons available over the counter.
Time Feb 6, 1989 p. 25

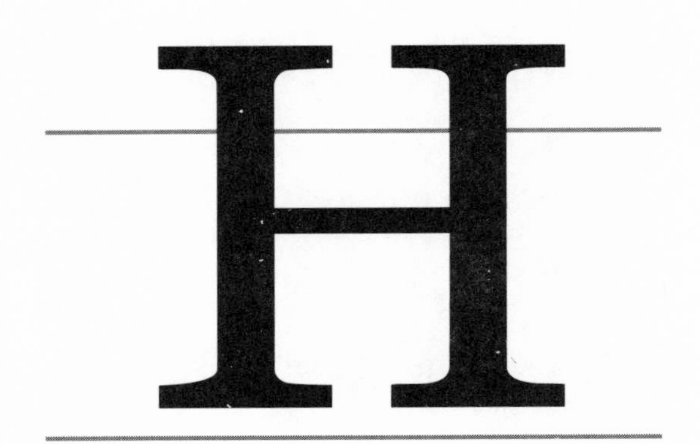

HAITI - ECONOMIC ASSISTANCE

Major countries supplying aid to Haiti, 1987. Countries. Suspend aid in 1988?
Black Enterprise Apr 1988 p. 26

HALLOWEEN

Halloween at the polls. Best-selling political masks since 1970. Hottest costumes this year.
U.S. News Oct 31, 1988 p. 82

HARVARD UNIVERSITY

The skyrocketing costs of college. Annual expenses. 1990 estimates. Percent increase from 1980.
Newsweek Feb 5, 1990 p. 6

HAZARDOUS WASTE

A trail of defense pollution. Generating hundreds of thousands of tons of hazardous waste yearly, the military is the nation's most pervasive and protected polluter. Some of the worst sites. Map.
Newsweek Aug 6, 1990 p. 21

Toxic waste in the balance. According to manufacturers' reports, here's how many millions of pounds of toxic waste moved across state borders in a year. 1987.
U.S. News Jun 11, 1990 p. 74

The DOE's [Department of Energy] expensive studies. Costs of RIFS work at 12 DOE sites. DOE site. Original estimate. Cost to date.
U.S. News Feb 19, 1990 p. 28

America's most hazardous waste sites (number per state). Map.
Scholastic Update Jan 12, 1990 p. 29

How a nuclear-weapons plant poisoned a community. The Feed Materials Production Center at Fernald, Ohio, has been contaminating the air, water and ground around it almost from the day its doors opened 37 years ago. Diagram.
U.S. News Oct 23, 1989 pp. 22–23

Waste not, want not. Hazardous-waste sites on the Superfund priority list. Toxic-waste sites per state. Map.
U.S. News Feb 6, 1989 pp. 48–49

Toxic trade. Known shipments [of hazardous waste]. Exported from - to - [countries]. Incompleted shipments [of hazardous waste]. Exported from - to - [countries].
U.S. News Nov 21, 1988 p. 55

Clamp down on toxic terrorists. Sierra Leone. Congo. Guinea. Nigeria.
Black Enterprise Nov 1988 p. 31

No more open pits. The days of outright dumping are gone. Waste-management companies now employ a range of sophisticated techniques.
Newsweek Oct 3, 1988 p. 39

HEAD-WOUNDS AND INJURIES

Causes of head injuries.
FDA Consumer Sep 1990 p. 11

Head-trauma cases [various statistics].
Newsweek Apr 9, 1990 p. 49

HEADACHES

Headache boom. Migraine headaches. Here is the rate per 1,000 [for migraine headaches]. 1981-1989.
U.S. News Jun 24, 1991 p. 71

The anatomy of a headache. Cause and effect. Transmitting pain. [The] brain. The serotonin connection. Diagrams.
U.S. News Jul 31, 1989 pp. 46–47

HEALTH

Fatty risks: How to estimate health hazards. Body-mass index (BMI) is a standard method doctors use to estimate degree of body fat.
U.S. News May 14, 1990 p. 58

Age 25. Age 45. Age 85. Max. heart rate. Lung capacity. Cholesterol level. Muscle strength. Kidney function.
Newsweek Mar 5, 1990 p. 44

The top 10 health frauds. FDA's list of the top 10 health frauds.
FDA Consumer Oct 1989 p. 29

HEALTH CARE

See MEDICAL CARE

HEALTH EDUCATION

Diet-health trends. Shoppers who "pay attention" to ingredient lists. Shoppers who use lists to "avoid" or limit.
FDA Consumer May 1988 pp. 7–8

HEALTH MAINTENANCE ORGANIZATIONS

The shift to HMO's. Percentage of insured in HMO's and PPO's [health-maintenance organizations and preferred provider organizations]. 1984-1997.
U.S. News Dec 25, 1989 p. 67

HEART

Two tests are better than one. Wave patterns. Electrocardiogram. Seismocardiogram.
Newsweek Sep 16, 1991 p. 69

HEART - DISEASES

Fatal attraction. The growth factor PDGF [platelet derived growth factor] plays a central role in atherosclerosis. Diagram.
U.S. News Sep 9, 1991 pp. 56–57

The state of black health. Tracking the silent killers. Mortality statistics by race. Cardiovascular disease. Heart disease. Stroke. Black. White. Men. Women.
Black Enterprise Jul 1991 p. 43

Heart. Blocked arteries. The usual way: Coronary bypass. The minimal way: Atherectomy. Procedures per year. Average cost. Hospital stay. Recovery time. Risk.
U.S. News May 20, 1991 p. 77

[Five ways of] keeping arteries open. Bypass surgery. Motorized scraper. Balloon angioplasty. Mesh metal stent. Drill-type scrapers.
FDA Consumer Apr 1991 pp. 24–25

Gap narrows between bypass and angioplasty. Bypass surgeries. Angioplasties. 1980-1988.
FDA Consumer Apr 1991 p. 26

Opening arteries. Tissue buildup is what blocks coronary arteries. Synthetic molecules might stop the process. Diagram.
Newsweek Jan 14, 1991 p. 57

Growing perceptions of diet-heart disease links. 1983, 1986, 1988.
FDA Consumer Mar 1989 p. 26

The arteries' heros and villains. Diagram.
Time Dec 12, 1988 pp. 62–63

Healthy artery. Diseased artery. Diagram.
Time Dec 12, 1988 pp. 64–65

Controversy over a common operation. Heart-bypass surgeries performed in the U.S. 1970-1986.
Newsweek Aug 1, 1988 p. 60

A hot heart fix-it. Number of procedures. Coronary bypasses. Angioplasties. 1970-1986. Cost. Time. Advantages. Disadvantages.
U.S. News Jul 25, 1988 p. 65

The spread of heartworm disease. 1960s. 1980s. Maps.
FDA Consumer Jun 1988 p. 27

Balloon angioplasty opens clogged coronary arteries by compressing the deposits of fat — called plaque — that block the flow of blood to the heart muscle. Diagram.
FDA Consumer May 1988 p. 26

Despite the growing popularity of balloon angioplasty, some 230,000 coronary bypass operations were performed in 1985. Diagram.
FDA Consumer May 1988 p. 27

HEART ATTACKS

The grim odds. [Selected statistics].
U.S. News Jul 22, 1991 p. 54

Lifesaving jolt. In five different places, survival rates for victims of sudden cardiac arrest improved dramatically when emergency medical teams began carrying defibrillators on all heart attack calls.
U.S. News Jul 22, 1991 p. 56

Improving the odds. [Percent who survive] with current clotbuster use. With optimum clotbuster use. With optimum clotbuster use plus prompt medical care.
U.S. News Mar 25, 1991 p. 69

The path to a heart attack. Diagram.
U.S. News Aug 6, 1990 pp. 58–59

How the risk grows. Deaths by heart attack per 100,000 people in each age group. Men. Women. Ages 35-85.
U.S. News Dec 18, 1989 p. 75

The heart-attack gap widens. Deaths from heart disease. Black males. Black females. White males. White females. 1970-1985.
Newsweek Oct 31, 1988 p. 66

Lowering the odds. If you [do various activities] your risk of a heart attack may drop [by this percent].
Business Week Feb 22, 1988 p. 169

HEART TRANSPLANTATION

See TRANSPLANTATION OF ORGANS, TISSUES, ETC.

HEPATITIS

America's hepatitis chart. New infections per year. Chronic carriers.
U.S. News Aug 13, 1990 p. 14

A year's cases: Estimated outcomes. Hepatitis B infections. Chart.
FDA Consumer May 1990 p. 16

HEREDITY

Genetic liabilities. Some of the most common hereditary disorders that strike Americans, usually incurring heavy medical costs. Lifetime rates.
U.S. News Jul 23, 1990 p. 58

HEROIN

Highs and lows: Counting costs in the drug market. Ice. Crank. Crack. Heroin. Cost. Duration. Effects. Side effects.
Newsweek Nov 27, 1989 p. 38

HIGH SCHOOL EQUIVALENCY EXAMINATIONS

Diplomas for dropouts. Here are the state-by-state results for high-school equivalency tests in 1989. Percentage who passed. Graduates over age 40.
U.S. News Jun 25, 1990 p. 66

HIGH SCHOOL GRADUATES

Diplomas aplenty. A record number of adults have high-school and college degrees. [Here are] the leading states. High-school graduates. College graduates.
U.S. News Dec 16, 1991 p. 95

The less educated. Median annual family income. By type of family head under 30. Dropout. High school grad. Some college. College grad. 1973, 1989.
Business Week Aug 19, 1991 p. 81

Diplomas for dropouts. Here are the state-by-state results for high-school equivalency tests in 1989. Percentage who passed. Graduates over age 40.
U.S. News Jun 25, 1990 p. 66

A larger crowd at the finish. The following percentages of students, graduated from public schools in 1987.
U.S. News Jun 12, 1989 p. 69

High school graduates as a percentage of total 18-24 year olds, selected years. Black. White. Total. 1976, 1981, 1985.
Black Enterprise Sep 1988 p. 43

HIGH SCHOOL STUDENTS

Children in the United States who are gifted. Average SAT score of college-bound high-school seniors. Who dropped out of high school. [Other selected statistics.]
U.S. News Nov 11, 1991 p. 14

A formula for failure. Math-achievement levels of high-school seniors. Below basic. Basic. Proficient. Advanced.
Newsweek Oct 14, 1991 p. 54

No pain, no gain. Time [spent] on homework [by] seventeen-year-old students. 1988. Chart.
Newsweek May 27, 1991 p. 62

The sporting life. Top ten sports for high school girls. 1971. 1986-87. Percent increase.
Ms. Aug 1988 p. 83

See Also DROPOUTS

HIGH SCHOOLS

Way off Broadway. Plays from the 1980s are edging into high-school repertoires. Here are the current favorites.
U.S. News Oct 8, 1990 p. 83

Big men on a bigger campus. Of the 9,820 public high schools in the U.S. with grades 9 through 12, those with the most students are: School [name and location]. Students. Teachers.
U.S. News Dec 11, 1989 p. 74

HISPANIC AMERICANS

In 20 years Hispanics will be the largest minority group. Blacks. Hispanics. 1980-2020.
Time Jul 29, 1991 p. 15

Young, diverse and growing. Hispanics by age. And by origin.
Newsweek Apr 9, 1990 p. 18

By 2056, whites may be a minority group. Non-Hispanic whites as a % of total population. 1920-2056.
Time Apr 9, 1990 p. 30

[Percent] increase in each population group (1980-1988). Asians and others. Hispanics. Blacks. Whites.
Time Apr 9, 1990 p. 30

Birth rates and immigration. Whites. Blacks. Hispanics. Asians and others.
Time Apr 9, 1990 p. 31

Hourly wages of U.S. Hispanic population, age 21-25, by years in the U.S.
U.S. News Sep 25, 1989 p. 31

U.S. Hispanics: Their numbers are taking off. Percent of total U.S. population. 1970-2000.
Business Week Feb 20, 1989 p. 21

Latin power: A demographic upheaval. Dade County's ethnic mix. Breakdown of the Latin population.
Newsweek Jan 25, 1988 p. 28

HISPANIC AMERICANS - ECONOMIC CONDITIONS

The poverty line. White. Black. Hispanic. Americans living below it.
Time Sep 30, 1991 p. 30

Devastating minorities. Median annual family income. By type of family head under 30. White. Black. Hispanic. 1973, 1989.
Business Week Aug 19, 1991 p. 81

Catching up, sort of. Percentage of households with incomes of $50,000 or more. Whites. Blacks. Hispanics. 1972-1988.
U.S. News Dec 11, 1989 p. 74

As Hispanic Americans grow wealthier, advertisers spend more to reach them. Total purchasing power. Ad spending aimed at Hispanics. 1983-1987.
Business Week Jun 6, 1988 p. 64

HISPANIC AMERICANS - EDUCATION

Percentage of Hispanic population with a high-school education, by year of immigration.
U.S. News Sep 25, 1989 p. 31

Reading scores of U.S. Hispanic population, age 21-25, by years in U.S.
U.S. News Sep 25, 1989 p. 31

HISPANIC AMERICANS - EMPLOYMENT

Formal training by sex and race of trainee. Share of training. Share of the work force. Males. Females. Whites. Blacks. Hispanics. [By] age.
Black Enterprise Aug 1989 p. 41

HOBBIES

Amateur artistry. Expressing one's artistic tendencies has become much more popular. 1975, 1987.
U.S. News Mar 5, 1990 p. 72

HOLIDAYS

Any cause for celebration. Unusual holidays states will celebrate in 1989. Map.
U.S. News Dec 19, 1988 p. 70

HOME HEATING OIL

Price spike. Retail gasoline. Retail home-heating oil (cents per gallon). 1989.
U.S. News Jan 22, 1990 p. 49

HOME SHOPPING

TV shopping: More viewers, less choice. [By network]. 1986, 1990.
Business Week Oct 22, 1990 p. 70

HOME-BASED BUSINESSES

The shortest commute. Americans who work at home. 1980-2000.
U.S. News Dec 25, 1989 p. 67

Home is where the work is. Growth of full-time home-based businesses. 1980-1995 [est.].
U.S. News Dec 26, 1988 p. 120

For more workers, there's no place like home. White-collar workers operating out of home offices. Full-time. Part-time. PCs used in home offices.
Business Week Oct 10, 1988 p. 105

HOMELESS

Overcrowding has come down but more 'couch people' go to shelters. Black. Hispanic. White. 1978, 1983, 1989.
U.S. News Dec 23, 1991 p. 31

The changing homeless population. Percentage of homeless that are: Families with children, severely mentally ill, substance abusers.
U.S. News Jan 15, 1990 p. 27

Homelessness. Estimated homeless population for 29 cities. Map.
Scholastic Update Jan 12, 1990 p. 28

A nationwide problem. Homeless people. New York, Los Angeles, Minneapolis, Phoenix, Washington D.C., Chicago. Map.
Scholastic Update Feb 10, 1989 p. 8

Four causes of homelessness. More people live in poverty. Number of Americans below the poverty line. 1969-1986.
Scholastic Update Feb 10, 1989 p. 12

Four causes of homelessness. Federal aid hasn't kept up. Percent rise in federal budget for social welfare programs. 1960-1985.
Scholastic Update Feb 10, 1989 p. 12

Four causes of homelessness. Less low-income housing. Annual budget of Department of Housing and Urban Development (HUD). 1981-1987.
Scholastic Update Feb 10, 1989 p. 13

Four causes of homelessness. Release of mentally ill. Total days spent by all patients in state and county mental hospitals. 1969-1986.
Scholastic Update Feb 10, 1989 p. 13

Who the homeless are. [Percentage that are] alcohol abusers. Drug abusers. Mentally ill. Members of homeless families.
U.S. News Feb 29, 1988 p. 33

HOMES

See HOUSING

HOMOCIDE

See MURDER AND MURDER RATES

HOMOSEXUALITY

Tiny bundle of neurons that might be linked to homosexuality. Diagram of brain.
Newsweek Sep 9, 1991 p. 52

HORMONES

Effects of human-growth hormone injections in men over 60.
Time Jul 16, 1990 p. 54

HORMONES, FEMALE

Sugar and spice and everything nice. The female sex hormones, especially estrogen, shape the brain as well as the body. Brain. Cardiovascular. Immune system. Fat.
Newsweek May 28, 1990 p. 59

HORMONES, MALE

Snakes and snails and puppy dog tails. The male sex hormone testosterone strengthens muscles but threatens the heart. Senses. Bones. Heart. Blood.
Newsweek May 28, 1990 p. 58

HOSPITALS

The best of the best [listed by medical specialty]. To identify the nation's best hospitals, *U.S. News* solicited opinions from 1,501 physicians in 15 specialties.
U.S. News Aug 5, 1991 pp. 38–66

Hospital debt is rising. Debt per hospital bed. Patients admitted per hospital. Operating profit margin. 1984-1990.
Business Week Aug 27, 1990 p. 67

It's no vacation. The average hospital stay in days. 1970, 1979, 1988.
U.S. News Apr 30, 1990 p. 56

A hospital stay. Cost represents expense of typical stay and excludes physician charges. 1965, 1975, 1980, 1988.
U.S. News Apr 30, 1990 p. 60

Hospitals in black communities. Hospitals serving black communities (with more than 500 beds). Number of hospitals serving black communities, selected years.
Black Enterprise Jul 1988 p. 41

Where hospitals get their money. Community-hospital revenue. Percentage of hospital expenses that go for charity care and bad debt.
U.S. News May 16, 1988 p. 26

Hospitals are suffering as patients dwindle and outpatient treatment soars. Community-hospital closures. Days medicare patients spent in short-stay hospitals. Medicare payments to hospitals. [To] doctors.
U.S. News Apr 11, 1988 pp. 50–51

HOSPITALS, PSYCHIATRIC

Shrinkage growth. Types of psychiatric hospitals. Private for profit. State and county. 1984, 1988.
Newsweek Nov 4, 1991 p. 51

HOSTAGES - LEBANON

Who holds whom—and why. Since 1984, a variety of Shi'ite fundamentalist groups have kidnapped more than 30 Westerners in Lebanon. [List of groups and hostages they hold]. 1984-1991.
Time Aug 19, 1991 pp. 28–29

HOUSEWORK

Home, sweep home. Proportion of household tasks done by women and men. Meals. Bills. Cleaning. Outdoor chores/repairs. 1965-1985.
Ms. Apr 1989 p. 83

Who does the work? Time spent on housework and childcare for married couples with children. 1969-1983.
Ms. Feb 1988 p. 19

HOUSING - AFFORDABILITY

Home ownership will stay strong. Number of households owning homes. 1970-2000.
Business Week Jul 16, 1990 p. 22

Median income ratios of owners and renters to all households: 1950-1985. Owners. All households. Renters.
Black Enterprise Jan 1990 p. 45

The squeeze on first-time buyers. Percentage of married couples with own home (by age). 1974-1988.
U.S. News Dec 11, 1989 p. 56

More-affordable houses. Housing-affordability index. 1980-1988. Percentage of population who own their home, by age. 1980-1987.
U.S. News Aug 29, 1988 p. 102

A dream deferred. Percent owning homes, by age. 1979-1987.
Newsweek Apr 11, 1988 p. 65

HOUSING - BUYING AND SELLING

State by state. Where disclosure is mandatory. Where disclosure is voluntary.
U.S. News Sep 16, 1991 p. 62

The rebound in home buying. Sales of new single-family homes. 1990, 1991.
Business Week Aug 12, 1991 p. 18

New single family home sales (in thousands). 1989-1991. New home sales rate. Median home prices. New. Existing. Average 30-year fixed mortgage rate. Mortgage delinquency rate.
U.S. News Jun 10, 1991 p. 14

Housing hopes. Housing starts. New-home sales. Existing-home sales. 1990-1991.
U.S. News May 27, 1991 pp. 48–49

A deeper slump in homebuilding. Sales of new single-family homes. 1986-1991.
Business Week Jan 14, 1991 p. 121

Sales of existing one-family homes. Median sale price of existing one-family homes. [Office] vacancy rate. 1986-1991.
Business Week Dec 31, 1990 p. 139

Home sales by region. Median price on existing, single-family homes. 1981-1989.
Black Enterprise Oct 1990 p. 68

The newest boom towns. Even in the midst of a real-estate recession housing values in a number of cities are soaring. The five hottest metropolitan areas in the last year.
Newsweek Oct 1, 1990 p. 49

[Ten] coolest markets [selected cities]. 1989 1st qtr. 1990 1st qtr. % change.
U.S. News Apr 9, 1990 p. 94

Big gains on real estate are over. Annual returns including rents and appreciation. Increase in median price of existing single-family home. 1984-1989.
Business Week Jan 22, 1990 p. 61

The office glut goes on. Home sales slide, but prices keep rising. Sales of existing one-family homes. Median sale price of existing one-family homes. 1985-1989.
Business Week Dec 25, 1989 p. 140

Such a deal. Potential first-time home buyers. 1970-2000.
Newsweek Dec 25, 1989 p. 55

For-sale signs of the times. Housing sales dropped in 38 states. Here's the number of houses sold per state. States where sales improved. States where sales dropped. Houses sold.
U.S. News Oct 9, 1989 p. 81

A slump in new households will dampen housing growth. Millions of households, average annual increase. Real annual growth of residential investment. 1970s, 1980s, 1990s.
Business Week Jan 23, 1989 p. 104

Office vacancies remain high. Values lag behind inflation. Home prices keep rising, although far from evenly. Median sale price of existing one-family homes. Northeast. Midwest. 1983-1988.
Business Week Dec 26, 1988 p. 172

Home buyers are starting to show their age. Median age of: All home buyers. First-time buyers. Repurchasers. 1977, 1987.
Business Week Jul 11, 1988 p. 18

See Also HOUSING - COSTS

HOUSING - CONSTRUCTION

Fewer houses built. Housing starts, in millions. 1988-1991.
Time Dec 9, 1991 p. 22

Housing starts. 1989-1991.
Time Nov 25, 1991 p. 64

Hard-hat blues. Unemployment rate. Construction industry. U.S. total 1990, 1991.
U.S. News Aug 26, 1991 p. 45

A sharp slowing in the growth of households. Annual rise in number of U.S. households. 1980-1990.
Business Week Mar 25, 1991 p. 18

Homes become more affordable but housing starts remain weak. Housing affordability index. Housing starts, annual rate.
Business Week Mar 25, 1991 p. 23

Housing hits a nine-year low. Housing starts. 1990-1991.
Business Week Mar 4, 1991 p. 14

A deeper slump in homebuilding. Sales of new single-family homes. 1986-1991.
Business Week Jan 14, 1991 p. 121

How bad will it get? Construction spending. 1990.
Newsweek Jan 14, 1991 p. 33

The housing market: No sign of a bottom yet. Housing starts. Millions of units, seasonally adjusted annual rate. 1984-1990.
Business Week Dec 10, 1990 p. 18

Another slow year for builders. Starts for single-family homes. 1985-1990.
Business Week Jan 8, 1990 p. 122

Rules of the road. An elevated highway consists of many parts. It must be designed so that a quake does not undermine one segment, bringing the entire structure down. Diagram.
Newsweek Oct 30, 1989 p. 35

Four causes of homelessness. Less low-income housing. Annual budget of Department of Housing and Urban Development (HUD). 1981-1987.
Scholastic Update Feb 10, 1989 p. 13

Housing's long decline. Starts. 1985-1988.
Business Week Dec 5, 1988 p. 21

Is housing headed for the cellar? Permits. 1983-1988.
Business Week May 30, 1988 p. 22

HOUSING - COSTS

Median selling price of single-family homes, U.S. West. Midwest. South. North East. National average. Nov. '89 - Nov. '90.
Black Enterprise Apr 1991 p. 29

Homes become more affordable but housing starts remain weak. Housing affordability index. Housing starts, annual rate.
Business Week Mar 25, 1991 p. 23

Starter homes. East. South. Midwest. West. Price range. Days on market, 1990. Change in days since 1989.
U.S. News Nov 12, 1990 p. 70

Executive homes. East. South. Midwest. West. Price range. Days on market, 1990. Change in days since 1989.
U.S. News Nov 12, 1990 p. 72

A sampling of housing prices (first quarter, 1990). Metro areas. Median-priced existing home. Median family income. Qualifying income.
Black Enterprise Oct 1990 p. 67

What can you afford? Income. Home price. Down payment. Mortgage.
Black Enterprise Oct 1990 p. 68

The most affordable first homes [by geographical region]. Average income 25-29 year-old. Median first home price. Down payment needed (20%). Monthly housing costs as a percentage of income.
U.S. News Aug 6, 1990 p. 41

Home. Average home price. Midwest. West. South. Northeast. 1989.
U.S. News Jul 30, 1990 p. 73

Real home prices went nowhere in the 1980s. New home prices adjusted for amenities and inflation. 1960-1990.
Business Week May 28, 1990 p. 18

No place like Houston. Here are the two-year projections for housing prices in the 50 largest metro areas.
U.S. News May 28, 1990 p. 75

This new house. Here's how the cost components break down for a four-bedroom house with 2 1/2 baths and a two-car garage, on a 7,800-square-foot lot.
U.S. News May 21, 1990 p. 78

Mortgage sampler. Here's how adjustable-rate and 30-year fixed-rate mortgages stack up in 20 metropolitan areas. 1990.
U.S. News Apr 9, 1990 p. 64

[Forty] more markets [selected cities]. 1989 1st qtr. 1990 1st qtr. % change.
U.S. News Apr 9, 1990 p. 94

[Ten] coolest markets [selected cities]. 1989 1st qtr. 1990 1st qtr. % change.
U.S. News Apr 9, 1990 p. 94

The monthly mortgage bite. Here are the top 25 markets ranked by affordability—the share of the average monthly household income in a given area spent on the monthly mortgage payment.
U.S. News Apr 9, 1990 p. 94

Cracking the country's hottest housing markets. Although the national median home price is $93,100, buyers can still break into some hot markets for less. 1990.
U.S. News Apr 9, 1990 p. 94

The monthly cost of owning. Payments on a 30-year, fixed-rate mortgage for a median-priced house. Monthly payment. U.S. median price. Interest rate. 1980-2000.
U.S. News Dec 25, 1989 p. 67

Houses on the rise. How the value of a hypothetical house worth $100,000 today will climb by 1999 at different appreciation rates.
U.S. News Dec 4, 1989 p. 71

Mortgage payment milestone. Average monthly mortgage payment for home buyers. Household income. Avg. monthly payment.
Black Enterprise May 1989 p. 51

Cashing in on rentals. Location. Home price. Monthly cost. Owning. Similar rental.
U.S. News Apr 17, 1989 p. 68

Houses keep getting costlier. Median sale price of existing single-family homes. 1969-1989.
Business Week Mar 27, 1989 p. 102

Location, location, location. [Percentage change in median price of existing single-family homes, 1988]. Map.
U.S. News Mar 6, 1989 pp. 46–47

Where the living is easy, and not so easy. Cities ranked according to median home prices during the last quarter of 1988. Lowest prices. Highest.
Time Feb 27, 1989 pp. 50–51

Out of reach. Price paid by typical first-time home buyers, yearly averages. 1978-1988.
Time Feb 27, 1989 p. 51

Who can afford it? Prices are up. Price of a starter home. So are down payments. But incomes lag. Income needed to qualify for mortgage. Median income of potential first-time home buyers. 1979-1988.
U.S. News Dec 19, 1988 p. 65

Home prices have been slipping in real terms. Average new home price adjusted for inflation and amenities. 1973-1988.
Business Week Oct 31, 1988 p. 20

Where will it stop? Median prices of single-family homes. New. Existing. 1970-1988.
U.S. News Sep 19, 1988 p. 63

[Twenty-five] markets and how they grew. Housing prices in 25 markets. Metropolitan area. [Cost of] existing house. New house.
U.S. News Sep 19, 1988 p. 64

Priced in, priced out. Housing costs as share of income [by metropolitan area]. Most-affordable markets. Least-affordable markets. U.S. average.
U.S. News Sep 19, 1988 p. 67

Home sweet home—what one costs. Where prices climbed most. Where prices dropped most [by U.S. city].
U.S. News Feb 29, 1988 p. 8

Six ways to meet a mortgage. Term to maturity. Total interest. Regular payment. Total annual payments.
U.S. News Feb 22, 1988 p. 91

See Also HOUSING - BUYING AND SELLING

HOUSING - FEDERAL AID

Federal housing dollars have been slashed. Net budget authority for housing aid. 1980-1989.
Business Week Jul 10, 1989 p. 75

Federal government operating subsidy for city public housing—1987. N.Y. S.F. L.A. K.C., Mo.
Black Enterprise Jan 1989 p. 15

HOUSING - GREAT BRITAIN

Percentage of home ownership. United Kingdom. White. Black. Asian.
Black Enterprise Dec 1989 p. 96

HUBBLE SPACE TELESCOPE

[Hubble Space Telescope] Problem. Solution. Diagram.
Time Jul 9, 1990 p. 43

Orbiting observer [Hubble Space Telescope]. Operations. Maintenance. Diagram
U.S. News May 15, 1989 pp. 54–55

HUMAN BODY

The body's peak performance. Endurance. Strength. Coordination. Twenties. Thirties. Forties. Fifties.
Time Apr 22, 1991 p. 80

HUMAN RIGHTS

Torture: A guide to human rights abuses around the world. Nations where torture is practiced. Map.
Scholastic Update Oct 21, 1988 pp. 8–9

HUNGARY - ECONOMIC CONDITIONS

Adding up the damage. Bulgaria. Czechoslovakia. Hungary. Poland. Romania. Real economic growth. Hard-currency balance. Oil imports.
U.S. News Nov 5, 1990 pp. 55–56

Troubled economies. Yugoslavia. Hungary. Poland. Change in real wages. Inflation. Gross domestic product per capita. 1984-1988.
Scholastic Update Oct 20, 1989 pp. 16–17

HUNGER

People who starve to death each year. U.S. expenditures for overseas food aid. U.S. children under 12 who are regularly hungry. [Other selected statistics.]
U.S. News Oct 21, 1991 p. 18

Hunger. Percentage of infants with low birth weight (1982-88) [by country].
Scholastic Update Jan 25, 1991 pp. 4–5

Children's crusade. Malnutrition of children under age 5 [by region].
Newsweek Oct 8, 1990 p. 48

Millions go hungry. People with "food insecurity" [by country].
Scholastic Update Mar 23, 1990 p. 13

Africa: A hungry continent. The hunger problems of six countries: Ethiopia, Sudan, Mozambique, Angola, Liberia and Zimbabwe. Map.
Scholastic Update Jan 27, 1989 pp. 4–5

HUNTING

The hunting toll on U.S. wildlife. Animals taken by licensed hunters in the 1988-89 season. Mammals. Birds.
U.S. News Feb 5, 1990 p. 33

HYPNOSIS

How hypnosis works. Diagram.
U.S. News Sep 23, 1991 p. 71

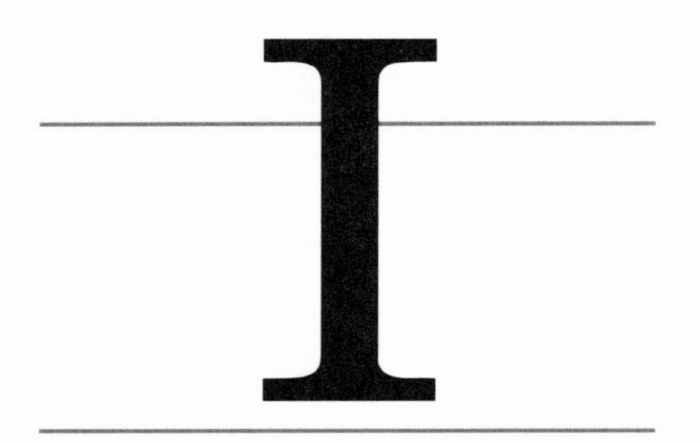

IBM

See INTERNATIONAL BUSINESS MACHINES CORPORATION

ILLEGAL ALIENS

See ALIENS, ILLEGAL

IMMIGRATION AND EMIGRATION - EUROPE

Cold shoulders. Immigrants accepted this year under regular immigration procedures. Refugees applying for asylum this year. Percent of refugees accepted. [By European country.]
Newsweek Dec 9, 1991 p. 36

IMMIGRATION AND EMIGRATION - SOVIET UNION

Freedom flights. Jewish emigration from U.S.S.R. 1977-1989.
U.S. News Sep 18, 1989 p. 12

IMMIGRATION AND EMIGRATION - UNITED STATES

One year's new arrivals [in California]. Where they came from in 1989.
Time Nov 18, 1991 p. 68

Many Soviets are pouring into the U.S. 1987-1991.
Business Week Nov 4, 1991 p. 95

Whom the U.S. lets in. Refugees the U.S. has admitted each year since 1980, and the region from which they have come. No. of refugees. 1980-1991.
Scholastic Update Oct 18, 1991 p. 12

How to get a green card. The new law raises the number of immigrants. Chart.
Time Oct 14, 1991 p. 27

Visas granted European immigrants. 1965, 1990. Immigration visas issued. 1980, 1990. Applicants on immigration waiting list. 1981, 1990. [Other selected statistics.]
U.S. News Oct 7, 1991 p. 12

The import of human capital. Asia. Europe. Mexico. West Indies. Africa. 1983-1987.
Black Enterprise Jun 1990 p. 59

Immigrants bring skills to the U.S. Immigrants entering in 1988. Total U.S. population. [Percent who are] college-educated. Engineers. Teachers. Doctors. Skilled blue-collar workers.
Business Week Oct 30, 1989 p. 128

Immigrants to the U.S. [By country of origin.] 1820-1988.
U.S. News Oct 23, 1989 p. 47

The new immigrants. All immigrants. Europe and Canada. 1900-1987.
Business Week Sep 25, 1989 p. 106

Where they come from [by country or region]. Asia. Mexico. Latin America. Caribbean. Europe. Canada. 1929, 1987.
Business Week Sep 25, 1989 p. 106

Coming to America. Immigrants admitted to the U.S. 1900-1990.
U.S. News Feb 13, 1989 p. 30

The march of immigrants. Percentage of legal immigrants to the U.S. from Asia, Latin America, Europe and Canada, other. 1955-1964, 1987.
U.S. News Feb 13, 1989 p. 30

Huddled masses. Applications for political asylum by Hispanics and Haitians now in the U.S. Country. Cases pending. As of Nov. 1, 1988.
Business Week Feb 6, 1989 p. 52

Will immigrants fill the job gap? Legal immigration. 1967-1986.
Business Week Sep 19, 1988 p. 120

The feds gear up. Immigration & naturalization service budget. Immigration & naturalization service staff. 1986-1989.
Business Week May 16, 1988 p. 37

The U.S. immigration story: Today and yesterday. Legal immigration in 1986. Total immigrants legally entering U.S.
Scholastic Update May 6, 1988 p. 4

160 years of U.S. immigration. Legal immigration to the U.S. by decade and source. 1831-1990.
Scholastic Update May 6, 1988 p. 4

New faces. Immigrants to U.S. [by country]. From new places. Immigrants to U.S. from top five Asian countries. 1976-1986.
U.S. News Apr 25, 1988 p. 42

IMMUNE SYSTEM

Combat zone. When a foreign invader enters the body, the immune system initiates a strategic defense. Diagram.
U.S. News Jul 2, 1990 pp. 50–51

Biological warfare. [How the immune system works.] The body commands a wide assortment of defenders to eliminate the danger and guard against repeat invasions. Diagram.
Time May 23, 1988 pp. 58–59

IMPORTS AND EXPORTS

Back on top. Leading exporters in 1989 [in billions of dollars].
U.S. News Dec 3, 1990 p. 65

The U.S. trails Japan in capital spending and high-tech exports. U.S. Japan. W. Germany. 1970-1986.
Business Week Jun 15, 1990 p. 45

Imports: Not that big in the U.S. economy. Ratio of imports to gross domestic product. Japan. U.S. Australia. France. Canada. Britain. W. Germany.
Business Week May 7, 1990 p. 24

The Pacific threat to U.S. high-tech trade. Total U.S. high-tech balance. High-tech balance with Japan and Asian tigers. 1980-1989.
Business Week Apr 16, 1990 p. 18

The new superpower. Exports. United Germany. United States. Japan. Soviet Union.
Newsweek Feb 26, 1990 p. 18

A Third World export pattern. U.S. Malaysia. G-7 trading partners. U.S.S.R. Composition of exports (1985).
U.S. News Nov 20, 1989 p. 32

Sizing up world markets. Gross domestic product. Exports. Imports. [By country or region.]
U.S. News Jun 20, 1988 p. 52

Trade. Increase in exports. 1980-86. Market share. 1986. U.S. Japan. China. Hong Kong. Singapore. S. Korea. Taiwan.
Newsweek Feb 22, 1988 p. 45

See Also INTERNATIONAL TRADE

IMPORTS AND EXPORTS - CANADA

Top traders. Share of Canadian exports bought by [various countries]. Share of U.S. exports bought by [various countries].
U.S. News Feb 5, 1990 p. 54

Good neighbor and best trader. Top U.S. imports from Canada. Top U.S. exports to Canada. U.S. trade with major partners.
U.S. News Jun 27, 1988 pp. 38–39

IMPORTS AND EXPORTS - CARIBBEAN REGION

Going in the wrong direction. U.S. imports from CBI [Caribbean Basin Initiative] countries. 1983-1986.
Black Enterprise May 1988 p. 45

IMPORTS AND EXPORTS - CHINA

What China ships to the U.S. Toys and games. Clothing. Footwear. Petroleum. Export value 1990.
Business Week Jul 29, 1991 p. 38

China's 10-year economic explosion. Exports and imports have been surging. Exports. Imports. Billions of dollars. 1979-1988.
Business Week Jun 19, 1989 p. 33

IMPORTS AND EXPORTS - EUROPE

Why Europe is in dollar shock. Dollars per German mark. Merchandise exports to 23 European countries. U.S.-European trade balance. 1985-1990.
Business Week Mar 4, 1991 p. 36

What's bothering Europe. Exports are slowing. Current account balance. 1986-1991.
Business Week Oct 15, 1990 p. 47

IMPORTS AND EXPORTS - GERMANY

Doing well, doing good. How West Germany compares with the U.S. on Soviet trade—and aid. Imports. Exports. Aid.
Time Jul 16, 1990 p. 17

IMPORTS AND EXPORTS - IRAN

A busy two-way street. Imports from Iran. Exports to Iran. Oil imports from Iran. U.S. Western Europe. Japan.
Newsweek Mar 6, 1989 p. 33

IMPORTS AND EXPORTS - IRAQ

The big exporters to Iraq last year. U.S. West Germany. Turkey. Britain. Romania. Japan. France.
Business Week Sep 17, 1990 p. 22

Iraq's hunger for food imports. Imports as a share of total consumption.
Business Week Aug 27, 1990 p. 26

IMPORTS AND EXPORTS - JAPAN

The U.S. trails Japan in capital spending and high-tech exports. U.S. Japan. W. Germany. 1970-1986.
Business Week Jun 15, 1990 p. 45

Not an easy sell. Japanese imports. World imports. Imports from the U.S. 1984-1988.
Newsweek Feb 13, 1989 p. 49

A yen for Western goods. American products that command a substantial share of the Japanese market. Brand. Product. Japanese market share. Annual sales in Japan.
U.S. News Dec 5, 1988 p. 93

Foreign cars Japan likes. Number of cars imported to Japan [by country and make of car].
U.S. News Aug 22, 1988 p. 65

IMPORTS AND EXPORTS - KOREA

Korea's exports are flattening. Exports. GNP growth. Average monthly earnings. Percent change. 1986-1990.
Business Week Dec 3, 1990 p. 57

Korea's export drive stalls out. Footwear. Autos. Televisions. Videocassette recorders. Home appliances. Total exports. 1988, 1989.
Business Week Apr 30, 1990 p. 40

Sticker shock in South Korea. Why a $10,000 imported car costs so much in Korea.
Business Week Feb 15, 1988 p. 47

IMPORTS AND EXPORTS - MEXICO

Can linking quite different nations provide a dramatic lift to trade? Mexico. U.S.
Business Week May 27, 1991 p. 33

Bumper crop from Mexico. Mexico's food exports to the U.S. 1985-1990.
Business Week Feb 25, 1991 p. 71

U.S.-Mexico trade is soaring. Mexican exports to the U.S. U.S. exports to Mexico. 1986-1990. What Mexico buys from the U.S. What the U.S. buys from Mexico.
Business Week Nov 12, 1990 p. 104

IMPORTS AND EXPORTS - NORTH AMERICA

North American trade patterns. [From and to] Canada. U.S. Mexico. Rest of the world. Export. Import.
Black Enterprise Sep 1991 p. 47

IMPORTS AND EXPORTS - SOVIET UNION

Shopping centers. The U.S.S.R. imports mostly from Eastern Europe and the Third World. Goods. Value (rubles). Percentage from West. Primary source.
U.S. News Nov 20, 1989 p. 35

Here today, gone tomorrow. U.S. exports. U.S. imports from U.S.S.R. 1976-1987.
U.S. News Apr 4, 1988 p. 51

IMPORTS AND EXPORTS - UNITED STATES

The trade gap is much less gaping. Imports. Exports. 1988-1991.
Business Week Sep 2, 1991 p. 18

The export boom's surprising leaders. Percent change in exports. 1990, 1991.
Business Week Aug 12, 1991 p. 16

Detroit tries again. The Big Three's auto exports to Japan. GM. Ford. Chrysler. 1979-1990.
Business Week Jul 22, 1991 p. 82

America heads toward a trade surplus. U.S. trade in goods and services. 1975-1991. The scorecard for major trading partners and for some key industries.
Business Week Jun 3, 1991 p. 25

U.S. exports: Growth is still slowing. Real exports. Current dollar exports. 1984-1991.
Business Week Mar 11, 1991 p. 18

Overseas blues. Real growth in total [U.S.] exports. 1980-1990.
Newsweek Feb 25, 1991 p. 65

Where you stand in today's economy. If you work in the exporting sector. Percent of private nonagricultural employment. Average weekly wages 1990 dollars. 1979-1990.
Business Week Dec 17, 1990 p. 67

Where you stand in today's economy. If you're in an industry hurt by imports. Percent of private nonagricultural employment. Average weekly wages 1990 dollars. 1979-1990.
Business Week Dec 17, 1990 p. 67

Vital exports. Exports as a percentage of real GNP (in 1982 dollars). 1980-1990.
U.S. News Dec 3, 1990 p. 65

As the U.S. economy slows, exports' role is growing. Growth in gross national product. Exports. All else. Billions of 1982 dollars. 1986-1990.
Business Week Nov 26, 1990 p. 51

Steady exports [U.S.]. Annual percentage change. 1989-1991.
U.S. News Nov 26, 1990 p. 57

Shrinking imports [U.S.]. Annual percentage change. 1989-1991.
U.S. News Nov 26, 1990 p. 57

Rising exports. Real exports. 1980-1990.
U.S. News Oct 29, 1990 p. 88

Higher inflation. Import prices excluding oil. 1980-1990.
U.S. News Oct 29, 1990 p. 88

America's dependence on export growth. Export growth as a percent of GNP growth. 1986-1989.
Business Week Apr 30, 1990 p. 25

High-tech imports keep gaining ground. Share of U.S. domestic market. Information technology. All other manufactured imports. 1980-1989.
Business Week Feb 5, 1990 p. 18

The world wants U.S. chemicals. Exports of chemicals. 1985-1990.
Business Week Jan 8, 1990 p. 78

U.S. farmers cash in overseas. Exports of grain and feeds. 1985-1990.
Business Week Jan 8, 1990 p. 105

Mixed performance. Resurgent exports. U.S. exports' share of total OECD [Organization for Economic Cooperation and Development] trade. 1980-1988.
U.S. News Dec 25, 1989 p. 42

A smarter bet? U.S. exports 1988. To European community. To Japan.
Newsweek Nov 6, 1989 p. 60

What the U.S. sells abroad. These 35 products make up more than 60 percent of total U.S. shipments in 1988.
U.S. News Aug 21, 1989 p. 43

What the U.S. buys from abroad. These 35 items accounted for nearly two thirds of total imports in 1988.
U.S. News Aug 21, 1989 p. 44

You can't just buy American. 1988 U.S. food imports. Value. Major countries of origin.
Newsweek Mar 27, 1989 p. 19

Energy *deja vu*. Net U.S. oil imports. 1977-1988.
U.S. News Mar 27, 1989 p. 55

Food from abroad. American agricultural imports total $21 billion. Percentages coming from [various countries].
U.S. News Mar 27, 1989 p. 56

Outward bound. Growth in U.S. exports and imports over previous year. Exports. Imports. 1987, 1988.
U.S. News Nov 14, 1988 p. 82

As steel imports drop. Imports as a percentage of U.S. market. 1981-1988.
Business Week Sep 26, 1988 p. 49

Trade woes. Total U.S. imports and exports in billions of U.S. dollars. Imports. Exports. 1970-1987.
Scholastic Update Sep 9, 1988 p. 2

Exports growth: 12-month average. Monthly. 1986-1988.
Black Enterprise Sep 1988 p. 41

Rising U.S. food imports. 1973-1986.
FDA Consumer Sep 1988 p. 14

How well the states sell. Exports in 1987 [by state]. Amount exported. Per worker.
U.S. News Jun 13, 1988 p. 71

Foreign imports have been booming. 1980-1987.
Black Enterprise Jun 1988 p. 260

Narrowing gap. Imports. Trade deficit. Exports. 1987-1988.
Time May 30, 1988 p. 44

Manufacturing a trade gap. U.S. imports of machinery and plant equipment. 1982-1987. U.S. imports. 1982. U.S. imports. 1987. U.S. trade balance. 1983-1987.
U.S. News May 16, 1988 pp. 38–39

How five U.S. industries are losing their home market. Steel. Autos. Machine tools. Textiles and clothing. Consumer electronics. 1970-1987.
Time May 9, 1988 pp. 62–63

Grain sales are growing faster. Consumption of American grain. Corn. Wheat. Soybeans. U.S. Export. 1985-1988.
Business Week Mar 28, 1988 p. 81

The fallout from protectionist measures. Annual cost to consumers. Total cost. Cost per job saved.
U.S. News Mar 21, 1988 p. 51

The dollar's delayed impact on exports. Trade-weighted value of the U.S dollar vs. a basket of currencies. Volume of U.S. manufactured exports. 1978-1988.
Business Week Feb 29, 1988 p. 62

Some sectors enjoy export surpluses but others remain huge losers. Ten key industries in U.S. foreign trade, their estimated trade balance, and some of the companies in each category.
Business Week Feb 29, 1988 pp. 64–65

The wild swings in U.S. exports. Annual change in merchandise exports. 1971-1988.
Business Week Feb 1, 1988 p. 25

U.S. merchandise trade. Imports. Exports. 1981-1987.
Time Jan 25, 1988 pp. 48–49

Chemical exports rise again. The value of products sold abroad by U.S. chemical processing companies. 1983-1988.
Business Week Jan 11, 1988 p. 78

IMPOTENCE

Four methods that restore potency. A suction device. Rerouting an artery. Inflatable implant. Semirigid implant. Diagrams.
U.S. News Jan 16, 1989 p. 65

Bedroom blues. Percentage of men in different age groups who are impotent.
U.S. News Jan 16, 1989 p. 66

IN VITRO FERTILIZATION

See FERTILIZATION IN VITRO, HUMAN

INCENTIVES IN INDUSTRY

Why companies establish incentive pay programs [various reasons].
Black Enterprise Jun 1990 p. 246

Incentive pay takes off. Bonuses. Options and other long-term incentives. Percent of senior executives' base salary.
Business Week Jan 8, 1990 p. 26

INCOME

Money income, by race, 1988-1991. White. Black. Other races.
Black Enterprise Jan 1991 p. 48

A persistent black income deficit. Received. Parity. 1988, 1989, 1990.
Black Enterprise Jun 1990 p. 224

The middle class shrinks. Earnings. 1973-1987.
Business Week Sep 25, 1989 p. 95

See Also WAGES AND SALARIES

INCOME - DISPOSABLE

Consumers' income growth is sluggish. Annual rate of change in real disposal income. 1969-1989.
Business Week Nov 11, 1991 p. 134

With income growth dragging. Real disposable personal income per capita. 1987-1991.
Business Week Sep 30, 1991 p. 22

Disposable personal income. 1990.
Newsweek Apr 1, 1991 p. 38

Consumers will have less to spend. Growth in disposable income. 1986-1991.
Business Week Jan 14, 1991 p. 85

How bad will it get? Disposable personal income. 1990.
Newsweek Jan 14, 1991 p. 32

The savings rate is in the cellar. Savings as a percent of disposable personal income. 1966-1988.
Business Week Nov 21, 1988 p. 24

INCOME - FAMILY

Shrinking pocketbooks. Median family income. 1980-1990.
U.S. News Dec 9, 1991 p. 65

Incomes have become more unequal. Growth in average family income, after inflation, 1977-1992. Pretax. Aftertax.
Business Week Nov 18, 1991 p. 85

In poverty: 34 million Americans. Median family income. Inflation-adjusted dollars. Current dollars. 1980-1990.
U.S. News Oct 7, 1991 p. 12

Incomes have dropped. Median annual family income. By age of family head. 1973, 1989.
Business Week Aug 19, 1991 p. 81

Devastating minorities. Median annual family income. By type of family head under 30. White. Black. Hispanic. 1973, 1989.
Business Week Aug 19, 1991 p. 81

Payday blues. Median family income. In 1989 dollars. 1970-1989.
U.S. News Oct 29, 1990 p. 107

Median income by race. 1960-1987. White family income. Black family income.
Black Enterprise Jan 1990 p. 58

Median family incomes, by race and family type, 1987. Black. White. Black/white ratio.
Black Enterprise Nov 1989 p. 59

Median family income. 1950, 1970, 1985. Black. White.
Black Enterprise Oct 1989 p. 26

Earnings flatten. Median family income. Males. Females. 1960-1987.
Business Week Sep 25, 1989 p. 95

Suburbs outearn cities. Suburbs. Central cities. Real family income. 1959-1989.
Business Week Sep 25, 1989 p. 95

Increasing incomes. Family income in 1987 dollars. Median income. Poverty rate. 1970-1987.
Newsweek Sep 19, 1988 p. 49

Progress is a sometime thing. Median family income (adjusted for inflation). 1947-1988.
U.S. News Aug 29, 1988 p. 40

What families earn. Number of families. Median income.
Ms. Apr 1988 p. 81

Median family income by family type and employment status of householder, by race: 1985. White. Black. Black/white ratio.
Black Enterprise Jan 1988 p. 50

Median family income, by race and origin: 1986, 1985, and 1982. White. Black. Hispanic.
Black Enterprise Jan 1988 p. 52

INCOME - HOUSEHOLD

Median household income of householder [renter]: 1988 and 1989. Hispanic origin. White. Black. Asian or Pacific Islander.
Black Enterprise Jan 1991 p. 49

Median household income of homeowner: 1988 and 1989. Hispanic origin. White. Black. Asian or Pacific Islander.
Black Enterprise Jan 1991 p. 49

More downscale consumers. Share of households with annual income of $15,000 and below. 1970-2000.
Business Week Oct 8, 1990 p. 62

Median income ratios of owners and renters to all households: 1950-1985. Owners. All households. Renters.
Black Enterprise Jan 1990 p. 45

Americans: Who are we? How much we earn. Yearly household income. Percent of households earning [by various yearly incomes]. Chart.
Scholastic Update Dec 2, 1988 p. 5

Making it in Grand Forks. Metropolitan areas projected to have highest after-tax household income in 1991. Average income.
U.S. News Jan 11, 1988 p. 63

INCOME - MEDIAN

Income. Living in poverty. Unemployment. Median income. Blacks. Whites. 1966, 1987.
Scholastic Update Apr 7, 1989 p. 21

INCOME - PER CAPITA

Comparing state stats. Per capita income. Poverty rate. Job growth. Federal taxes paid per person. [By state.]
Scholastic Update Jan 11, 1991 pp. 26–27

Comparing state stats. Per capita income. Poverty rate. Job growth. Federal taxes paid per person. [By state.]
Scholastic Update Jan 12, 1990 pp. 30–31

Americans are getting richer. Income per capita. Net worth per capita. 1979-1988.
U.S. News Nov 20, 1989 p. 73

Black per capita income rises, still below that of whites. Overall. Whites. Blacks. 1967, 1987.
Black Enterprise Nov 1989 p. 59

U.S. Almanac: Comparing state stats. Per capita income. Poverty rate. Job growth. U.S. taxes paid per person. U.S. spending per capita. [By state.]
Scholastic Update Dec 2, 1988 pp. 18–19

Progress is a sometime thing. Per capita income (adjusted for inflation). 1947-1988.
U.S. News Aug 29, 1988 p. 40

INCOME - PERSONAL

A dramatic change in fortunes. Personal income growth. New England. U.S. Rocky Mountains. 1986-1990.
Business Week Mar 19, 1990 p. 20

INCOME DISTRIBUTION

The squeeze is on. Most families are middle income. Distribution of families, by income level, 1990.
Newsweek Nov 4, 1991 p. 24

A majority of money goes to the affluent. Distribution of total family income, 1990.
Newsweek Nov 4, 1991 p. 24

The gap has widened. Percent change in distribution of total family income, 1975-1990.
Newsweek Nov 4, 1991 p. 24

The less educated. Median annual family income. By type of family head under 30. Dropout. High school grad. Some college. College grad. 1973, 1989.
Business Week Aug 19, 1991 p. 81

The rich get richer. Percentage change by income (1977-1990). Real income. Federal tax rate. [From poorest 20% to richest 1%.]
U.S. News Aug 13, 1990 p. 49

Income. Percentage of households that make [various income levels]. Age of head of household. 1990.
U.S. News Jul 30, 1990 p. 71

Catching up, sort of. Percentage of households with incomes of $50,000 or more. Whites. Blacks. Hispanics. 1972-1988.
U.S. News Dec 11, 1989 p. 74

Haves vs. have-nots. The portion of total household income held by the richest 20% and the poorest 20% of each country. Brazil. Mexico. Argentina. U.S. Japan.
Time Nov 6, 1989 p. 65

Rich vs. poor: The gap widens. Percent change in household income from 1979 to 1987. Married couples with children. Single mothers with children. Heads of household under 25. Heads of household over 65. Blacks.
Business Week Apr 17, 1989 pp. 78–79

The widening gap between the rich and poor. Percentage of black families receiving income in each range. 1970, 1986.
Black Enterprise Jan 1989 p. 51

The widening income gap. Average income of U.S. population, by income quintile. 1955-1986.
U.S. News Aug 29, 1988 p. 60

INCOME TAX

Tax changes widened the gap. [Selected statistics.]
U.S. News Nov 18, 1991 p. 35

A race against time. Total personal tax rate. 1962-1991.
Business Week Oct 25, 1991 p. 133

Bad timing. Taxes raised for 1991. Income taxes. Gasoline. Alcohol. Tobacco.
U.S. News May 27, 1991 p. 56

How the cookie crumbles. Annual income. Effective tax rates. Present. Proposed. Average change.
Newsweek Nov 5, 1990 p. 22

Bearing the burden. Percent change in federal income taxes [by income].
Time Nov 5, 1990 p. 28

Moving the bubble. Marginal tax rates for a couple filing jointly with two children. Current tax law. Proposed tax law.
Time Nov 5, 1990 p. 28

Exploiting the 'fairness' issue. Pecent increase 1980-1990. At pre-tax income. Of taxes paid.
Newsweek Oct 22, 1990 p. 22

1990 rates on joint returns, four exemptions. Marginal tax rate. "The bubble." Effective tax rate (percent of income paid in taxes). Taxable income before exemptions.
Time Oct 22, 1990 p. 28

Does a tax increase mean slower growth? [Changes in federal tax laws]. Percentage change in real GNP. 1980-1989.
U.S. News Sep 17, 1990 p. 30

The rich get richer. Percentage change by income (1977-1990). Real income. Federal tax rate. [From poorest 20% to richest 1%.]
U.S. News Aug 13, 1990 p. 49

Where the cap pinches. In 1987, residents of these states paid the highest average state and local taxes per household [state and amount paid].
U.S. News Aug 13, 1990 p. 50

The state tax man cometh. The average American spends 2 hours and 45 minutes of every 8-hour workday earning enough to pay taxes. The time spent each day working for the state and local tax man. 1980-1989.
U.S. News Jun 11, 1990 p. 38

The great tax shuffle. Here's the score card for a two-income family of four in five different pay brackets. Total wages. 1980-1990 taxes. Income. Social Security. Total.
Newsweek Mar 12, 1990 p. 70

Closing the loopholes. Number of new registrations for tax shelters including oil and gas and real estate partnerships since the 1986 Tax Reform Act. 1986-1989.
U.S. News Mar 12, 1990 p. 62

Can you beat the clock? Time the IRS says it typically takes to do a return. Keeping records. Reading about the rules. Preparing the form.
U.S. News Mar 12, 1990 p. 77

Who scores from a tax cut on capital gains. Average tax reduction. Percentage of benefits.
Business Week Mar 5, 1990 p. 26

The falling tax burden for rich and poor. High income. Low income. Entire population. 1966-1988.
Business Week Jan 15, 1990 p. 16

The income tax rates for 1989. Single income. Head of household. Married couple (filing jointly). Married couple filing separately. [By] income level.
Black Enterprise Dec 1989 p. 72

How America compares with its peers. Maximum tax rates for capital gains on equities. Short-term tax rate. Long-term tax rate. Period to qualify for long-term gains treatment. [Selected countries.]
U.S. News Aug 7, 1989 p. 42

Uncle Sam's record bite. April tax revenues. 1983-1989.
Business Week Jun 5, 1989 p. 32

Pretend it's a game of 'Beat the Clock.' Average time needed to complete various tax forms.
U.S. News Mar 27, 1989 p. 81

States that tax fairly. State income taxes ranked by progressivity index. Joint filers. Single filers.
U.S. News May 9, 1988 p. 84

A taxing day. How much of an 8-hour workday is spent earning enough money to pay taxes (in hours and minutes). Federal. Local. 1955-1988.
U.S. News May 9, 1988 p. 84

Diminishing returns. Percentage of federal income-tax returns filed by [given dates]. Percentage of returns with an extension. 1985-1988.
U.S. News Apr 4, 1988 p. 73

Where the living is taxing. Total taxes collected per person in [12 countries].
U.S. News Jan 11, 1988 p. 63

See Also TAXATION

INCOME TAX - AUDITING

Overtaxed auditors. Nearly a million 1988 individual tax returns were audited. The message for taxpayers: The odds vary from state to state. 1988.
U.S. News Apr 16, 1990 p. 62

The IRS on the prowl. Chance of audit. [By income group and type of return.] 1981, 1989.
U.S. News Mar 12, 1990 pp. 80–81

Number of Americans audited by the IRS. 1984-1987.
Black Enterprise Apr 1988 p. 39

INCOME TAX - CORPORATE

Back where they began. Corporate income-tax receipts as a percentage of total federal revenue. 1980-1991.
U.S. News Sep 26, 1988 p. 42

What relief? Proposed 1981 corporate tax cuts. Actual changes. 1982-1991.
U.S. News Sep 26, 1988 p. 43

INCOME TAX - DEDUCTIONS

Take a peek at your peers' deductions. Average itemized deductions by income range. Medical. Taxes. Charity. Interest.
Business Week Mar 11, 1991 p. 110

What a child care savings plan saves. Net saving above tax credit. One child. Two children. Adjusted gross income (thousands of dollars).
Business Week Nov 19, 1990 p. 177

INDIA

An explosive political landscape. Diverse religious and ethnic groups. Recent history of violence. 1983-1991. Map.
Newsweek Jun 3, 1991 p. 30

One nation—or many? Dominant religion. Election violence. Religious clashes. Separatist rebellions. Map.
U.S. News Jun 3, 1991 pp. 40–41

India's complex society. % of total population. The government job market [quotas].
Time Oct 15, 1990 p. 63

INDUSTRIAL RESEARCH
See RESEARCH AND DEVELOPMENT

INDUSTRY
Percentage change in capital spending for heavy industry. 1984-1991. Spending decline in heavy industry. Decline in overall new-construction. Office vacancy rates.
U.S. News Jun 17, 1991 p. 12

A spotlight on 1990 industry profits. Aftertax profits. Industries with the biggest dollar change. Winners and losers [by industry] in the year's profit race.
Business Week Mar 18, 1991 p. 54

A spotlight on 1989 industry profits. Industries with the biggest dollar change in 1989. Profit race. The best. The worst. [By industry.]
Business Week Mar 19, 1990 p. 66

Capacity utilization. The average is only edging up but for some industries, rates are soaring. Industry. 1987, 1988.
Business Week May 2, 1988 p. 31

INFANT MORTALITY
Infancy in death's shadow. Infant mortality rate of selected countries.
Black Enterprise May 1990 p. 25

Infant deaths. Drug addiction among pregnant women is driving up the U.S. infant mortality rate. Deaths in first year of life, per 1,000 live births [by city].
Newsweek Oct 16, 1989 p. 10

Growing up against the odds. [Black infants' health and mortality.]
Newsweek Sep 11, 1989 p. 31

Survival odds. 22 of the countries with lower infant-mortality rates than the U.S.
U.S. News Aug 8, 1988 p. 10

Many births, many deaths. Countries with highest birth rates. Fertility rate. Infant-mortality rate. Countries with lowest birth rates.
U.S. News May 23, 1988 p. 76

INFANTS, PREMATURE
Pushing back the envelope. Developmental range of newborns. 24 weeks to 40 weeks. Chart.
Newsweek May 16, 1988 p. 64

The $300,000 baby. Breakdown of a four-month premature baby's expenses.
Newsweek May 16, 1988 p. 67

INFLATION (FINANCIAL)
British price problems. Inflation rate. U.S. France. Britain. 1988-1991.
U.S. News Jul 22, 1991 p. 47

INFLATION (FINANCIAL) - CHINA
Annual inflation rate [China]. 1977-1989.
U.S. News Jun 5, 1989 p. 23

INFLATION (FINANCIAL) - EUROPE
What's bothering Europe. Inflation is heating up. Consumer prices. Percent change. 1986-1991.
Business Week Oct 15, 1990 p. 47

INFLATION (FINANCIAL) - UNITED STATES
Driving down inflation. Economic growth. Commodities Futures Price index. Money supply.
U.S. News Oct 7, 1991 p. 58

Turning back the clock? Inflation rate. 1914-1990.
U.S. News Oct 7, 1991 p. 60

Mild price increases. U.S. consumer price inflation. 1978-1991.
U.S. News Aug 12, 1991 p. 39

Shrinking fast. Inflation rate. Value of $1000 after: 5 yrs. 10 yrs. 15 yrs. 20 yrs.
Newsweek Apr 8, 1991 p. 46

Who says inflation is rising? Year-to-year change in consumer price index excluding food and energy. 1979-1990.
Business Week May 21, 1990 p. 31

The pressures under inflation. Excluding food and energy. Producer prices. Consumer prices. 1988-1990.
Business Week Apr 2, 1990 p. 27

Core inflation eases a tad. Consumer price inflation. Wage and benefit growth. 1988, 1989.
Business Week Feb 12, 1990 p. 22

The sting of cupid's arrow. The cost of celebrating Valentine's Day. Chocolate. Roses. Valentine card. 1950-1987.
U.S. News Feb 15, 1988 p. 88

The tooth fairy digs deeper. Money left by the tooth fairy for the first tooth lost. Amounts left. Average. Inflation adjusted. 1900-1987.
U.S. News Feb 1, 1988 p. 73

INFLUENZA

See FLU (DISEASE)

INHERITANCE

Why the inheritance boom is for real. Who has the wealth. Where the wealth is.
U.S. News May 7, 1990 p. 33

INJURIES, OCCUPATIONAL

The rising cost of railroad injuries. Year. Railroad employees. Injuries. Payouts. 1982-1988.
Business Week Nov 6, 1989 p. 93

INSECT CONTROL

So many plants, so little time. California. [Area] currently infested with whiteflies. Plagues of the past. Date, insect. Eradication cost.
Newsweek Nov 25, 1991 p. 63

INSURANCE

The rising cost of coverage. Total premiums paid [for all types of insurance]. 1982-1989.
Newsweek Apr 23, 1990 p. 47

What consumers around the world spend on insurance [by country]. Per capita premiums. Property/casualty insurance. Life insurance.
Business Week Mar 6, 1989 p. 85

Risky business. Non-traditional insurance. Amount spent, in billions of dollars. 1984-1988.
Newsweek Mar 7, 1988 p. 75

INSURANCE, AUTOMOBILE

The 10 most expensive states to insure your car [by state and cost of premium].
Time Sep 30, 1991 p. 52

Accelerating costs. Auto insurance. Claims paid out. Premiums. 1980-1990.
U.S. News Jul 2, 1990 p. 46

Tall tolls. States with the highest average premiums for private passenger car.
U.S. News Feb 12, 1990 p. 13

Out of control. Auto insurance costs, annual average increase nationwide. 1982-1987.
Business Week Nov 28, 1988 p. 38

Cost of coverage. Average 1986 premiums for private passenger cars. Ten highest [states]. Ten lowest.
Time Nov 28, 1988 p. 75

Climbing rates. Average annual automobile-policy premiums. 1981-1988.
U.S. News Apr 4, 1988 p. 68

Cars that get stolen. Average theft payments for 1987 cars. Highest claims. Lowest claims.
U.S. News Feb 1, 1988 p. 73

INSURANCE, HEALTH

A year's worth of health care. Annual premiums. Medical. Dental. Traditional plan. HMO. Point-of-service plan.
U.S. News Nov 4, 1991 pp. 78–79

Unhealthy benefits. Percentage of companies offering health insurance [by number of employees].
U.S. News Jun 3, 1991 p. 59

Living dangerously. States with the highest percentages of uninsured in 1988.
U.S. News Jan 14, 1991 p. 62

The smaller the firm. Percent of employees with company-sponsored health insurance, by size of firm. Number of employees in company. Percent with insurance.
Business Week Nov 26, 1990 p. 187

Long-term care. Annual premiums for long-term-care insurance, by age at the time of purchase. Policy A. Policy B. Policy C. 1989.
U.S. News Jul 30, 1990 p. 74

Health insurance coverage among the civilian population under age 65, by source of coverage and poverty status, 1986.
Black Enterprise Mar 1990 p. 39

Steep bill. Business expenditures for health care. 1980-1987.
Newsweek Aug 28, 1989 p. 46

Say goodbye to the age of free rides. Types of group health plans. 1984, 1988.
Newsweek Jan 30, 1989 p. 47

Shifting burden. Spending, personal health-care. Direct patient payments. Third-party payments. 1967, 1987.
Newsweek Jan 30, 1989 p. 50

A quick guide to policy quality [long-term-care insurance]. Daily benefit. Cost-of-care adjustment. Level premium. Guaranteed renewable for life. Maximum benefit period. Waiting period.
U.S. News Jan 23, 1989 pp. 56–57

Who pays for care. Unlike many countries, the United States relies on a patchwork of providers. Source of coverage. Population covered in millions.
Newsweek Aug 22, 1988 p. 53

INSURANCE INDUSTRY

A mounting tab. Payment to policy holders of failed property and casualty companies. 1980-1989.
Newsweek Mar 25, 1991 p. 44

The insurance industry's vicious cycle. Net income before taxes for property/casualty insurers. 1982-1988.
Business Week Apr 11, 1988 p. 60

A new kind of risk. Problem [insurance] companies: Life/health. Property. Insolvencies. 1981-1987.
Newsweek Mar 14, 1988 p. 40

INSURANCE, LIABILITY

Premium punishment. Nonmedical professional liability insurance. 1978-1987.
U.S. News Aug 8, 1988 p. 64

INSURANCE, LIFE

Life insurance. Median yearly household premium for life insurance. Husband and wife. Single. Two incomes. One income. 1989.
U.S. News Jul 30, 1990 p. 73

Comparison of types of life insurance. Policy type. Cost. Return in cash value. Control over investment mix. Fees and service charges. Flexibility. Tax treatment.
Black Enterprise Nov 1989 p. 64

Saving in whole-life policy vs. the "buy term and invest the difference" concept. $100,000 whole-life policy for a 35-year old male non-smoker.
Black Enterprise Nov 1989 p. 68

Survival of the fittest: How insurers see it. Annual deaths per 1,000 insured persons. 1958. Current. Smoker. Nonsmoker.
Business Week May 30, 1988 p. 105

INSURANCE, UNEMPLOYMENT

Increasing joblessness. Unemployment benefits, initial claims, monthly average. 1991.
U.S. News Dec 9, 1991 p. 65

Unemployed workers receiving jobless benefits. 1961-1990.
U.S. News Sep 2, 1991 p. 13

The young collect less unemployment. Percent of jobless filing for unemployment insurance. All ages. Under 35. 1979, 1990.
Business Week Aug 19, 1991 p. 85

The frayed safety net of jobless insurance. Share of unemployed receiving unemployment benefits. 1970-1991.
Business Week May 6, 1991 p. 22

The state of the insurance funds. Unemployment compensation reserves. Net reserves. Weeks of cash on hand. Best shape. Worst shape. [By state.]
U.S. News Jan 21, 1991 p. 61

Jobless claims are trending higher. Initial unemployment insurance claims. Eight-week moving average. 1986-1987.
Business Week Feb 1, 1988 p. 18

INSURANCE, WORKERS' COMPENSATION

The compensation explosion. Total cost of workers' compensation. 1977-1991.
U.S. News Jul 29, 1991 p. 26

The priciest states for worker's comp. Employer 1988 payment for $100 in payroll. National average.
Business Week Jul 22, 1991 p. 23

INTEGRATION, RACIAL

See BLACKS - INTEGRATION

INTELLIGENCE

Sibling smarts. When the Mensa society, a group open to anyone who scores in the top 2 percent on a standardized IQ test, surveyed its members, here is how they fell in their family's lineup.
U.S. News May 22, 1989 p. 81

INTELLIGENCE SERVICE

The U.S. intelligence maze. This chart offers a glimpse of the dozens of intelligence organizations in the United States. Military. Civilian.
U.S. News Jun 3, 1991 pp. 24–25

INTEREST (ECONOMICS)

Intense interest. Corporations. Net interest payments as a percentage of cash flow. 1975-1990.
U.S. News Nov 19, 1990 p. 56

Interest eats up cash. Interest payments as a percentage of cash flow. 1972-1988.
Business Week Nov 7, 1988 p. 141

INTEREST RATES

Defying gravity. Credit cards. New cars. Consumer finance rate. Fed discount rate. 1988-1991.
Time Oct 14, 1991 p. 44

Borrowers aren't enjoying the full benefit of fed rate cuts. Prime rate as a percentage of federal funds rate. 1989-1991.
Business Week Sep 30, 1991 p. 21

Reading the rates. Interest rates. 30-year AAA corporate bonds. 3-month commercial paper. 1990.
Newsweek May 13, 1991 p. 48

Real interest rates have been falling. 30-year treasury bond rate. 1984-1991.
Business Week Apr 1, 1991 p. 64

Who's got low rates. Short-term interest rates. U.S. Japan. 1985-1990.
U.S. News Sep 24, 1990 p. 69

Interest rates' gentle tumble. Federal funds rate. 1988-1990.
U.S. News Jul 30, 1990 p. 17

Will the real cost of money stay high? Global. U.S. 1965-1990.
Business Week Apr 16, 1990 p. 83

The case for lower interest rates. Percentage change in consumer prices. 1988-1990. Percentage change in household debt. Percentage change in total nonfinancial debt. 1980-1989.
U.S. News Oct 23, 1989 p. 56

INTERNAL REVENUE SERVICE

Overtaxed auditors. Nearly a million 1988 individual tax returns were audited. The message for taxpayers: The odds vary from state to state. 1988.
U.S. News Apr 16, 1990 p. 62

Can you beat the clock? Time the IRS says it typically takes to do a return. Keeping records. Reading about the rules. Preparing the form.
U.S. News Mar 12, 1990 p. 77

The IRS on the prowl. Chance of audit. [By income group and type of return.] 1981, 1989.
U.S. News Mar 12, 1990 pp. 80–81

Number of Americans audited by the IRS. 1984-1987.
Black Enterprise Apr 1988 p. 39

INTERNATIONAL BUSINESS MACHINES CORPORATION

OS/2's battle from behind. OS/2. MS-DOS. Windows. Thousands of units. 1988-1990.
Business Week Apr 22, 1991 p. 33

Computer cash. U.S. market share for large multi-user computers.
Newsweek Sep 17, 1990 p. 50

IBM is still the [computer] market leader. Market share.
Business Week Jun 27, 1988 p. 30

How currency gains are masking IBM's European problem. Sales are rising in dollar terms but not in most local currencies. France. West Germany. Italy. Britain. 1982-1987.
Business Week Jun 20, 1988 p. 97

INTERNATIONAL TRADE

Competing on wages. Hourly compensation costs in manufacturing, U.S. dollars. Germany. U.S. Japan. 1985-1989.
U.S. News Jun 10, 1991 p. 57

The better balance in trade. Current-account surplus or deficit (in billions of dollars). Japan. Germany. South Korea. U.S. 1980-1991.
U.S. News Jul 23, 1990 p. 56

The new superpower. Balance of trade. Japan. United Germany. Soviet Union. United States.
Newsweek Feb 26, 1990 p. 19

Pacific overtures. Free-for-all. Economic muscle in the Pacific Rim. Economic growth rate. Per capita income. Economic strength. [By country.]
U.S. News Nov 20, 1989 p. 65

How companies stack up globally [leading international companies]. Sales. Profits. Share-price gain.
Business Week Jul 18, 1988 p. 137

A rush to the altar. International joint ventures. U.S.-Europe. Europe-Japan. U.S.-Japan. 1979-1987.
U.S. News Jun 20, 1988 p. 49

Sizing up world markets. Gross domestic product. Exports. Imports. [By country or region.]
U.S. News Jun 20, 1988 p. 52

See Also IMPORTS AND EXPORTS

INTERNATIONAL TRADE - EUROPE

The twelve as traders equal EEC [by country]. Year of entry. Per capita GDP. Trade within EEC. EEC trade as share of total trade. Trade with U.S.
U.S. News Feb 29, 1988 p. 40

Doing well by doing good. Eastern Europe's two-way trade with the European Economic Community, 1986. Six Warsaw Pact countries.
U.S. News Feb 15, 1988 p. 41

INTERNATIONAL TRADE - JAPAN

Japan's trade surplus is expanding globally. Europe. Southeast Asia. U.S. 1986-1991.
Business Week Jul 1, 1991 p. 42

Japan's investment outflow dwarfs its trade surplus. Net investment overseas. Trade surplus. 1983-1989.
Business Week Apr 9, 1990 p. 16

Japan's swelling import bill and shrinking trade surplus. 1980-1987.
U.S. News Feb 26, 1990 p. 49

INTERNATIONAL TRADE - NORTH AMERICA

The shape of North America Inc. The new North America. Population. GNP. Total trade. Canada. U.S. Mexico.
Business Week Nov 12, 1990 p. 103

INTERNATIONAL TRADE - SOVIET UNION

Trade proposals. Soviet trade with U.S. Soviet trade with E.C. 1988.
Time Dec 18, 1989 p. 36

INTERNATIONAL TRADE - UNITED STATES

The top contributors to U.S. foreign earnings. Industry. 1990 sales.
Business Week Jul 15, 1991 p. 34

Closing the gap. U.S. merchandise trade deficit, percentage of GNP. 1980-1991.
U.S. News Jun 10, 1991 p. 57

America heads toward a trade surplus. U.S. trade in goods and services. 1975-1991. The scorecard for major trading partners and for some key industries.
Business Week Jun 3, 1991 p. 25

Economic outlook. Why the war will widen the trade deficit. U.S. trade deficit as a percentage of GNP. 1986-1990. Change in U.S. foreign assets, as a share of GNP. 1980-1990.
U.S. News Feb 18, 1991 p. 54

Trade rebound. Percentage change in dollar value of world trade. 1980-1989.
U.S. News Dec 3, 1990 p. 65

A vanishing trade deficit. 1985-1990.
U.S. News Sep 24, 1990 p. 69

Winds of change. Foreign investment. Export value index. 1981-1987.
Newsweek Jul 23, 1990 p. 27

Foreigners are doing more of the buying. Overseas sales for U.S. equipment producer. 1980-1989.
U.S. News Apr 23, 1990 p. 57

The budget gap is narrowing, but the trade deficit is due to widen. Unified budget deficit. Merchandise trade deficit. 1985-1989.
Business Week Dec 25, 1989 p. 85

Recipe for red ink. U.S. trade balance by sector. Agriculture. Manufacturing. Oil. Other. Total trade balance. 1980, 1988.
U.S. News May 22, 1989 p. 52

Over a barrel. Total trade deficit and percentage devoted to energy. 1986-1988.
U.S. News Mar 27, 1989 p. 55

The U.S. trade gap stays stubbornly wide. Imports. Trade deficit. Exports. 1980-1988.
Business Week Feb 27, 1989 pp. 86–87

How U.S. trade has changed. U.S. trade balances with major foreign markets.
Business Week Feb 27, 1989 p. 92

America the needy. U.S. current-account balance. 1980-1987. Net U.S. international investment. 1980-1991.
U.S. News Jun 13, 1988 p. 48

Narrowing gap. Imports. Trade deficit. Exports. 1987-1988.
Time May 30, 1988 p. 44

How five U.S. industries are losing their home market. Steel. Autos. Machine tools. Textiles and clothing. Consumer electronics. 1970-1987.
Time May 9, 1988 pp. 62–63

The trade deficit. 1980-1987.
Black Enterprise May 1988 p. 55

The fallout from protectionist measures. Annual cost to consumers. Total cost. Cost per job saved.
U.S. News Mar 21, 1988 p. 51

U.S. merchandise trade. Imports. Exports. 1981-1987.
Time Jan 25, 1988 pp. 48–49

INTERNATIONAL TRADE WITH CANADA

The free-trade agreement. [U.S. and Canada]. Merchandise exports. Direct investment.
Time Dec 5, 1988 p. 40

INTERNATIONAL TRADE WITH CHINA

China widens the gap. Trade surplus with the U.S. 1985-1990.
Business Week Apr 22, 1991 p. 46

End of an era? Total trade with China. 1980-1988.
Newsweek Jun 19, 1989 p. 26

INTERNATIONAL TRADE WITH EUROPE

Reversal of fortune. U.S. trade balance with the European Community. 1985-1991.
U.S. News Jun 24, 1991 p. 48

Why Europe is in dollar shock. Dollars per German mark. Merchandise exports to 23 European countries. U.S.-European trade balance. 1985-1990.
Business Week Mar 4, 1991 p. 36

INTERNATIONAL TRADE WITH JAPAN

The trade gap: Autos take a growing bite. U.S.-Japanese trade deficit. Vehicles and parts. Remainder. 1987-1990.
Business Week Nov 18, 1991 p. 109

Trade wins. Japanese trade surplus with United States. Japanese trade surplus with Asia. Japanese trade surplus with Europe.
U.S. News Nov 18, 1991 p. 78

A dwindling trade gap. U.S. trade deficit with Japan. 1986-1990.
Business Week Apr 8, 1991 p. 24

Gaining fast. Japan's share of the U.S. car market. Transplants. Imports. 1986-1990.
Newsweek Apr 8, 1991 p. 42

The Japanese edge in laptops. 1990 U.S. market share. Japan. U.S. Other.
Business Week Mar 18, 1991 p. 121

U.S. trade deficit with Japan, in billions of dollars. 1985-1989.
Time Jul 9, 1990 p. 18

Japan's growing market share. Total U.S. market for new machine tools valued at $2500 or more. 1985, 1989.
Business Week Jun 4, 1990 p. 64

A paradox of trade patterns. Japan's share of U.S. imports. Japan's share of U.S. exports. Japan's share of the U.S. manufacturing-trade deficit. 1987-1989.
U.S. News May 21, 1990 p. 55

Promises, promises. [What] Japan agreed to [in a trade agreement]. [What the] United States agreed to.
Time Apr 16, 1990 p. 43

Japan's growing share of the U.S. trade deficit. 1983-1989.
Business Week Jan 15, 1990 p. 38

Japan's surplus with the U.S. Total U.S. trade deficit. Japan's share. 1985-1989.
Business Week Sep 4, 1989 p. 47

INTERNATIONAL TRADE WITH MEXICO

Can linking quite different nations provide a dramatic lift to trade? Mexico. U.S.
Business Week May 27, 1991 p. 33

Free trade with Mexico will boost America's economy. U.S. non-oil trade surplus with Mexico. Real GNP growth rates. U.S. Mexico. Growth of foreign investment in Mexico.
U.S. News May 13, 1991 p. 58

INTERNATIONAL TRADE WITH PACIFIC RIM COUNTRIES

The Pacific threat to U.S. high-tech trade. Total U.S. high-tech balance. High-tech balance with Japan and Asian tigers. 1980-1989.
Business Week Apr 16, 1990 p. 18

INTERNATIONAL TRADE WITH THE SOVIET UNION

Fickle trade winds. U.S.-U.S.S.R. bilateral trade. 1970-1987.
Newsweek Jun 13, 1988 p. 47

Here today, gone tomorrow. U.S. exports. U.S. imports from U.S.S.R. 1976-1987.
U.S. News Apr 4, 1988 p. 51

INTERVIEWING

Career quiz. Job seekers in certain trades must prove their proficiency. These test questions might be asked of applicants in the following occupations [various occupations].
U.S. News Apr 2, 1990 p. 68

INVESTMENT TRUSTS

Money-market trade-off. Money-market funds. Average annual yield. 1979-1989.
U.S. News Dec 4, 1989 p. 70

INVESTMENTS

Private placements shrink. Privately placed securities. 1986-1991.
Business Week Aug 12, 1991 p. 21

Small investors are helping out. Mutual fund net new sales and net exchanges. Taxable money market funds. Total bond and income funds. Total equity funds.
Business Week Aug 12, 1991 p. 21

Reaping the dividends. Net U.S. investment income. 1980-1990.
U.S. News Jun 10, 1991 p. 57

A year to forget. [Type of] investment. 1990 return.
Newsweek Jan 14, 1991 p. 38

Black vs. white consumers. Types of financial vehicles owned. Blacks. Whites.
Black Enterprise Oct 1990 p. 45

How the competition stacks up. Investment. Japan. West Germany. U.S. Investment in plant and equipment. 1980-1990. Nondefense R&D expenditures. 1980-1988. Public investment. 1967-1985.
U.S. News Jul 16, 1990 p. 25

The case for holding on. Holding period [of stocks]. Your chance of earning. Chance of losing.
Newsweek Jun 4, 1990 p. 63

Best and worst investments of the '80s. Type of asset. Total return. Annual rate of return.
U.S. News Dec 4, 1989 p. 70

Rising risks. Junk bond market. 1985-1989.
Newsweek Sep 25, 1989 p. 32

Loading up on fees. Value of $10,000 investment with 10% annual return. With 6.5% front-end load. With 1% annual fee.
U.S. News Feb 6, 1989 p. 64

Final tally [various investments]. An investor's score card for 1988. Value at end of 1987. 1988 high, low. Value at end of 1988. Change 1987-88.
U.S. News Jan 23, 1989 p. 66

Where do the pension funds invest and how big is the pension nest egg. 1981-1987.
Black Enterprise Apr 1988 p. 52

Setting up a low-risk portfolio. Common stocks. Bonds. Money market. Gold stocks.
Black Enterprise Feb 1988 p. 63

Black stock ownership, 1982-1987. Total corporate stock outstanding. Amount. Percent of total.
Black Enterprise Feb 1988 p. 73

INVESTMENTS, AMERICAN

U.S. assets abroad vs. foreign assets in the U.S. 1980-1988.
Scholastic Update Sep 7, 1990 p. 21

Who owns what. U.S. direct investment as a share of foreign economy. Foreign direct investment as a share of the U.S. economy. [Various countries.] 1988.
U.S. News Feb 19, 1990 p. 53

International investment position of the United States in constant (1982) dollars. 1970-1986. U.S. assets abroad. Foreign assets in U.S.
Black Enterprise Jun 1989 p. 95

U.S. capital outflows and inflows as a percent of gross national product. 1960-1986.
Black Enterprise Jun 1989 p. 95

Two-way street. Foreign direct investment in the U.S. U.S. direct investment abroad. 1988.
U.S. News May 29, 1989 p. 46

U.S. corporate divestment. Total companies with direct investment in South Africa: U.S. Non-U.S. 1984-1988.
Black Enterprise Mar 1989 p. 41

A bigger U.S. stake in Japan. Capital spending in Japan by U.S. companies. 1985-1989.
Business Week Dec 19, 1988 p. 44

The steady globalization of business. U.S. direct investment overseas. Foreign direct investment in the U.S. International phone traffic. Equipment market. 1985-1989.
Business Week Mar 21, 1988 p. 140

INVESTMENTS, FOREIGN

High stakes. Japanese direct investment in the United States. 1980-1990.
Newsweek Nov 11, 1991 p. 48

Foreigners strike gold. Non-U.S. corporations own or control most of the biggest gold mines in the United States. The 15 top foreign-owned mines.
U.S. News Oct 28, 1991 p. 46

Economic outlook. Why foreigners aren't buying up America anymore. Percentage of U.S. manufacturing controlled by foreign firms. 1977-1988. Inflows of foreign direct investment. 1989-1990. Foreign direct investment. 1985-1990.
U.S. News Apr 29, 1991 p. 61

Kuwait's holdings. U.S. Europe. Asia. Company. Business. Ownership. Estimated value.
Time Dec 24, 1990 p. 30

A sampler of the largest foreign employers in the U.S. Foreign parent. Home office. U.S. subsidiary. U.S. employment.
Business Week Dec 17, 1990 p. 81

Who owns U.S.? Total value of direct investment [by country]. Recent acquisition.
Time Dec 10, 1990 p. 23

Foreign investors return to Latin America. Foreign direct investment. 1984-1991.
U.S. News Nov 19, 1990 p. 64

Returns recover on U.S. foreign direct investments in Latin America. Annual rate of return. 1984-1988.
U.S. News Nov 19, 1990 p. 64

Today treasuries, tomorrow Rockefeller Center? Real estate. Manufacturing plants. Takeovers.
Newsweek Oct 15, 1990 p. 55

Japanese plants in North America. U.S. plants. [Comparison of] assembly time per car. Worker training. Defects per 100 cars. Inventory supply. Work teams.
Business Week Oct 8, 1990 p. 83

U.S. assets abroad vs. foreign assets in the U.S. 1980-1988.
Scholastic Update Sep 7, 1990 p. 21

Winds of change. Foreign investment. Export value index. 1981-1987.
Newsweek Jul 23, 1990 p. 27

The shrinking role of overseas capital. Foreign investment as a percentage of U.S. GNP. 1983-1989.
U.S. News Apr 16, 1990 p. 45

Who owns what. U.S. direct investment as a share of foreign economy. Foreign direct investment as a share of the U.S. economy. [Various countries.] 1988.
U.S. News Feb 19, 1990 p. 53

Taking stock. Foreign investment [by various countries] in U.S. manufacturing. 1980, 1988.
U.S. News Jan 15, 1990 p. 44

Buying binge. Japanese investment in U.S. real estate. 1985-1988.
Newsweek Nov 13, 1989 p. 63

Japanese auto plants in the U.S. and Canada. Parent. Production started. Planned yearly capacity. Planned employment. Map.
Business Week Aug 14, 1989 pp. 74–75

Japanese plants take a bigger slice. Cars made in America. Percent of domestically produced cars. 1984-1989.
Business Week Aug 14, 1989 p. 78

It [Japan] adds to its U.S. base. Direct investment in the U.S. 1985-1989.
Business Week Aug 7, 1989 p. 46

International investment position of the United States in constant (1982) dollars. 1970-1986. U.S. assets abroad. Foreign assets in U.S.
Black Enterprise Jun 1989 p. 95

Inward and outward foreign direct investment for selected countries, 1967 and 1987. U.S. Europe. Japan. Canada. Developing countries. All countries.
Black Enterprise Jun 1989 p. 95

Outside influence. The percentage of total sales, manufacturing employment and assets from foreign-owned firms. U.S. Britain. W. Germany. Japan. 1977, 1986.
U.S. News May 29, 1989 p. 45

Two-way street. Foreign direct investment in the U.S. U.S. direct investment abroad. 1988.
U.S. News May 29, 1989 p. 46

Japan: Now the top foreign investor in the U.S. Japan. Britain. Netherlands. Direct investment. 1986-1988.
Business Week May 1, 1989 p. 20

U.S. corporate divestment. Total companies with direct investment in South Africa: U.S. Non-U.S. 1984-1988.
Black Enterprise Mar 1989 p. 41

Comparing the competition. Investments. Exports. Imports. Gross domestic product.
Newsweek Feb 27, 1989 p. 15

Who's buying America. Top 10 purchases of U.S. companies by foreign firms in 1988.
U.S. News Jun 27, 1988 p. 64

The growing foreign stake in the U.S. Foreign direct investment as share of domestic capital stock. Manufacturing. Non-manufacturing. 1970-1987.
Business Week May 9, 1988 p. 36

A new land of the rising sun. Japan comes to Tennessee. Companies owned by Japanese have established 60 operations in the Volunteer State.
U.S. News May 9, 1988 pp. 44–45

Foreign nations with highest direct investments in U.S. Netherlands. Britain. Canada. Japan. Switzerland. West Germany. 1980-1987.
U.S. News May 9, 1988 p. 59

Cars they make here. Automobiles and pickup trucks manufactured in U.S. by Japanese companies. 1983-1987.
U.S. News May 9, 1988 p. 59

Buying buildings. Japanese investments in U.S. real estate. Office. Hotel. Residential. Other. 1985-1987.
U.S. News May 9, 1988 p. 59

Top cities. Japanese investments in U.S. real estate [by selected cities] through 1987. Map.
U.S. News May 9, 1988 p. 59

Factories & workers. Japanese-owned factories in U.S. [Number of American] employes.
U.S. News May 9, 1988 p. 59

Portfolios. Japanese holdings in U.S. stocks, bonds and treasury securities. 1985-1987.
U.S. News May 9, 1988 p. 59

The steady globalization of business. U.S. direct investment overseas. Foreign direct investment in the U.S. International phone traffic. Equipment market. 1985-1989.
Business Week Mar 21, 1988 p. 140

What they've bought. Total foreign-held assets in the U.S. Foreign direct investment in the U.S. 1982-1987.
U.S. News Feb 22, 1988 p. 68

Where Japan is leaving its mark. Number of Japanese factories in U.S. Number of workers. Japanese-owned factories. 1983-1987. Map.
U.S. News Feb 1, 1988 p. 73

INVESTMENTS, JAPANESE

Investing in the neighborhood. Japanese direct investment. [In] Taiwan. Hong Kong. Thailand. Malaysia. Singapore. Indonesia. 1988, 1989, 1990.
Newsweek Aug 19, 1991 p. 32

As Japan's surpluses fall it's investing less overseas. Current account surplus. Net capital outflow. Stocks, bonds, and loans. Direct investment. 1986-1990.
Business Week Nov 12, 1990 p. 46

Japan's investment outflow dwarfs its trade surplus. Net investment overseas. Trade surplus. 1983-1989.
Business Week Apr 9, 1990 p. 16

Japan's investment in Asia is growing [by country]. 1984, 1988.
Business Week Apr 10, 1989 p. 44

IRAN-IRAQI WAR, 1979-1988

The bloody cost of getting nowhere. Total number of war dead. Farthest penetration into enemy territory by troops of Iraq, Iran. Map.
Newsweek Aug 1, 1988 p. 28

Armed to kill. Estimated military spending. Iraq. Iran. 1980-1986. Value of military equipment sent to Iraq and Iran (1979-1986). [By country.]
U.S. News Aug 1, 1988 p. 30

The seesaw war. Territory won and lost. By Iran. By Iraq. [Chronology Sept. 1980 - April 1988.] Map.
U.S. News Aug 1, 1988 p. 30

IRANIAN AIR DOWNING, 1988

Overwhelmed by their own data? Aegis combat information center. Diagram.
Newsweek Aug 15, 1988 p. 20

High technology, high tension. Sequence of events. Chart. Map.
Newsweek Jul 18, 1988 p. 20

IRAQ

Iraq. Cities that have reported hostilities. Kurdish concentrations. Allied controlled. Map.
Time Apr 1, 1991 p. 35

Across the great divides. Iraqi insurgents. Anti-regime clashes. Map.
Newsweek Mar 25, 1991 p. 19

Cities that have reported hostilities. Sunni Arab. Sunni Kurd. Shiite Arab.
Time Mar 18, 1991 p. 56

Kuwait. Iraq. Saudi Arabia. Population 1988. Area. GNP. Oil production. Oil reserves. Troops. Tanks. Aircraft. Ships. [Map of oil pipelines.]
Time Aug 13, 1990 p. 17

IRAQ - MILITARY STRENGTH

Iraq's lethal plants. Iraqi nuclear facilities. Map.
Newsweek Oct 7, 1991 p. 33

The evidence: How Saddam planned to build a bomb. Evidence uncovered by U.N. inspectors. Plutonium. Uranium. The bomb. Equipment. Delivery. Chart.
Newsweek Oct 7, 1991 pp. 34–35

Iraqi military strength. Troops. Tanks. Armored vehicles. Artillery. Armed helicopters. Combat aircraft. Ships. August 1990 - March 1991.
U.S. News Apr 15, 1991 p. 30

What ever happened to the Republican Guard. Chemical weapons. Air defenses. The front line.
Time Mar 11, 1991 p. 19

What's left of Iraq's army. Tanks. Artillery. Aircraft. Before Jan. 16. Destroyed. Number remaining.
Time Mar 11, 1991 p. 36

Armed to the teeth. Active forces. Combat aircraft. Combat ships. Tanks. Iraq. U.S. and multi-national forces.
Newsweek Oct 29, 1990 p. 30

Iraq. Ballistic missile sites. Nuclear plants. Chemical plants. Airfields. Map.
Business Week Sep 3, 1990 p. 28

The military matchup. Active armed forces. Main battle tanks. Combat aircraft. Surface-to-surface missiles. Iraq. Israel.
Newsweek Aug 13, 1990 p. 21

A sampling of the weapons delivered to Iraq in the 1980s [from other countries].
U.S. News Jun 4, 1990 p. 37

Iraq's missiles. Missile. Range. Payload. Status.
U.S. News Jun 4, 1990 p. 38

Iraqi weapons-procurement trail. Saddam Hussein has assembled a worldwide procurement network that uses banks, companies and inconsistent Western export laws to obtain technology that can be used to make weapons.
U.S. News Jun 4, 1990 p. 44

Iraq's atomic future. If it chooses, there are two main routes Iraq can use to produce plutonium or highly enriched uranium for an atomic bomb.
U.S. News Jun 4, 1990 p. 50

Iraq's military might. Missiles. Range. Nuclear weapons. Chemical weapons.
U.S. News Apr 9, 1990 p. 34

IRISH AMERICANS

St. Patrick's Day bookkeeping. Where the Irish live [by city].
U.S. News Mar 14, 1988 p. 76

ISRAEL

Tough times for new arrivals. Soviet immigration [to Israel]. Unemployment rate. 1990-1995.
U.S. News Oct 14, 1991 p. 50

Israel's vital statistics. Population. GNP. Gross external debt. Unemployment. Inflation. Trade balance.
Business Week Oct 7, 1991 p. 49

Facts on the ground. Estimated population of the West Bank and Gaza. Jewish. Palestinian. 1967-1987. Settlements founded between 1967 and mid-1977. Settlements as of April, 1987. Map.
U.S. News Feb 22, 1988 p. 53

ISRAEL - MILITARY STRENGTH

The military matchup. Active armed forces. Main battle tanks. Combat aircraft. Surface-to-surface missiles. Iraq. Israel.
Newsweek Aug 13, 1990 p. 21

Quality versus quantity. Country. Population in millions. Armed forces: Regular, reserves. Tanks. Combat aircraft.
Newsweek May 1, 1989 p. 44

Arab-Israeli military balance. Army. Air Force and Air Defense. Navy. Map.
U.S. News Apr 4, 1988 pp. 42–43

ISRAEL - POLITICS AND GOVERNMENT

The ultra-Orthodox make their political mark. Knesset seats. Labor and other center-left parties. Religious parties. Likud and other right-wing parties. 1984, 1988. Breakdown of religious parties.
Newsweek Nov 14, 1988 p. 33

ISRAELI-ARAB RELATIONS

Palestinians comprise [percentage of labor force and various jobs in Israel].
Time Nov 26, 1990 p. 47

Israel. Iraq. Total armed forces. Reserves. Tanks. Artillery. Aircraft. Attack helicopters.
Time Oct 8, 1990 p. 34

Israel and the Middle East. [Profiles of] Israel. Egypt. Lebanon. Jordan. Syria. Saudi Arabia. Map.
Scholastic Update Apr 22, 1988 pp. 2–3

Facts on the ground. Estimated population of the West Bank and Gaza. Jewish. Palestinian. 1967-1987. Settlements founded between 1967 and mid-1977. Settlements as of April, 1987. Map.
U.S. News Feb 22, 1988 p. 53

Challenging the Jewish state. Israel's population. Jews. Arabs in Israel. Arabs in occupied territories. 1948-2000. Map.
Newsweek Jan 4, 1988 p. 27

ISRAELI-ARAB RELATIONS - TERRITORIAL QUESTION

Strategic terrain, precious water. The West Bank [gives] the region great military importance and is crucial to Israel's water supply. The struggle for land. Map.
U.S. News Dec 16, 1991 p. 60

The lines that divide. Israeli occupied territories. Israeli annexed territories. Map.
Newsweek Nov 11, 1991 pp. 32–33

Set in stone. Jewish settlements on the West Bank. Pre-Camp David settlements. Post-Camp David settlements.
Newsweek Oct 28, 1991 p. 37

The world's hot spot. Israel and the Occupied Territories. Population of West Bank. Population of Gaza Strip. Pre-1967-1991. A country-by-country look at the volatile region. Maps.
Scholastic Update Oct 4, 1991 pp. 4–5

Disputed lands, divided peoples. What will be on the table. Syria. Lebanon. Jordan and the Palestinians. The hard parts. Settlements. Water. Jerusalem. Map.
U.S. News Aug 12, 1991 pp. 18–19

Areas occupied by Israel. Map.
Time Mar 18, 1991 p. 38

A new climate? Israel. Israeli military administration. Demilitarized zone. Major Jewish settlement areas. Map.
Newsweek Dec 26, 1988 p. 21

Israel and the occupied territories. Land conquered by Israel in 1967 war. Israeli civilian settlements. Israeli military settlements. Palestinian refugee camps. Major cities.
Scholastic Update Apr 22, 1988 pp. 2–3

ITALY - ECONOMIC CONDITIONS

Italy's growth is tailing off. Italy. France. Germany. Real GDP growth rates. Annual consumer price increases, 1990. Budget deficit as percent of GDP, 1990.
Business Week Jun 11, 1990 p. 34

IVORY TRADING

From the bloody hands of poachers into the stashes of smugglers, ivory moves across Africa under the noses of often corrupt officals. 1979, 1989. Map
Time Oct 16, 1989 p. 68

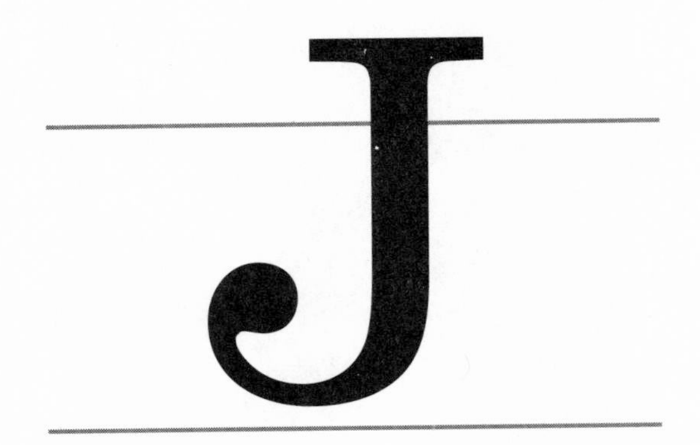

JAPAN

The jump in land prices. Index of land prices for Japan's six largest cities. Commercial. Residential. 1985-1990.
Business Week Nov 26, 1990 p. 72

JAPAN - ARMED FORCES

Japan's self-defense forces.
Time Oct 29, 1990 p. 54

JAPAN - ECONOMIC CONDITIONS

Asia ascending. Japan. U.S.A. China. Per capita GNP. 1991, 2041.
Newsweek Dec 9, 1991 p. 6

Japan plays catch-up. GNP per capita. U.S. Japan. 1967-1987. Percent change in GNP. 1971-1990.
Scholastic Update Nov 15, 1991 p. 13

As Japan's growth slows down. Increase in GNP. Capital spending takes a big hit. 1988-1992.
Business Week Sep 16, 1991 p. 47

Investing in the neighborhood. Japanese direct investment. [In] Taiwan. Hong Kong. Thailand. Malaysia. Singapore. Indonesia. 1988, 1989, 1990.
Newsweek Aug 19, 1991 p. 32

Japan's trade surplus is expanding globally. Europe. Southeast Asia. U.S. 1986-1991.
Business Week Jul 1, 1991 p. 42

Japan's bursting bubble. [Value of] modest home. Office building. Imperial Palace. 1986, 1990.
Newsweek Apr 1, 1991 p. 44

What's bothering Japan. As interest rates rise. Official discount rate. Percent. 1986-1990.
Business Week Dec 17, 1990 p. 43

What's bothering Japan. Construction is falling. Housing starts. Millions of units. 1986-1990.
Business Week Dec 17, 1990 p. 43

What's bothering Japan. Profits are dropping. Percent change in corporate profits. 1986-1990.
Business Week Dec 17, 1990 p. 43

What's bothering Japan. Growth is slowing. Percent change in gross national product. 1986-1991.
Business Week Dec 17, 1990 p. 43

Tougher times in Tokyo. Percentage change in Japanese investment. Percentage change in Japanese auto sales. 1989, 1990.
U.S. News Oct 22, 1990 p. 55

The painful consequences of Japan's high interest rates. Quarterly averages for 10-year government bonds (adjusted for inflation). Japan. U.S. 1988-1990.
U.S. News Sep 17, 1990 p. 42

Where the big economies stand. United States. Japan. West Germany. Economic growth. Inflation. Cost of credit. 1987-1990.
Business Week Aug 20, 1990 p. 29

Shop till you drop. Consumer spending. Adjusted for inflation, in trillions of yen. 1980-1990.
Newsweek Aug 13, 1990 p. 47

The postwar shocks for Japan's economy. Japan has maintained steady growth despite economic shocks and a volatile stock market. 1965-1990. Chart.
Business Week Apr 23, 1990 p. 49

Why the Japanese market is feeling so sick. Interest rates are climbing. The yen is falling. The land boom is over. 1985-1989.
Business Week Apr 23, 1990 p. 51

Japan topped Silicon Valley by outinvesting the U.S. Share of world chip production. Average capital spending as share of chip revenues. 1979-1989.
Business Week Feb 5, 1990 p. 55

Japan keeps its engine humming. Capital spending. 1985-1989.
Business Week Aug 7, 1989 p. 46

Comparing the competition. Investments. Exports. Imports. Gross domestic product.
Newsweek Feb 27, 1989 p. 15

Sharing the burden. Expenditures of development aid 1977-89. Geographical distribution of development aid 1986-87. Japan. United States.
Newsweek Feb 6, 1989 p. 36

Salaried Japanese keep falling behind the average. Financial assets. Average of all households. Households headed by salaried workers. 1983-1987.
Business Week Sep 12, 1988 p. 51

JAPAN - EMPLOYMENT

Behaviors and savers. Work week. Labor disputes. Unemployment rate. Female work force. Savings rate. Industrial robots.
Newsweek Feb 27, 1989 p. 17

JAPAN - POLITICS AND GOVERNMENT

The economics of Japanese politics. Where money comes from. Where the money goes. Chart.
Time Apr 24, 1989 p. 39

JAPAN - SOCIAL CONDITIONS

Cramped consumers. Population density. Living space. Average rent. Food. Electricity. Persons per passenger car.
Newsweek Feb 27, 1989 p. 20

JAPAN AND THE UNITED STATES

A cheap currency. Yen per dollar. 1985-1990.
U.S. News Sep 24, 1990 p. 69

The painful consequences of Japan's high interest rates. Quarterly averages for 10-year government bonds (adjusted for inflation). Japan. U.S. 1988-1990.
U.S. News Sep 17, 1990 p. 42

How the competition stacks up. Labor. Japan. West Germany. U.S. Productivity index. 1980-1990. Unit labor costs. 1980-1990. Scientists and engineers engaged in R&D. 1980-1987.
U.S. News Jul 16, 1990 pp. 24–25

Biggest banks. Home base of the world's top 100 banks. U.S. Japan. West Germany.
U.S. News Jul 16, 1990 p. 24

Bright ideas. Percentage of total U.S. patents granted. U.S. Japan. West Germany. 1975, 1989.
U.S. News Jul 16, 1990 p. 24

How the competition stacks up. Capital. Japan. West Germany. U.S. Cost of capital. 1980-1988. Real long-term interest rates. 1975-1990.
U.S. News Jul 16, 1990 p. 24

How the competition stacks up. Investment. Japan. West Germany. U.S. Investment in plant and equipment. 1980-1990. Nondefense R&D expenditures. 1980-1988. Public investment. 1967-1985.
U.S. News Jul 16, 1990 p. 25

Gray power. Percent of population now over 65. U.S. Japan. West Germany. World average.
U.S. News Jul 16, 1990 p. 25

Time clock. Average number of working hours per year. U.S. Japan. West Germany.
U.S. News Jul 16, 1990 p. 25

Striking out. Number of workdays lost to labor disputes (in thousands). U.S. Japan. West Germany.
U.S. News Jul 16, 1990 p. 26

The U.S. trails Japan in capital spending and high-tech exports. U.S. Japan. W. Germany. 1970-1986.
Business Week Jun 15, 1990 p. 45

Promises, promises. [What] Japan agreed to [in a trade agreement]. [What the] United States agreed to.
Time Apr 16, 1990 p. 43

Japan topped Silicon Valley by outinvesting the U.S. Share of world chip production. Average capital spending as share of chip revenues. 1979-1989.
Business Week Feb 5, 1990 p. 55

Comparing two superpowers. Japan. U.S. Education. Economy. Quality of life. Map.
Scholastic Update Dec 8, 1989 pp. 4–5

Japan moves ahead. National and global markets. Europe. Japan. U.S. World. 1980-1989.
U.S. News Dec 4, 1989 p. 76

We'll take Manhattan. Goods cost more in Tokyo than in New York City. Product. Maker. Tokyo. New York.
U.S. News Nov 20, 1989 p. 13

[Japan keeps] its surplus high. Trade surplus with the U.S. 1985-1989.
Business Week Aug 7, 1989 p. 46

Comparing the competition. Investments. Exports. Imports. Gross domestic product.
Newsweek Feb 27, 1989 p. 15

Sharing the burden. Expenditures of development aid 1977-89. Geographical distribution of development aid 1986-87. Japan. United States.
Newsweek Feb 6, 1989 p. 36

Superconductors. Japan grabs the lead in research. Japan. U.S. Launches work on applications. Area of superconductor development. Percent of Japanese companies involved.
Business Week Sep 19, 1988 p. 76

Where the money goes. [Japanese money spent to influence the U.S.].
Business Week Jul 11, 1988 p. 64

Japanese-endowed chairs at MIT.
Business Week Jul 11, 1988 p. 70

Personal savings. Japan's frugality is one reason that it has so much money to invest. [Savings] as a percentage of disposable personal income. Japan. U.S. 1980-86.
Newsweek Feb 22, 1988 p. 46

Where Japan is leaving its mark. Number of Japanese factories in U.S. Number of workers. Japanese-owned factories. 1983-1987. Map.
U.S. News Feb 1, 1988 p. 73

See Also INTERNATIONAL TRADE WITH JAPAN

JAPANESE LANGUAGE

Lip service. U.S. college students enrolling in Japanese-language courses. 1977-1987.
U.S. News Mar 28, 1988 p. 41

JESUITS

Where the Jesuits are [by geographic area]. Map.
Time Dec 3, 1990 p. 89

JEWS

After the Holocaust. A half century after Nazi Germany mounted its war on the Jews of Eastern Europe, the Soviet Union alone contains a substantial Jewish population [Jewish population by country.]
Newsweek May 7, 1990 p. 38

Open gates. Jewish emigration from the U.S.S.R. 1974-1989.
Newsweek Sep 25, 1989 p. 52

Freedom flights. Jewish emigration from U.S.S.R. 1977-1989.
U.S. News Sep 18, 1989 p. 12

JOB HUNTING

Jobs on the rise. Fastest-growing occupations. Jobs in decline. Shrinking occupations. 1986-2000.
U.S. News Apr 25, 1988 pp. 60–61

A geographical guide to the jobs of the future. Regional job growth. U.S. metropolitan areas with slowest projected employment growth. U.S. metropolitan areas with greatest projected job growth. 1987-2000. Map.
U.S. News Apr 25, 1988 p. 62

Starting over. Average length of a job search by a midlevel executive trying to find a new job. By age. By salary. By position. Average new salary.
U.S. News Apr 18, 1988 p. 83

JOB SATISFACTION

Job satisfaction: Made in the U.S.A. U.S. Canada. European Community. Japan.
Business Week Oct 28, 1991 p. 45

Grumbling in the office. Employees are unhappier now than they were 16 years ago. The following percentage of 2,000 employees surveyed in 1973 and 1989 expressed complete satisfaction with [various conditions].
U.S. News Sep 18, 1989 p. 77

JOBS

Struggling to bounce back. Nonfarm payrolls (in millions of workers). 1989-1991.
Business Week Dec 23, 1991 p. 25

More work in services. Distribution of men under 30 by longest job held during year. Services. Retail trade. Manufacturing. 1973, 1989.
Business Week Aug 19, 1991 p. 80

Whites are still faring much better. Working as: Managers/professionals. Administrators. Service personnel. Laborers. Whites. Blacks.
Business Week Jul 8, 1991 p. 53

The changing work force. Percent of work force. Women. Men. Hispanics. Asians. Blacks. 1966, 1980, 1990.
Business Week Jul 8, 1991 p. 62

Job machine. Percentage of new jobs created by: Firms with fewer than 100 employees. Firms with more than 10,000 employees. 1986-1990.
U.S. News Jun 3, 1991 p. 52

Where the jobs are. Fastest employment growth rates 1990-91 [by state]. Slowest employment growth rates 1990-91 [by state].
U.S. News Feb 18, 1991 p. 47

Civilian labor force and employment/unemployment, by race, 1989-1991.
Black Enterprise Jan 1991 p. 48

Where you stand in today's economy. If you work in the exporting sector. Percent of private nonagricultural employment. Average weekly wages 1990 dollars. 1979-1990.
Business Week Dec 17, 1990 p. 67

Where you stand in today's economy. If you're in an industry hurt by imports. Percent of private nonagricultural employment. Average weekly wages 1990 dollars. 1979-1990.
Business Week Dec 17, 1990 p. 67

Where you stand in today's economy. If your job is in the domestic sector. Percent of private nonagricultural employment. Average weekly wages 1990 dollars. 1979-1990.
Business Week Dec 17, 1990 p. 67

A sampler of the largest foreign employers in the U.S. Foreign parent. Home office. U.S. subsidiary. U.S. employment.
Business Week Dec 17, 1990 p. 81

Employment status of civilian labor force by age, sex and race. Employed—age 45-64. Unemployed—age 45-64. Black males. White males. Black females. White females. 1989.
Black Enterprise Dec 1990 p. 41

How companies are addressing cultural diversity in the workplace. % of total companies with diversity programs. % of companies with management involvement in diversity programs.
Black Enterprise Nov 1990 p. 54

Career quiz. Job seekers in certain trades must prove their proficiency. These test questions might be asked of applicants in the following occupations [various occupations].
U.S. News Apr 2, 1990 p. 68

The services keep hiring. Number of jobs in: Services. Manufacturing. 1988-1990
U.S. News Mar 26, 1990 p. 51

Job skills: A widening gap. Percent of students and jobs available each level of verbal skill. Employers' current needs. Employers' future needs. Students' current level.
Black Enterprise Dec 1989 p. 53

Selected white-collar occupations filled by blacks, 1940-1980. Managerial. Professional and technical. Number (percent).
Black Enterprise Nov 1989 p. 58

City jobs. White males still hold most top municipal jobs. Here are the percentages of female and black leaders by position.
U.S. News Oct 9, 1989 p. 81

Percentage of women, blacks and Hispanics in technical occupations. All workers. All technical workers. Technicians and related support.
Black Enterprise Aug 1989 p. 41

Why Johnny must read. Average time spent reading for the job. Occupation. Avg. time. Type of material.
U.S. News Jun 26, 1989 p. 46

Jobs. Careers of black workers. White collar. Farm. Blue collar. Household help. Other. 1965, 1985.
Scholastic Update Apr 7, 1989 p. 21

The young and the restless. Jobs in which workers stay the longest. Jobs in which workers stay the shortest. Median tenure (in years). Median age.
U.S. News Jan 23, 1989 p. 66

Job growth in the 1980s. Percentage increase in jobs, 1950-1987 [by U.S. region]. Map.
Scholastic Update Dec 2, 1988 p. 9

Work hard and play hard. Average number of working hours and vacation days in major (international) cities. [City.] Hours worked per week. Vacation days per year.
U.S. News Nov 14, 1988 p. 82

Minorities are stuck in the wrong jobs. Percent of jobs held in 1986 by: Blacks. Hispanics. Percent change in demand for jobs. 1986-2000. Too few in fast-growing jobs. Too many in slow-growing jobs. [By occupation.]
Business Week Sep 19, 1988 p. 108

Where we work. Employment by major job groups. Male. Female.
Scholastic Update Sep 9, 1988 p. 2

Employment boom. Where jobs have been won and lost, by industry. 1988,1980.
U.S. News Aug 29, 1988 p. 102

Growing work force. Number of employed workers. Percentage of U.S. population employed. 1980-1988.
U.S. News Aug 29, 1988 p. 103

Government jobs. Number of federal employes. Number of state and local employes. Percent of total labor force. 1987, 1960.
U.S. News Aug 22, 1988 p. 65

Job growth has slowed in Massachusetts. Annual percentage increase in total nonagricultural employment. Mass. U.S. 1983-1988.
Business Week Aug 1, 1988 p. 51

The emerging job market. All men. All women. Black men. Black women. Share of current jobs. Implied share of new jobs. Share of labor force growth.
Ms. Jul 1988 p. 77

Future jobs will require more education. Current jobs. New jobs. Median years of school [required].
Ms. Jul 1988 p. 79

Traveling on the job. Distances walked in the workplace [by occupation] during a workday. Calories burned daily. Days needed to lose 1 lb.
U.S. News May 9, 1988 p. 84

Starting over. Average length of a job search by a midlevel executive trying to find a new job. By age. By salary. By position. Average new salary.
U.S. News Apr 18, 1988 p. 83

Is Uncle Sam a cheapskate? What different jobs pay in the private sector and the federal government.
U.S. News Mar 14, 1988 p. 76

Which jobs are killers. Annual number of deaths per 1,000 workers. Blue-collar and service workers. White-collar and clerical workers.
U.S. News Jan 18, 1988 p. 75

See Also WAGES AND SALARIES

JOBS - EUROPE

Europe's job rolls are swelling. New jobs in EC nations. 1985-1989.
Business Week Jan 16, 1989 p. 46

JOBS - FUTURE

Jobs for the year 2000. Here are the 20 occupations projected to have the greatest percentage increase in the next 10 years.
U.S. News Jan 29, 1990 p. 62

Job skills: A widening gap. Percent of students and jobs available each level of verbal skill. Employers' current needs. Employers' future needs. Students' current level.
Black Enterprise Dec 1989 p. 53

Jobs of the future. Occupations with the largest number of new jobs (total jobs added 1986-2000).
U.S. News Jun 26, 1989 p. 45

A degree of difficulty. Tomorrow's jobs will require more education. Years of schooling needed to perform job. Current jobs. Future jobs.
Black Enterprise Feb 1989 p. 70

White males now dominate the job market. Composition of the labor force, 1985. But they will play a smaller role in the future. New entrants to the labor force, 1985-2000.
Business Week Sep 19, 1988 pp. 102–103

Industries projected to generate the largest number of wage and salary jobs, by the year 2000.
Black Enterprise Aug 1988 p. 43

A jobs boom. Employment in the nonprofit sector. 1972-1995.
U.S. News Jul 25, 1988 p. 64

Jobs on the rise. Fastest-growing occupations. Jobs in decline. Shrinking occupations. 1986-2000.
U.S. News Apr 25, 1988 pp. 60–61

A geographical guide to the jobs of the future. Regional job growth. U.S. metropolitan areas with slowest projected employment growth. U.S. metropolitan areas with greatest projected job growth. 1987-2000. Map.
U.S. News Apr 25, 1988 p. 62

The fastest-growing jobs, 1986-2000. Occupation. Number. % rise.
Black Enterprise Feb 1988 p. 166

[The fastest-growing jobs, 1986-2000] will continue to be service-oriented. Employment opportunities. Service 1986 vs. manufacturing 2000.
Black Enterprise Feb 1988 p. 168

[The fastest-growing jobs, 1986-2000] will continue to be service-oriented which could spell trouble for unskilled minorities. Black men. Black women. Hispanics.
Black Enterprise Feb 1988 p. 170

JOBS, TEMPORARY

Temporary workers: The boom fizzles. 1985-1989.
Business Week Dec 11, 1989 p. 62

Temporary workers. Growth of the temporary help services industry. 1983-1986. Total industry payroll and percentage of professional temporary workers. 1985, 1986. Career field breakdown of temporary help services industry. 1985, 1986.
Black Enterprise Apr 1988 p. 47

JOINT VENTURES

A rush to the altar. International joint ventures. U.S.-Europe. Europe-Japan. U.S.-Japan. 1979-1987.
U.S. News Jun 20, 1988 p. 49

JORDAN

Quality versus quantity. Country. Population in millions. Armed forces: Regular, reserves. Tanks. Combat aircraft.
Newsweek May 1, 1989 p. 44

JUDGES

On the bench (state and federal court judges by race, 1985).
Black Enterprise Dec 1988 p. 47

JUMPING

Well, it's leap year, right? Record jumps [animals and humans]. Longest. Highest.
U.S. News Feb 29, 1988 p. 75

JUNK BONDS

[Junk bond chronology]. 1969-1989.
Time Feb 26, 1990 p. 48

Who owns junk bonds [various groups].
Time Feb 26, 1990 p. 49

Who has been issuing junk bonds and
who has been buying them. 1988.
Business Week Oct 2, 1989 pp. 92–93

Rising risks. Junk bond market.
1985-1989.
Newsweek Sep 25, 1989 p. 32

The market Milken built. Percentage of
junk bonds owned by [various investors].
U.S. News Jan 9, 1989 p. 47

More junk bonds are being used.
Low-quality corporate debt (junk bonds)
issued. 1980-1987.
U.S. News Nov 7, 1988 p. 62

More junk bonds are being used. To fund
more and larger mergers. Value of
mergers and acquisitions. 1980-1987.
U.S. News Nov 7, 1988 p. 62

[More junk bonds are being used].
Resulting in a shrinking stock market.
Equity issues. 1980-1988.
U.S. News Nov 7, 1988 p. 63

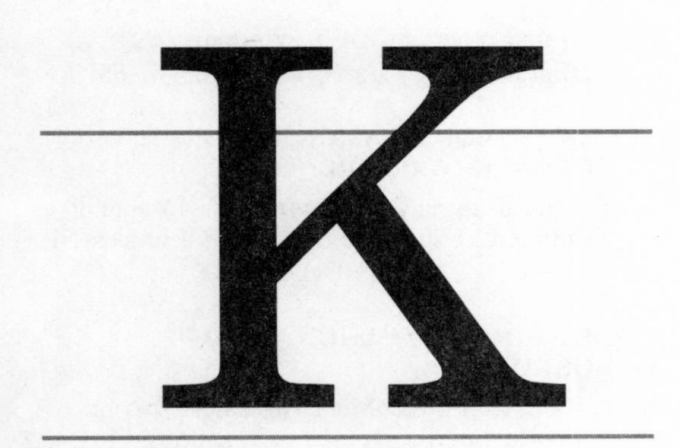

KAZAKHSTAN

Kazakhstan. Population. Languages. [Natural resources.] Map.
U.S. News Sep 23, 1991 p. 46

KIDNAPPING

Where ransom insurance is purchased [by geographic region].
Business Week Mar 19, 1990 p. 100

Where are they? Runaways. Abducted by relatives. Abducted by strangers. Found dead. Located alive. Still missing.
Newsweek Nov 27, 1989 p. 95

KIDNEY STONES

Sorting out the alternatives. How kidney stones are vanquished. Diagram.
U.S. News May 2, 1988 pp. 68–69

KOREA

Advantage: North. Weaponry [and] troops. North Korea. South Korea.
Newsweek Apr 29, 1991 p. 39

KOREA - ECONOMIC CONDITIONS

Korea's exports are flattening. Exports. GNP growth. Average monthly earnings. Percent change. 1986-1990.
Business Week Dec 3, 1990 p. 57

The miracle loses steam. South Korean economy. 1987-1989.
U.S. News Sep 18, 1989 p. 55

The shifting patterns of Korea's wealth. Trade surplus with the U.S. Overall trade surplus. Per capita income. 1984-1988.
Business Week Sep 5, 1988 p. 46

KURDS

[Kurdish] refugee centers in northern Iraq. Staging areas. Map.
Time Apr 29, 1991 p. 41

Mountains of death. As many as 2 million Kurds may be displaced. Map.
Newsweek Apr 22, 1991 p. 23

Kurds: A history of betrayal. 1920-1988. Chronology.
Newsweek Apr 15, 1991 p. 24

Forced march. Kurdish regions. Iraq. Turkey. Iran. Map.
Newsweek Apr 15, 1991 p. 25

Who are the Kurds? A people apart. Years of defeat. Where they live. Oil.
Time Apr 15, 1991 p. 27

Iraq. Cities that have reported hostilities. Kurdish concentrations. Allied controlled. Map.
Time Apr 1, 1991 p. 35

KUWAIT - ECONOMIC CONDITIONS

Kuwait's tab for the war. Total. Kuwait's financial reserves.
Business Week Mar 11, 1991 p. 34

Kuwait's holdings. U.S. Europe. Asia. Company. Business. Ownership. Estimated value.
Time Dec 24, 1990 p. 30

Kuwait's global empire. U.S. Europe. Asia. Other. Company. Industry. Percent owned. Value.
Business Week Oct 1, 1990 p. 51

Kuwait's shifting income stream. Foreign investment revenues. Oil-export revenues. 1984-1988.
Business Week Mar 7, 1988 p. 98

KUWAIT - OIL WELL FIRES

Snuffin' and cappin' in the desert. 520 wells still ablaze in Kuwait. [Three-step plan to extinguish them.] Diagram.
Newsweek Mar 25, 1991 p. 30

[Kuwait and Kuwait City]. Refineries. Oil fields. Pipelines. Roads. Area of smoke. Map.
Time Mar 4, 1991 p. 35

"Little Fortress." Kuwait. Oil refinery. Oilfield. Oil pipeline. Map.
Newsweek Aug 13, 1990 p. 24

Kuwait. Iraq. Saudi Arabia. Population 1988. Area. GNP. Oil production. Oil reserves. Troops. Tanks. Aircraft. Ships. [Map of oil pipelines.]
Time Aug 13, 1990 p. 17

KUWAIT AND THE UNITED STATES

Rebuilding the desert emirate [Kuwait]. Oil refining. Telecommunications and computers. Motor vehicles. Aerospace manufacturing. Medical supplies. Construction.
U.S. News Feb 4, 1991 pp. 58–59

KUWAIT-IRAQI INVASION, 1990

Before and after the invasion. U.S. oil imports.
Time Feb 25, 1991 p. 64

Possible targets. Major allied ground forces in the area [by country]. Iraqi forces. What the U.S. might do. What Saddam might do. Map.
Time Dec 10, 1990 p. 28

Let's make a deal. The Bush administration and its allies struck some costly and, at times, distasteful bargains in their resolution endorsing the use of force against Iraq [by country].
U.S. News Dec 10, 1990 pp. 28–29

Policing the embargo. Map.
U.S. News Oct 8, 1990 p. 46

Inside Iraq. Key installations. Oil. Chemicals. Missiles. The military. Hostages held. Map.
Time Sep 3, 1990 p. 27

Standoff in the desert. Military forces in the Middle East [by country]. Strategic targets inside Iraq. Possible attack and counterattack plans. Map.
U.S. News Aug 20, 1990 pp. 20–21

KUWAIT-IRAQI INVASION, 1990 - AMERICAN INVOLVEMENT

[Chronology of the] total U.S. forces deployed. Aug. 2, 1990 - Feb. 1, 1991. Chart.
Time Dec 31, 1990 p. 26

Quick reaction. Gulf. Vietnam.
Newsweek Nov 5, 1990 p. 33

KUWAIT-IRAQI INVASION, 1990 - CHEMICAL WARFARE

What Iraq may use. Infection. Effects if inhaled. Estimated death rate if untreated.
Time Dec 10, 1990 p. 39

KUWAIT-IRAQI INVASION, 1990 - HOSTAGES

Caught in the conflict. [Westerners] in greatest peril. Denied exit. Allowed to leave. Citizens of [various countries] in Kuwait. In Iraq.
Time Aug 27, 1990 p. 25

KUWAIT-IRAQI INVASION, 1990 - MILITARY STRENGTH

Iraq. U.S. Allies. [Troops, tanks, planes, ships].
Time Nov 12, 1990 p. 29

Defense in depth. Defensive fortification. Overhead view. Republican Guards. Iraqi defensive positions. Iraqi force. Multinational force. Map. Diagram.
U.S. News Nov 12, 1990 pp. 30–31

Armed to the teeth. Active forces. Combat aircraft. Combat ships. Tanks. Iraq. U.S. and multi-national forces.
Newsweek Oct 29, 1990 p. 30

Israel. Iraq. Total armed forces. Reserves. Tanks. Artillery. Aircraft. Attack helicopters.
Time Oct 8, 1990 p. 34

Iraq's total forces. Iraq's forces in Kuwait. Allied forces in the Persian Gulf now.
Time Oct 1, 1990 p. 51

The forces build up. Iraq [forces]. Saudi Arabia [forces]. U.S. military in the area. International forces. Map.
Time Aug 27, 1990 p. 19

[Current deployment of] U.S. ships. Combat aircraft. Ground troops. Saudi forces. Iraqi forces. Map.
Time Aug 20, 1990 p. 30

Kuwait. Iraq. Saudi Arabia. Population 1988. Area. GNP. Oil production. Oil reserves. Troops. Tanks. Aircraft. Ships. [Map of oil pipelines.]
Time Aug 13, 1990 p. 17

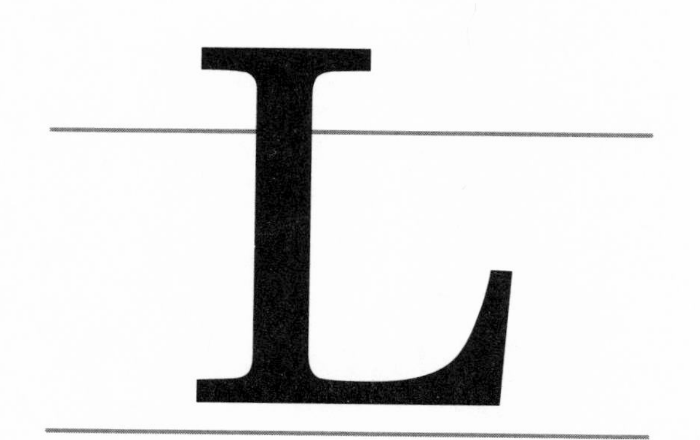

LABOR

More and more workers are 'professionals.' Professional and technical workers. 1950-1988.
Business Week Apr 2, 1990 p. 23

LABOR PRODUCTIVITY

See PRODUCTIVITY, INDUSTRIAL

LABOR SUPPLY - CHINA

Percentage of the labor force employed in agriculture, industry, services. China. U.S.
U.S. News Jun 5, 1989 p. 23

LABOR SUPPLY - UNITED STATES

U.S. population, civilian labor force, employment, and unemployment; By race, 1987-1990.
Black Enterprise Jan 1990 p. 60

Percentage of the labor force employed in agriculture, industry, services. China. U.S.
U.S. News Jun 5, 1989 p. 23

We need every person in the country. While jobs will increase, the number of new workers, aged 16 to 24, will decrease. 1980-1995.
Black Enterprise Sep 1988 p. 15

LABOR UNIONS

Union workers lag. Changes in compensation. Nonunion. Union. 1981-1990.
U.S. News Nov 5, 1990 p. 58

The dwindling might of U.S. unions. Share of work force that belongs to unions. 1950-1989.
Business Week Sep 10, 1990 p. 26

Disunited front. Percentage of union members. Total workers. Manufacturing. Service industries. 1983, 1989.
U.S. News Mar 19, 1990 p. 37

Union membership. 1980-1988.
Newsweek Mar 20, 1989 p. 24

LABORATORY ANIMALS

The toll on the animal kingdom [used in laboratory experiments].
Newsweek Dec 26, 1988 p. 51

The price of doing business. The cost of lab animals varies widely. The more exotic or complex the genetic makeup, the more expensive the animal.
Newsweek Dec 26, 1988 p. 57

The ark is overpopulated [dogs and cats in laboratory experiments].
Newsweek Dec 26, 1988 p. 59

Instead of men, guinea pigs. Researchers test new drugs and procedures on animals whose biological characteristics resemble those of humans. Animal. Some systems and structures similar to humans.
Newsweek May 23, 1988 p. 60

See Also ANIMAL EXPERIMENTATION

LANGUAGE AND LANGUAGES

Scrabble games sold since 1931. Estimated vocabulary of world's smartest people. Languages with the most words. [Other selected statistics.]
U.S. News Sep 30, 1991 p. 15

Spreading the word. [How] the ancestral tongue from which most modern European languages are descended was spread throughout the Continent. Map.
U.S. News Nov 5, 1990 p. 62

All in the family. A family tree for the human species. Genetic groups. Major language families.
U.S. News Nov 5, 1990 p. 65

Which second language? Languages likely to be helpful in the '90s.
U.S. News Dec 25, 1989 p. 66

LAS VEGAS (NEVADA)

Las Vegas [various statistics].
Newsweek Oct 29, 1990 p. 83

LATIN AMERICA

The epidemic [cholera]. Epicenter, widespread infection. Fewer cases. At risk. Map.
Newsweek May 6, 1991 p. 44

Death in the time of cholera. Countries with cholera, date of first case and number of reported cases. Spread of cholera. Map.
Time May 6, 1991 p. 58

LATIN AMERICA - ECONOMIC CONDITIONS

Latin America rebounds. Growth rates of real GDP per capita. 1965-2000.
U.S. News Dec 10, 1990 p. 60

Foreign investors return to Latin America. Foreign direct investment. 1984-1991.
U.S. News Nov 19, 1990 p. 64

Exports rebound [Latin America]. Merchandise exports. 1984-1989.
U.S. News Nov 19, 1990 p. 64

Returns recover on U.S. foreign direct investments in Latin America. Annual rate of return. 1984-1988.
U.S. News Nov 19, 1990 p. 64

Chile breaks from the pack. Percentage change in — Chile, Latin America. Exports (1980-1989). Inflation (1988, 1989). Foreign debt outstanding (1986-1989). Per capita gross domestic production (1981-1989).
U.S. News Mar 19, 1990 p. 44

How Latin America's economies are sliding backward. Growth. Inflation. Investment. Debt. Argentina. Brazil. Mexico. 1980-1989.
Business Week Jul 10, 1989 p. 45

Heap of trouble. Foreign debt [Latin American countries] in billions. 1987.
Time Jan 9, 1989 p. 33

The 'Silent Revolution' of South America. Inflation in percent. GDP growth in percent. Cars per 1000 population. 1987. [Selected countries.]
Newsweek Aug 22, 1988 p. 36

LATIN AMERICA AND THE UNITED STATES

Long-term U.S. military interventions in Latin America. [History of U.S. troops in Latin America.] Chart.
U.S. News Jan 8, 1990 p. 21

Key milestones in U.S.-Latin American relations. 1823-1981.
Scholastic Update Mar 11, 1988 p. 11

LAW SCHOOLS

Top 25 law schools. 25 law schools with the highest scores in the *U.S. News* survey.
U.S. News Apr 29, 1991 p. 74

Moot court (law school enrollment by race, 1985-86).
Black Enterprise Dec 1988 p. 47

LAWSUITS

More cases. U.S. federal district court filings. 1960-1990.
Time Aug 26, 1991 p. 54

More money. Average punitive damages awarded by jury. 1965-69, 1980-84.
Time Aug 26, 1991 p. 54

Two actual cases. One was tried by a jury; the other was mediated.
U.S. News Aug 26, 1991 p. 18

Lawsuits hit the U.S. economy hardest. Ratio of tort costs to gross domestic product. U.S. Canada. France. West Germany. Britain. Japan.
Business Week Nov 6, 1989 p. 34

LAWYERS

More lawyers. Number per 100,000 population (1989). Japan. Germany. Britain. U.S.
Time Aug 26, 1991 p. 55

The bar's swelling ranks. Lawyers. 1970-2000.
Business Week Aug 6, 1990 p. 57

Where J.D.'s meet M.D.'s. States with the most lawyers. States with the fewest lawyers. States with the most doctors. States with the fewest doctors.
U.S. News Feb 13, 1989 p. 84

Blacks and the law. Although the number of black lawyers has increased by almost 600% over the past 17 years, they only constitute 4% of the total number of lawyers in the United States.
Black Enterprise Dec 1988 p. 47

Legal eagles (number of lawyers by race, 1980).
Black Enterprise Dec 1988 p. 47

LAYOFFS

Big layoffs. Recently announced plans for staff reductions [by company].
Business Week Dec 23, 1991 p. 25

Pink slips. Layoffs announced by major U.S. corporations in 1991. Company. Number of jobs. Percent of work force.
Time Dec 23, 1991 p. 61

Who's gone for good. Number of net layoffs. Manufacturing. Construction. Federal government. Retail. Printing and publishing. Finance. The percent who won't reclaim those jobs.
Time Sep 9, 1991 p. 55

LEAD POISONING

Lead poisoning. Solutions. Diagram.
U.S. News Jul 29, 1991 p. 52

A silent hazard. [Selected statistics.]
Newsweek Jul 15, 1991 p. 43

Where lead is found in the home. Diagram.
Newsweek Jul 15, 1991 p. 44

Lead and money. [Selected statistics on lead paint in homes.]
Newsweek Jul 15, 1991 p. 48

The perils of lead. Pregnancy. Reproduction. Diagram.
Newsweek Apr 1, 1991 p. 56

Children and lead don't mix. Sources of lead that children can ingest. Source. Where it's frequently found or comes from. Number of children exposed.
U.S. News Oct 3, 1988 p. 71

LEARNING DISABILITIES

Labels come, labels go. Ratio of students labeled learning disabled to students labeled mentally retarded. 1977-1987.
U.S. News Mar 13, 1989 p. 61

LEASE AND RENTAL SERVICES

Rent-a-boom. Value of equipment leased. 1978-1989.
U.S. News Aug 14, 1989 p. 45

America's most wanted. Rapidly changing technologies and the tax laws are prompting businesses to lease everything from automobiles to satellites. Percentage of all equipment leased.
U.S. News Aug 14, 1989 p. 46

LEISURE

Simple pleasures. 2,000 adults were asked how they relaxed. Here is the percentage who said they [various responses].
U.S. News Jan 29, 1990 p. 62

Less time for play. Median number of hours available for leisure each week. 1973-1987.
U.S. News Aug 29, 1988 p. 103

Americans and the arts: VCR's take off. Decrease in leisure time. 1973-1987. Change in performing-arts audiences and other leisure-time activities.
Newsweek Mar 28, 1988 p. 69

LEVERAGED BUYOUTS

LBO activity has slowed and so has LBO fund-raising but many players have new money to spend. 1986-1989.
Business Week Feb 12, 1990 p. 62

A leveraged buyout in action. Raising the cash. Buying & reshaping. Making it work. Chart.
Time Dec 5, 1988 p. 68

Corporate America's big bet: More debt, less equity. Debt. Equity. 1983-1988.
Business Week Nov 7, 1988 pp. 138–139

Corporate debt rises to historic highs. Ratio of debt to net worth. 1972-1987.
Business Week Nov 7, 1988 p. 140

An off-the-charts deal. The top LBOs. Cash value of the deal, in billions of dollars.
Newsweek Nov 7, 1988 p. 77

LIABILITY

See DAMAGES (LEGAL)

LIBRARIES

Adults in the United States who used a library in 1990. Most popular reference books. Adults who didn't read a single book in 1990. Average time spent per day reading a book. Watching TV.
U.S. News Apr 15, 1991 p. 12

Next: Books on videotape? Public-library systems in largest cities. Yearly circulation per person. Items in the stacks. Items per person.
U.S. News Sep 19, 1988 p. 72

LIFE EXPECTANCY

Life expectancy. White females. Black females. White males. Black males. 1980-1990.
Time Sep 16, 1991 p. 51

Life expectancy by country, 1988.
Bull Atomic Sci Jun 1991 p. 15

Growing gap. Life expectancy at birth. Whites. Blacks. 1980-1988.
Time Dec 10, 1990 p. 78

The graying of the numbers. Life expectancy has increased the demands on Social Security [various statistics].
Newsweek May 7, 1990 p. 56

Black males and life expectancy. Who lives longer? White females. Black females. White males. Black males. 1950-1985.
Black Enterprise May 1990 p. 41

Projected life expectancy. [People] born in 1990. Your life expectancy at birth. Your life expectancy today is. Men. Women.
Newsweek Mar 5, 1990 p. 46

Life goes on. Expected life span, U.S. and Japanese people. Men. Women. 1960-2000.
U.S. News Dec 25, 1989 p. 67

Mapping out a long life. Countries where people live the longest. Countries where people die the earliest. Age. Both sexes. Male. Female.
U.S. News Nov 21, 1988 p. 81

Matters of life and death. Life expectancy at birth. Women. Men. 1945-1984.
U.S. News Aug 8, 1988 p. 53

How much time you've got left. Expected life remaining.
U.S. News Jan 25, 1988 p. 77

Get married, stop smoking. Estimated time lost [life expectancy] because of [various factors].
U.S. News Jan 25, 1988 p. 77

LIGHTNING

Fatal attraction. Lightning strikes the earth 100 times each second. Here is the number of lightning deaths in the U.S. 1940-1989.
U.S. News Jun 4, 1990 p. 78

Striking facts. Lightning deaths (1959-87). Where most people have been killed. Monthly death toll. States with most lightning deaths. Map.
U.S. News Sep 12, 1988 p. 71

LITERACY

Children's crusade. Education. Illiteracy. Developed countries. Nondeveloped countries. Global total.
Newsweek Oct 8, 1990 p. 48

Why Johnny must read. Average time spent reading for the job. Occupation. Avg. time. Type of material.
U.S. News Jun 26, 1989 p. 46

Putting the poor and minorities at a disadvantage. Reading scores. Dropout rate. White. Black. Hispanics. Poor.
Black Enterprise Feb 1989 p. 70

LITHUANIA

[What] the Soviet Union relies on Lithuania for. [What] Lithuania relies on the Soviet Union for.
Time Apr 30, 1990 p. 45

LOANS

Auto loans. Consumer finance rate. 1989-1991.
Time Nov 25, 1991 p. 64

Defying gravity. Credit cards. New cars. Consumer finance rate. Fed discount rate. 1988-1991.
Time Oct 14, 1991 p. 44

Borrowing has slowed sharply. Long-term borrowing by nonfinancial businesses. 1986-1991.
Business Week Aug 12, 1991 p. 20

Stopgap solution. Monthly changes. Outstanding non-financial commercial paper. Commercial and industrial loans at large commercial banks. 1990.
U.S. News Nov 19, 1990 p. 58

How the competition stacks up. Capital. Japan. West Germany. U.S. Cost of capital. 1980-1988. Real long-term interest rates. 1975-1990.
U.S. News Jul 16, 1990 p. 24

Lending by commercial banks. Percentage change in growth of personal loans. 1988-1990.
U.S. News Apr 23, 1990 p. 52

LONGEVITY

See LIFE EXPECTANCY

LOTTERIES

Annual lottery prizes. In billions of dollars, fiscal years. Number of states with lotteries. 1970-1991.
Time Nov 4, 1991 p. 81

Loot for lotto. On average, residents of states with lotteries spent this amount on tickets last year.
U.S. News May 22, 1989 p. 81

How lotteries boost state coffers. Gross sales of lottery tickets. Net revenue. 1978-1988.
Business Week Apr 10, 1989 p. 57

That's the ticket. Chances of winning states' "pick six" lotteries. State. Odds on a $1 bet.
U.S. News Nov 21, 1988 p. 81

What's in a game? Average percentage of gross lottery revenue that goes to: Prizes. State government. Administrative costs. Total revenue. Revenue going to state government. Dedication of net income [by state].
U.S. News Sep 19, 1988 p. 24

LUMBER INDUSTRY

See FOREST PRODUCTS INDUSTRY

LYME DISEASE

Reported U.S. Lyme disease cases. Map.
FDA Consumer Oct 1991 p. 5

The proliferation of Lyme disease. Reported cases. Highest incidence. Light to moderate. None. Map.
Newsweek May 22, 1989 p. 69

[Map of U.S. showing spread of Lyme disease.]
FDA Consumer Jul 1988 p. 23

MACHINE INDUSTRY

A lower trade deficit in machine tools. U.S. machine-tool trade deficit. 1983-1988.
Business Week Jan 11, 1988 p. 91

MAGAZINES

Magazines for all time. The oldest continuously published magazines in the United States.
U.S. News Oct 16, 1989 p. 125

MAGLEV TRAINS

How maglev works. Japanese and German versions of magnetically levitated high-speed vehicles. Magnetic repulsion (Japanese maglev). Magnetic attraction (German maglev).
U.S. News Jul 23, 1990 p. 52

MAIL-ORDER BUSINESS

Catalogs also feel the pinch. Percentage growth in U.S. mail-order dollar sales. Total mail-order sales. 1980-1990.
Business Week Nov 26, 1990 p. 140

More shopping from home. Mail orders as a share of total general merchandise sales. 1985-1990.
Business Week Jan 8, 1990 p. 85

MAMMOGRAM

Flirting with trouble. Share of all women surveyed who have had a mammogram. Reasons cited by women age 45 and older for not having had a mammogram during the last two years.
U.S. News Jul 11, 1988 p. 54

MAN - ORIGIN

The birth of modern humans. The "Multiregional model" and the "Out of Africa model." Map. Diagram.
U.S. News Sep 16, 1991 pp. 56–57

MAN, PREHISTORIC

Evolution's long march. Scientists now believe that the split between the ancestral hominid and chimpanzee was relatively recent—no more than 8 million years ago. [Evolution dateline]. Ancestral chimpanzee [to] modern homo sapiens. Chronological chart.
Newsweek Jan 11, 1988 p. 48

MANAGEMENT

Projections for management occupations, 1986 to 2000. Occupation. Number. Percent increases.
Black Enterprise Jul 1989 p. 31

MANUFACTURERS AND MANUFACTURING

Holding on in the heartland. Industrial production index, total industry. Average weekly earnings, manufacturing. Average weekly hours, manufacturing. 1990, 1991.
U.S. News Dec 9, 1991 pp. 58–59

Feeding the bottom line. Japan's productivity has pulled ahead. Output per hour in manufacturing. Overhead as a percent of manufacturing cost. U.S. Japan. Germany. 1986-1990.
Business Week Oct 25, 1991 p. 66

The productivity gap. Manufacturing productivity. Service productivity. 1980-1991.
U.S. News Jul 22, 1991 p. 44

This time the downturn isn't killing the Midwest. Manufacturing activity in the Midwest. 1979-1991.
Business Week Apr 29, 1991 p. 22

Falling employment. U.S. manufacturing employes. 1989, 1990.
U.S. News Dec 24, 1990 p. 51

The manufacturing revolution. Many U.S. companies have overhauled their factories in recent years and are now producing goods in dramatically different ways. New-age production. Old style production. Diagram.
U.S. News Oct 22, 1990 pp. 52–53

Manufacturing productivity. Factory output per hour. 1985-1990.
U.S. News Sep 17, 1990 p. 47

The profit potential. Manufacturers' profits. Factory production. Factory employment. 1989, 1990.
U.S. News Aug 6, 1990 p. 45

Unequal opportunities despite productivity gains. U.S. manufacturing productivity compared with major industrial competitors. 1980-1988.
U.S. News Dec 25, 1989 p. 41

Unequal opportunities. Higher capital costs. Rates on borrowing for equipment and machinery. U.S. Britain. Japan. West Germany. 1988, 1980.
U.S. News Dec 25, 1989 p. 41

Coming full circle. U.S. share of world manufacturing output. 1880-1986.
U.S. News Dec 26, 1988 p. 82

U.S. manufacturing: A far weaker picture. Average annual growth rate. Output. Productivity. 1973-1987.
Business Week Dec 12, 1988 p. 98

How the U.S. ranks in the productivity race. Output per hour in manufacturing. 1979-87, 1987. U.S. Japan. France. West Germany. Britain. Percent increase, annual rate.
Business Week Jul 25, 1988 p. 14

Manufacturing: Leaner and meaner. Manufacturing employment. [Manufacturing] output. 1979-1988.
Business Week Jul 25, 1988 p. 28

Manufacturing: The key to growth. The plunging dollar is boosting U.S. exports and spurring industrial output. 1980-1987.
Business Week Jan 11, 1988 pp. 70–71

MARIJUANA

Percent of high school students who have used these drugs in the past 30 days. Marijuana. Cocaine. 1979-1989.
Time Feb 26, 1990 p. 38

Domestic dope. Estimated quantities of marijuana from and available for use in the U.S. Quantity (metric tons). Percentage of total supply. 1985-1988.
U.S. News Nov 6, 1989 p. 28

MARINE CORPS

By the numbers. [Organization of] the Marines and the U.S. Army.
Newsweek Feb 18, 1991 p. 41

Arms and the men. An American Marine and his equipment. Diagram.
Newsweek Feb 18, 1991 p. 42

Who packs the most punch? Marine Battalion [vs.] Army Division. Personnel. Tanks. Armored vehicles. Artillery. Attack helicopters. Attack aircraft.
Time May 21, 1990 p. 28

MARRIAGE

[Percentage] of U.S. women who conceived out of wedlock but married before the child was born. 1960-1989.
Time Dec 16, 1991 p. 39

Percent of married couples with wives who work. 1960-1988.
Time Sep 3, 1990 p. 83

Second chances. Marriage rates per 1,000 divorced men and women. 1986.
U.S. News Jan 29, 1990 p. 50

Wedding bells. The percentage who say they would like to wed at the following ages. Male. Female.
U.S. News Sep 11, 1989 p. 87

Matrimonial misses. How long marriages last. Years of marriage. Percentage of couples who will divorce in this or later years. Average expected length of marriage.
U.S. News Feb 27, 1989 p. 75

Where love lasts best. Marriage and divorce rates [by state] per 1,000 people for 1987.
U.S. News May 30, 1988 p. 72

Later knot tying. Median age at first marriage. Women. Men. 1900-1987.
U.S. News May 30, 1988 p. 72

June's still the champ. Marriages during 1987 [by month].
U.S. News May 30, 1988 p. 72

U.S. in focus: Comparing the states. Percent population change. Metropolitan areas. Divorces. Marriages. [By state.]
Scholastic Update Jan 15, 1988 pp. 16–17

MASS MEDIA

Final delivery. [List of] recently deceased major newspapers.
U.S. News Feb 4, 1991 p. 71

Global media heavies. Sales. [Top five companies]. 1989.
Business Week Nov 12, 1990 p. 72

Where we get our news. TV is now America's number one source of news. Primary news sources: 1959-1988. (Percent of Americans who named each medium as a top source). Television. Newspapers. Radio. Magazines.
Scholastic Update Sep 8, 1989 p. 6

Plugging into the airwaves. Minority ownership of radio and television stations, 1979-1986.
Black Enterprise May 1988 p. 36

MASSACHUSETTS

Job growth has slowed in Massachusetts. Annual percentage increase in total nonagricultural employment. Mass. U.S. 1983-1988.
Business Week Aug 1, 1988 p. 51

MASTER OF BUSINESS ADMINISTRATION DEGREE

The pay gap between men and women. Annual salary after graduation. Discrepancy in pay scales. By school. By industry.
Business Week Oct 29, 1990 p. 57

Choosy grads. Percentages of recent M.B.A. graduates who say they would not work for these employers.
U.S. News Oct 22, 1990 p. 82

MATHEMATICS

A formula for failure. Math-achievement levels of high-school seniors. Below basic. Basic. Proficient. Advanced.
Newsweek Oct 14, 1991 p. 54

MAXWELL, ROBERT

Robert Maxwell Group. Maxwell Communication. Mirror Group Newspapers. Debt. Revenue.
Time Dec 16, 1991 p. 54

MAYORS

A mayoral pay scale. What mayors make in the nation's largest cities.
U.S. News Mar 6, 1989 p. 66

Blacks and politics. Black mayors of cities with populations over 100,000. Name. Term expires. City. Percentage of blacks.
Black Enterprise Mar 1988 p. 37

MEAT

U.S. per capita red meat consumption. Poultry consumption. Fish consumption.
U.S. News Apr 22, 1991 p. 8

Trimming the beef. Annual U.S. per capita beef consumption. 1974-1990.
U.S. News Jul 2, 1990 p. 44

Where's the fat? TV ads claim that pork, "the other white meat," is about as low in fat as chicken and turkey. Here's the breakdown for a 3-ounce serving of: Pork. Veal. Turkey. Chicken. Fat. Cholesterol. Calories.
U.S. News Jun 4, 1990 p. 78

MEDICAID

Spending on Medicaid and prisons is up. Outlays as a percentage of total state spending. Medicaid. Corrections. 1987-1989.
Business Week Jun 5, 1989 p. 88

MEDICAL CARE

Fading fast. Funds appropriated for NHSC [National Health Service Corps] scholarship awards. 1980-1989.
Black Enterprise Sep 1990 p. 17

Who will care for us? Aging America. As a percent of total population. Children under age 17. Adults over age 65. 1980-2040.
Newsweek Jul 16, 1990 p. 53

Blood tests doctors often order. Complete blood count. Fats. Other common tests.
U.S. News Jun 18, 1990 p. 72

The growing menu of health benefits. Number of state-mandated medical services. 1975-1989.
Business Week May 21, 1990 p. 47

Annual health costs per capita. Health costs as a % of GNP. Life expectancy. Infant mortality. % of population not covered by health insurance. [Comparing] U.S. Canada. Britain.
Time May 7, 1990 p. 50

Places you'll get cut. Differences in the relative frequency with which [surgeons] perform common surgical procedures. Northeast/Mid-Atlantic. Midwest. South. West. 1987.
U.S. News Apr 30, 1990 p. 56

It's no vacation. The average hospital stay in days. 1970, 1979, 1988.
U.S. News Apr 30, 1990 p. 56

Surgery's hot list. Most-common inpatient surgical procedures.
U.S. News Apr 30, 1990 p. 57

Who will look after mama? Need. Nursing homes. Cost. At home. Caregivers. Insurance.
Newsweek Mar 12, 1990 p. 73

What's up, Doc? The number of yearly physician visits per person in the U.S. for all ages: Female. Male. 1970-1987.
U.S. News Sep 18, 1989 p. 77

Say goodbye to the age of free rides. Types of group health plans. 1984, 1988.
Newsweek Jan 30, 1989 p. 47

Still more coverage. A sampling of state laws passed in 1987 requiring new or expanded health-care coverage.
U.S. News Mar 28, 1988 p. 61

MEDICAL CARE - SOVIET UNION

Medicine & health. Infant deaths per 1,000 births. Official life expectancy at birth. Kidney transplants. Medical supplies. CAT scanners. Rural hospitals.
U.S. News Apr 3, 1989 p. 43

MEDICAL CARE, COST OF

Feverish growth. Percent change since 1980. Total health expenditures. Consumer price index. Federal education expenditures. 1980-1991.
Time Nov 25, 1991 pp. 34–35

A year's worth of health care. Annual premiums. Medical. Dental. Traditional plan. HMO. Point-of-service plan.
U.S. News Nov 4, 1991 pp. 78–79

Which covers what. What a Medicare policy will cover. What a Medicare policy will not cover. What a "core" medigap policy will cover. What a "core" medigap policy will not cover. What a long-term-care policy will cover.
U.S. News Jul 1, 1991 pp. 60–61

A guide to the new medigap policies. Annual cost estimates and suggestions for people with various needs.
U.S. News Jul 1, 1991 p. 62

Unhealthy benefits. Percentage of companies offering health insurance [by number of employees].
U.S. News Jun 3, 1991 p. 59

Medicare's rising construction bills. U.S. government payments. 1985-1991.
Business Week Apr 22, 1991 p. 69

No Rx yet for health costs. Federal government Medicare outlays. 1986-1991.
Business Week Jan 14, 1991 p. 90

Cutting the cost of health care. To reduce the high costs of health care, companies are looking at the following solutions.
Black Enterprise Nov 1990 p. 81

Or pay later. The cost of prenatal care for a healthy pregnant woman. The cost of medical care for an extremely premature baby.
Time Oct 8, 1990 p. 45

Annual health costs per capita. Health costs as a % of GNP. Life expectancy. Infant mortality. % of population not covered by health insurance. [Comparing] U.S. Canada. Britain.
Time May 7, 1990 p. 50

A hospital stay. Cost represents expense of typical stay and excludes physician charges. 1965, 1975, 1980, 1988.
U.S. News Apr 30, 1990 p. 60

Are social trends affecting health costs? Percent of: GNP spent on health. Adults who are unmarried. Children living with one parent. Households made up of one person. 1970, 1988.
Business Week Mar 5, 1990 p. 20

Health costs keep climbing. National health spending as a share of GNP. 1985-1990.
Business Week Jan 8, 1990 p. 87

As health care costs shoot up. Cumulative percent change since 1980. Medical care. All consumer prices. 1980-1989.
Business Week Nov 20, 1989 p. 111

[Medical costs] eat up more of national income. Health spending as percent of national income. 1980-1988.
Business Week Nov 20, 1989 p. 111

Individuals foot more of the bill. Individuals' out-of-pocket spending as percent of total spending on health. 1980-1988.
Business Week Nov 20, 1989 p. 111

Paying your own way. Americans spent $123 billion, or an average of $488 apiece, out of their own pockets for health care in 1987. Per person, here's how it was spent:
U.S. News Jul 31, 1989 p. 66

The economic dilemma. Health care costs percent increase. Consumer price index percent increase. Percent of gross national product. 1985-2000.
Business Week Feb 6, 1989 p. 76

A soaring health bill. Total health care expenditures. Percent of GNP. 1965-1987.
Newsweek Jan 30, 1989 p. 45

Shifting burden. Spending, personal health-care. Direct patient payments. Third-party payments. 1967, 1987.
Newsweek Jan 30, 1989 p. 50

Bed and breakfast. Hospital-room costs [by state] for an average stay.
U.S. News Jan 30, 1989 p. 81

Picking up the tab. Payments to nursing homes. Out of pocket. Medicaid. Federal programs. Medicare. Private insurance plans.
U.S. News Jan 23, 1989 p. 58

The rising cost of getting well. U.S. health care outlays. 1984-1989.
Business Week Jan 9, 1989 p. 82

Medicare's bite. Per-year cost of coverage. 1983-1993.
U.S. News Jun 6, 1988 p. 13

No limits in sight. Health-care spending. Median age in U.S. 1965-2000.
U.S. News Feb 22, 1988 p. 78

Home vs. hospital. The cost of care [various kinds].
U.S. News Jan 25, 1988 p. 68

Health care costs outrun the rest. Percent increase from previous year. Consumer price index. Medical price index. 1983-1988.
Business Week Jan 11, 1988 p. 104

MEDICAL INSURANCE

See INSURANCE, HEALTH

MEDICAL RESEARCH

More conflicts of interest? Industry is paying for more of biomedical R&D. Sources of funding for total U.S. health research & development. NIH. Other public. Industry. Nonprofit. Total funding, 1980, 1989.
Business Week Nov 5, 1990 p. 148

AIDS ranks fourth in research dollars. Cancer. Genetic diseases. Heart disease. AIDS. Infectious diseases.
Business Week Sep 17, 1990 p. 97

MEDICAL SCHOOLS

Top 15 medical schools. 15 medical schools with the highest scores in the *U.S. News* survey.
U.S. News Apr 29, 1991 p. 88

Failing doctors. Science exam most medical students must pass. Failure rate. 1984-1990.
U.S. News Nov 12, 1990 p. 93

MEDICARE

Which covers what. What a Medicare policy will cover. What a Medicare policy will not cover. What a "core" medigap policy will cover. What a "core" medigap policy will not cover. What a long-term-care policy will cover.
U.S. News Jul 1, 1991 pp. 60–61

A guide to the new medigap policies. Annual cost estimates and suggestions for people with various needs.
U.S. News Jul 1, 1991 p. 62

Medicare's rising construction bills. U.S. government payments. 1985-1991.
Business Week Apr 22, 1991 p. 69

No Rx yet for health costs. Federal government Medicare outlays. 1986-1991.
Business Week Jan 14, 1991 p. 90

Out of bounds? Projected Medicare gap. Income of Medicare's Hospital Insurance Trust Fund. Medicare's payments for hospitalization. 1990-2000.
Newsweek Dec 26, 1988 p. 30

The high cost of health. Growth in Medicare costs. Payments to physicians. Hospital payments. GNP. Enrollees. 1977-1988.
U.S. News Sep 26, 1988 p. 25

A big gap in doctors' fees. Average Medicare charges for physicians. Large urban area. Small rural area. 1987.
U.S. News Sep 26, 1988 p. 25

Medicare into the red. Change in the hospital trust fund. 1987-1999.
U.S. News Aug 29, 1988 p. 60

Uncle Sam's soaring doctor's bill. Annual Medicare reimbursements to physicians. 1977-1986.
Business Week Jul 18, 1988 p. 77

Medicare's booster shot. Hospital. Doctors. Prescribed drugs. Nursing homes. Costs. Financed by. [Coverage provided by] current law. New law.
Time Jun 20, 1988 p. 20

Medicare's bite. Per-year cost of coverage. 1983-1993.
U.S. News Jun 6, 1988 p. 13

MEDICINE, ALTERNATIVE

How acupuncture works. Diagram.
U.S. News Sep 23, 1991 p. 70

How hypnosis works. Diagram.
U.S. News Sep 23, 1991 p. 71

How biofeedback works. Diagram.
U.S. News Sep 23, 1991 p. 73

MEMORY

How long-term memories are formed. Diagram.
Time Oct 28, 1991 p. 86

MENSA

Sibling smarts. When the Mensa society, a group open to anyone who scores in the top 2 percent on a standardized IQ test, surveyed its members, here is how they fell in their family's lineup.
U.S. News May 22, 1989 p. 81

MENTAL HEALTH CARE

Shrinkage growth. Types of psychiatric hospitals. Private for profit. State and county. 1984, 1988.
Newsweek Nov 4, 1991 p. 51

Four causes of homelessness. Release of mentally ill. Total days spent by all patients in state and county mental hospitals. 1969-1986.
Scholastic Update Feb 10, 1989 p. 13

Help with mental health. How the states rank in serving the seriously mentally ill. 1988 ranking. 1986 ranking. Psychiatrists/psychologists per 100,000 population.
U.S. News Dec 5, 1988 p. 93

MENTALLY HANDICAPPED

Labels come, labels go. Ratio of students labeled learning disabled to students labeled mentally retarded. 1977-1987.
U.S. News Mar 13, 1989 p. 61

MERGERS AND ACQUISITIONS

See CORPORATIONS - ACQUISITIONS AND MERGERS

METHANOL

Three ways to go. Electric. Compressed natural gas. Methanol. Diagrams.
Business Week Apr 8, 1991 pp. 56–57

MEXICO

Mexico turns the corner. Income per capita. 1979-1990.
U.S. News Jun 18, 1990 p. 46

Mexico-at-a-glance. Capital. Official language. Area. Population. Ethnic groups. Religion. Chief products. Federal government. Map.
Scholastic Update Nov 18, 1988 p. 7

MEXICO - ECONOMIC CONDITIONS

A tale of three nations. Comparative economic and social indicators. Population. GNP. Unemployment rate. Average hourly earnings. U.S. Mexico. Canada.
Black Enterprise Oct 1991 p. 27

Rising growth, falling inflation. Percentage change in real gross domestic product. Percentage change in consumer prices. 1980-1991.
U.S. News Jul 8, 1991 p. 39

The border is bursting with jobs. Workers in Mexico's Maquiladora plants. 1980-1990.
Business Week Nov 12, 1990 p. 105

Mexico's shaky foundation. The trade balance has turned negative. More Mexicans need jobs. National population. But inflation has been tamed. Change in the consumer price index.
Business Week Apr 3, 1989 pp. 52–53

Mexico's foreign debt. 1970-1988.
Scholastic Update Nov 18, 1988 p. 9

Some improvements but no growth. Exports. Debt. Gross domestic product. Prices. 1980-1988.
U.S. News Jul 4, 1988 p. 39

MEXICO - HISTORY

Mexico in focus [by region]. Map.
Scholastic Update Nov 18, 1988 pp. 6–7

MIAMI (FLORIDA)

Latin power: A demographic upheaval. Dade County's ethnic mix. Breakdown of the Latin population.
Newsweek Jan 25, 1988 p. 28

MICROCOMPUTERS

See COMPUTERS, PERSONAL

MICROSCOPE

How to see an atom. Atomic-force microscope. Diagram.
U.S. News May 9, 1988 p. 67

MICROWAVE COOKING

What makes the microwave run?
FDA Consumer Mar 1990 p. 20

MIDDLE CLASSES

The middle class shrinks. Earnings. 1973-1987.
Business Week Sep 25, 1989 p. 95

More families graduate to the upper class. Lower class. Middle class. Upper class. Percent of total U.S. families. 1969, 1986.
Business Week Aug 15, 1988 p. 34

MIDDLE EAST COUNTRIES

Mideast peace: The economic potential. Population. Gross domestic product per capita. Defense spending per capita. [By Mideast country.]
Business Week Nov 11, 1991 p. 39

The world's hot spot. Israel and the Occupied Territories. Population of West Bank. Population of Gaza Strip. Pre-1967-1991. A country-by-country look at the volatile region. Maps.
Scholastic Update Oct 4, 1991 pp. 4–5

Disputed lands, divided peoples. What will be on the table. Syria. Lebanon. Jordan and the Palestinians. The hard parts. Settlements. Water. Jerusalem. Map.
U.S. News Aug 12, 1991 pp. 18–19

Areas occupied by Israel. Map.
Time Mar 18, 1991 p. 38

The volatile Middle East. Capital city. Port. Oil field. Oil pipeline. Arab nation. Non-Arab Islamic nation. Jewish nation. Territory held by Israel. Map.
Scholastic Update Oct 5, 1990 p. 4

To have and have not. The Arab world is sharply divided between wealthy nations that have oil and poor ones that don't. Haves. Have-nots. GNP per capita. Population.
U.S. News Aug 20, 1990 p. 27

MIDDLE EAST COUNTRIES - MILITARY STRENGTH

Arms strength. Tanks. Artillery. Aircraft. Helicopters. [By country.]
U.S. News Aug 13, 1990 p. 23

Arab-Israeli military balance. Army. Air Force and Air Defense. Navy. Map.
U.S. News Apr 4, 1988 pp. 42–43

MIDWESTERN STATES - ECONOMIC CONDITIONS

This time the downturn isn't killing the Midwest. Manufacturing activity in the Midwest. 1979-1991.
Business Week Apr 29, 1991 p. 22

Midwest. Plain States. Employment. Population. Median home price. Annual rate of change from previous year. 1989-1990.
U.S. News Nov 13, 1989 p. 59

MILITARY ASSISTANCE, AMERICAN

U.S. nonnuclear defense missions. Worldwide commitments. U.S. defense spending—and missions. Map.
U.S. News Oct 14, 1991 pp. 28–29

Getting a helping hand: The top 10. Top recipients of direct U.S. economic and military aid, fiscal 1990 [by country].
Newsweek Apr 16, 1990 p. 23

What the U.S. has spent on Nicaragua. 1979-1989.
Time Mar 12, 1990 p. 12

U.S. foreign aid. [To] Israel. Egypt. Poland-Hungary. Pakistan. Turkey. Philippines. Panama. Map. Total. % of aid that is military.
Time Mar 12, 1990 p. 14

MILITARY ASSISTANCE, SOVIET

Look homeward, Mikhail. Per capita GNP. Soviet arms transfers. Soviet troops, advisers [for six countries receiving Soviet Military Aid].
U.S. News May 9, 1988 pp. 33–35

MILITARY BASES

Here a base, there a base. More than half of all Defense Dept. bases in the U.S. are situated in these 12 states. State. Installations.
Business Week Jul 25, 1988 p. 55

MILITARY PERSONNEL

Schwarzkopf's pension. Generals who became president. Former military personnel in the Senate. House. [Other selected statistics.]
U.S. News Aug 12, 1991 p. 8

The shrinking U.S. military. Budget. 1990, 1995. Personnel. 1987, 1995.
U.S. News Apr 15, 1991 p. 42

Blacks in uniform. Percent of blacks in: Forces. Desert Storm. Army. Navy. Marine Corps. Air Force.
Newsweek Mar 11, 1991 p. 54

The military. Who is really doing America's fighting. White. Black. Other. New enlistees. Active duty enlisted personnel. High school graduate or equivalent.
Business Week Feb 25, 1991 p. 35

Caught with their guard down. Total Armed Forces. Active Armed Forces. Reserves. National Guard.
Newsweek Feb 4, 1991 p. 15

Up in arms. Military personnel per 100,000 people. [By country.]
U.S. News Oct 1, 1990 p. 95

Strength in numbers. Active duty forces. World War I. World War II. Korean War. Vietnam War. 1990-1995.
Newsweek Mar 19, 1990 p. 21

Manpower: A glut at the top. Total military personnel. Highest ranking officers. 1945, 1988.
Newsweek Jan 23, 1989 p. 18

The fast track to general. The rapid road from West Point or an ROTC commission to a first star includes many stops for schooling. Chart.
U.S. News Apr 18, 1988 pp. 34–35

Conventional arms. Defense spending as percentage of gross domestic product. Military personnel as a percentage of total population [by country].
Newsweek Mar 7, 1988 p. 42

MILITARY SERVICE, COMPULSORY

What [mandatory and volunteer service] other nations do.
U.S. News Feb 13, 1989 p. 23

MILKEN, MICHAEL

[Junk bond chronology]. 1969-1989.
Time Feb 26, 1990 p. 48

MILLIONAIRES

The rich get richer. The number of millionaires in the U.S. has grown from about 600,000 in 1980 to an estimated 1.5 million today. 1980-1990.
U.S. News Jun 25, 1990 p. 52

MINIMUM WAGE

Minimum wage, minimal turnout. Hourly workers at or below $3.35 an hour. 1981-1988.
Business Week Mar 27, 1989 p. 35

MINORITY BUSINESSES

Amount of money awarded in minority contracts for selected cities, 1988. Cities. Amount. Percent of total.
Black Enterprise Apr 1989 p. 18

MISSILES

The path to a target [four-step diagram of how the Tomahawk cruise missile works].
Newsweek Feb 18, 1991 pp. 40–41

Patriot vs. Scud [how Patriot missiles track and destroy Scud missiles].
Newsweek Jan 28, 1991 p. 20

The Tomahawk. Diagram.
Newsweek Jan 28, 1991 p. 22

How MIRV [missile] works. Diagram.
U.S. News Jun 12, 1989 p. 35

The AMRAAM missile and its cousins. Sidewinder. Sparrow. AMRAAM. Cost. Diagram.
U.S. News May 1, 1989 pp. 34–35

Home-grown missile power. A sampling of surface-to-surface ballistic missiles, developed by [various countries].
U.S. News Jul 25, 1988 p. 36

The enlarging club. Third-world countries with missiles of at least 150-mile range. 1970-1988.
U.S. News Jul 25, 1988 p. 36

Who has the hardware. First flight or delivery. Missile. Source. Deployed. Tested. Developing. [By various countries.]
U.S. News Jul 25, 1988 p. 37

MISSING CHILDREN

Where are they? Runaways. Abducted by relatives. Abducted by strangers. Found dead. Located alive. Still missing.
Newsweek Nov 27, 1989 p. 95

MISSING IN ACTION

Missing in action. WWII. Korea. Vietnam. MIAs. As a percent of total deaths.
Newsweek Jul 29, 1991 p. 23

MISSISSIPPI RIVER VALLEY

Unemployment rate. United States. Delta region. 1987, 1988.
Black Enterprise Dec 1989 p. 20

MOMMY TRACK

See WORKING MOTHERS

MONEY - INTERNATIONAL ASPECTS

Dollar drops. Trade-weighted value of U.S. dollar. 1980-1991.
U.S. News Dec 31, 1990 p. 46

Strong currencies. Currency units per dollar. France. U.S. Germany. June 1989 - Nov. 1990.
U.S. News Dec 17, 1990 p. 73

Falling dollar. Dollar vs. major foreign currencies. 1989-1991.
U.S. News Nov 26, 1990 p. 57

Descending dollar. Value of dollar vs. 20 currencies. 1980-1990.
U.S. News Oct 29, 1990 p. 88

A cheap currency. Yen per dollar. 1985-1990.
U.S. News Sep 24, 1990 p. 69

Mixed performance. A weaker greenback. The dollar's value against the currencies of its major trading partners. 1980-1989.
U.S. News Dec 25, 1989 p. 42

How currency gains are masking IBM's European problem. Sales are rising in dollar terms but not in most local currencies. France. West Germany. Italy. Britain. 1982-1987.
Business Week Jun 20, 1988 p. 97

MONEY - SOVIET UNION

The sinking ruble. Dollars per ruble. Official. Black market. 1986-1989.
Business Week May 1, 1989 p. 41

MONEY LAUNDERING

Shelter from the financial storm. Here are six of the most popular financial havens, and the amount of money in their banks.
U.S. News Aug 19, 1991 pp. 58–59

MONEY MANAGEMENT

A fool and his money. When 1,000 adults were asked to review a list of items and activities and say which ones were a waste of money, here is what they responded.
U.S. News Nov 6, 1989 p. 108

MONEY SUPPLY

The downshift in global money growth. Money supply. Britain. Japan. U.S. 1988-1991.
Business Week Aug 26, 1991 p. 16

Money growth stalls. M3 [A broad money supply measure]. 1990, 1991.
Business Week Aug 12, 1991 p. 20

Private placements shrink. Privately placed securities. 1986-1991.
Business Week Aug 12, 1991 p. 21

Small investors are helping out. Mutual funds net new sales and net exchanges. Taxable money market funds. Total bond and income funds. Total equity funds.
Business Week Aug 12, 1991 p. 21

MORTGAGES

Mortgage rates. 30-year fixed. 1989-1991.
Time Nov 25, 1991 p. 64

The mortgage squeeze on minorities. Mortgage denial rates.
U.S. News Nov 4, 1991 p. 21

Going down. Mortgage rates. Annual averages. 30 year fixed. 1 year adjustable. 1984-1991.
Newsweek Sep 30, 1991 p. 45

What can you afford? Income. Home price. Down payment. Mortgage.
Black Enterprise Oct 1990 p. 68

How interest rates affect monthly payments. Mortgage [amount]. 8%. 10%. 12%.
Black Enterprise Oct 1990 p. 69

Mortgage. Average monthly mortgage payments by region. Household income. 1989.
U.S. News Jul 30, 1990 p. 73

Mortgage sampler. Here's how adjustable-rate and 30-year fixed-rate mortgages stack up in 20 metropolitan areas. 1990.
U.S. News Apr 9, 1990 p. 64

The monthly mortgage bite. Here are the top 25 markets ranked by affordability—the share of the average monthly household income in a given area spent on the monthly mortgage payment.
U.S. News Apr 9, 1990 p. 94

Where's the check? Percentage of overdue loans. Least prompt states. Most prompt states.
U.S. News Apr 17, 1989 p. 77

Conventional mortgages. Estimated in billions. Fixed. Adjustable. 1984-1987.
Black Enterprise Sep 1988 p. 31

Six ways to meet a mortgage. Term to maturity. Total interest. Regular payment. Total annual payments.
U.S. News Feb 22, 1988 p. 91

MORTGAGES - REFINANCING

When it pays to borrow all over again. New mortgage. New monthly payment. Time to recoup costs.
Business Week Mar 18, 1991 p. 132

The finances of refinancing. Original mortgage rate. Current monthly payments. Monthly savings at 9.5 percent. Time to recoup costs.
U.S. News Aug 7, 1989 p. 60

MOTHERS - EMPLOYMENT

See WORKING MOTHERS

MOTION PICTURE THEATERS

Movie screens are proliferating. Indoor screens. But not the moviegoers. Average ticket price. Number of admissions. 1985-1990.
Business Week Dec 3, 1990 p. 127

MOTION PICTURES

Year the first nickelodean opened. Weekly movie attendance. 1950-1990. Number of movie screens. Average movie ticket price. Money spent by Americans at movies. [Other selected statistics.]
U.S. News Aug 19, 1991 p. 10

A tale of two summers. Box-office gross, in millions. 1989 total as of 7/16. 1990 estimates as of 7/15.
Time Jul 30, 1990 p. 56

Born on Broadway. Here are the top-grossing movies of the '80s that were based on stage plays. U.S. box-office receipts.
U.S. News Jun 4, 1990 p. 78

Oscar faces the music. Here are the most recent Oscar-winning tunes and the movies in which they were performed. 1959-1988.
U.S. News Apr 2, 1990 p. 68

Blockbusters and bombs. Of the thousands of movies released in the U.S. during the 1980s, these were the most and least popular. Year distributed. Box-office gross.
U.S. News Jan 22, 1990 p. 67

Intermission required. The longest movie ever made, "The Cure for Insomnia," at 85 hours, was shown only once in full. Here are some other lengthy flicks.
U.S. News Aug 28, 1989 p. 102

Picture-perfect chiefs. These Presidents are portrayed most frequently in movies.
U.S. News Jul 10, 1989 p. 62

The envelope, please. All-time top 24 movies and their Academy Awards.
U.S. News Apr 11, 1988 p. 68

Where have all the G movies gone? Number of movies rated. G. PG. R. X. PG-13. 1970-1987.
U.S. News Feb 15, 1988 p. 88

Younger moviegoers. [Attendance by age.] 1985-1987.
U.S. News Feb 15, 1988 p. 88

MOTORCYCLES - MARKET SHARE

Honda no longer leads the pack by much. Percent share of U.S. motorcycle market. Honda. Yamaha. Kawasaki. Suzuki. Harley. 1985, 1987, 1989.
Business Week Sep 3, 1990 p. 74

MOVIES

See MOTION PICTURES

MOVING, HOUSEHOLD

Pack up the pooch. An employe asked to relocate can usually count on reimbursement for normal moving expenses, but what about extras? Percentage of companies that will pay [various extras].
U.S. News Jul 16, 1990 p. 66

Packed up and movin' on. As the '80s wore on, more people moved. Here's how long owners stayed put in 1988 and in 1980, by state.
U.S. News Feb 5, 1990 p. 74

Moving along. Why prospective home buyers want to leave their present homes.
U.S. News Feb 13, 1989 p. 84

On the move. Comparing the cost of various household moves by van lines and by rental trucks. Van line. Rental truck. Savings.
U.S. News Jul 25, 1988 p. 66

MUNICIPAL FINANCE

Harmful deficits. State and local budget balances, excluding social insurance funds. 1982-1991.
U.S. News Jul 8, 1991 p. 51

Boosting revenues. State and local government budget surplus/deficit. 1980-1991.
U.S. News May 27, 1991 p. 56

A tide of red ink swamps state and local budgets. Operating balance. 1985-1990.
Business Week Apr 15, 1991 p. 22

State and local budgets are deep in the red. Operating balance excludes social insurance funds. 1983-1990.
Business Week Jul 30, 1990 p. 12

State and local budgets head into a deeper rut. Operating budget. 1984-1988 (est).
Business Week Dec 26, 1988 p. 40

State and local budgets slide deeper into the red. Operating balance. 1984-1988.
Business Week May 2, 1988 p. 24

MUNICIPAL GOVERNMENT

City jobs. White males still hold most top municipal jobs. Here are the percentages of female and black leaders by position.
U.S. News Oct 9, 1989 p. 81

MURDER AND MURDER RATES

America's new murder capitals. City. Percent increase in murders, 1985-90.
Newsweek Jun 10, 1991 p. 17

U.S. murders by handguns. 1985-1989.
Time May 20, 1991 p. 26

Percentage of murders committed with handguns. Percentage of those arrested for murder who go to prison. 1990 increase in reports of violent crimes by region.
U.S. News May 6, 1991 p. 16

Kids who kill. Firearm murders committed by offenders under age 18. 1984, 1987, 1989.
U.S. News Apr 8, 1991 p. 26

The murder wave moves beyond big cities. Cities with new records for homicides. Homicides per 1000. 1989, 1990.
Business Week Jan 14, 1991 p. 42

Homicide rates. (Gun-related). Black males. Black females. White males. White females. 1984-1988.
Newsweek Dec 17, 1990 p. 33

Guns and crime. Cops under fire. Murders with guns. All firearms. Handguns. Knives or other cutting instruments were the next most common murder weapon. 1974-1989.
U.S. News Dec 3, 1990 p. 38

Femicide: Related circumstances [when women kill]. California, 1988. Rape. Drug-related. Gang related. Robbery, burglary. Argument. Other.
Ms. Sep 1990 p. 34

Murder: Relationship of victim to offender. Texas, 1988.
Ms. Sep 1990 p. 35

One measure of the war on drugs. How homicide totals changed during 1989 in 10 cities with the highest 1988 per capita murder rates. 1989 murders.
U.S. News Apr 16, 1990 p. 14

The short arm of the law. Change in homicide rate and police strength in largest U.S. cities since 1978.
U.S. News May 1, 1989 p. 76

A tour of the urban killing fields. Cities with the highest per capita murder rates. The biggest increases. 1988.
Newsweek Jan 16, 1989 p. 45

Uneven odds. Lifetime risk of being murdered. Black males. White males. Black females. White females.
U.S. News Aug 22, 1988 p. 54

See Also CRIME

MUSEUMS

See ART MUSEUMS

MUSIC, POPULAR

Oldest favorites. Of tunes played at least 3 million times on radio and TV, these are the oldest.
U.S. News Dec 10, 1990 p. 79

MUSIC RECORDING INDUSTRY

Paying the piper. Type of music purchased. Percentage of dollar value. 1985-1989.
Black Enterprise Dec 1991 p. 60

Where the growth is. Cassettes. CDs. 1986-1990.
Business Week May 27, 1991 p. 38

Songs Porter wrote. Number of the 100 most recorded songs between 1890 and 1954 written by Porter. Gershwin. Berlin. Gross revenue generated by the recording industry in 1940. In 1990.
U.S. News May 27, 1991 p. 14

CDs play taps for LPs. Manufacturers' shipments. CDs. LPs.
Business Week Dec 24, 1990 p. 30

Sounds and sales. Unit shipments in millions. Cassettes. LPs. CDs. 1983-1989.
Newsweek Jul 16, 1990 p. 46

How much the industry makes. Recording-industry profits (in millions). 1978-1989.
Scholastic Update May 18, 1990 p. 3

Who buys the product? Age of record buyers. 1988.
Scholastic Update May 18, 1990 p. 3

What do they buy? Type of music purchased. Format. % of market—1988.
Scholastic Update May 18, 1990 p. 3

Tops on the turntable. Most popular record albums since 1955.
U.S. News Apr 24, 1989 p. 80

The beat is back. Demand is growing again. Cassettes. LPs. CDs. U.S. unit sales (millions). The new mix is more profitable. Manufacturers' average revenue per sale. 1984-1988.
Business Week Aug 15, 1988 p. 89

MUSICAL INSTRUMENTS

Making sweet music. Instrument sales. Fretted instruments. Woodwinds. Pianos. Brass. String instruments. 1940-1987.
U.S. News Nov 14, 1988 p. 82

MUSICAL PITCH

Tuning up through history. The pitch of a concert A in hertz (Hz), or cycles per second.
U.S. News Jun 26, 1989 p. 56

MUSLIMS

A new frontier for Islam? Muslim population in Western Europe [by country]. 1970-1989.
U.S. News Aug 6, 1990 p. 34

MUTUAL FUNDS

See INVESTMENTS

NAMES, PERSONAL

The real first families. The *Mayflower* carried just over 100 passengers. Here are the surnames of those known to have made the trip.
U.S. News Nov 27, 1989 p. 91

NASA

See NATIONAL AERONAUTICS AND SPACE ADMINISTRATION

NATIONAL AERONAUTICS AND SPACE ADMINISTRATION

NASA's top projects. Share of total [1992 budget].
Business Week Jun 3, 1991 p. 122

NASA'S ups and downs [chronology]. 1958-1990.
Time Aug 6, 1990 pp. 26–27

[Hubble Space Telescope] Problem. Solution. Diagram.
Time Jul 9, 1990 p. 43

NATIONAL PARKS AND RESERVES

Forests that everyone owns. Timber harvested in national forests. 1950, 1990. National forests [states with]. States having no national forests.
Time Dec 9, 1991 p. 71

Number of National Park Service properties in 1916. In 1990. Visitors in 1916. In 1991. Amount allocated to the National Park Service in 1991.
U.S. News Jun 10, 1991 p. 12

The wilderness mosaic. Where the wild lands are. The 474 wilderness areas range over 44 states. Map.
U.S. News Jul 3, 1989 p. 20

Crowded at the parks. Annual number of visitors to national parks. Visits. Acres.
U.S. News Jun 20, 1988 p. 78

NATIONS

World Almanac. 1991-1992. Country. Area, population. Pop. annual rise. Capital. Major languages. Form of govt., head. Date of indep., % of pop. under 15 yrs. Literacy rate, life expectancy. [By country.]
Scholastic Update Oct 18, 1991 pp. 30–35

World Almanac. 1990-1991. Country. Area, population. Pop. annual rise. Capital. Major languages. Form of govt., head. Date of indep., % of pop. under 15 yrs. Literacy rate, life expectancy. [By country.]
Scholastic Update Nov 2, 1990 pp. 20–26

World Almanac. 1989-1990. Country. Area, population. Physical quality of life, pop. annual rise. Capital. Major languages. Form of govt., head, & freedom status. Date of indep., % of pop. under 15 yrs. Literacy rate, life expectancy. [By country.]
Scholastic Update Sep 22, 1989 pp. 19–24

1988-1989 update almanac. Country. Area, population. Capital. Major languages. Form of govt., head, & freedom status. Date of indep., % of pop. under 15 yrs. Literacy rate, life expectancy. Physical quality of life. Per cap. GNP.
Scholastic Update Sep 23, 1988 pp. 24–29

NATIVE AMERICANS

Native ground: Where America's Indians live. Federal Indian reservations. State Indian reservations. Other Indian groups. Map.
Scholastic Update May 26, 1989 pp. 6–7

NATIVE AMERICANS - EDUCATION

Native Americans: Facing hard times. Education. Percent of adults over age 25 who have graduated high school. Native Americans. U.S. average.
Scholastic Update May 26, 1989 p. 5

NATIVE AMERICANS - EMPLOYMENT

Native Americans: Facing hard times. Jobs. Percent unemployed. Native Americans. U.S. average.
Scholastic Update May 26, 1989 p. 5

NATIVE AMERICANS - SOCIAL CONDITIONS

Native Americans: Facing hard times. Health. Life expectancy (in years). Native Americans. U.S. average.
Scholastic Update May 26, 1989 p. 5

Native Americans: Facing hard times. Housing. Percent of homes that lack electricity. Native American. U.S. average.
Scholastic Update May 26, 1989 p. 5

NATURAL GAS

A look ahead for natural gas. U.S. natural-gas consumption. Natural-gas well completions. Wellhead gas prices. 1989-1995.
Business Week Jul 15, 1991 p. 35

More people are using natural gas. Percentage of new single-family homes using natural gas. 1980-1989.
U.S. News Jul 30, 1990 p. 38

It's cheaper. Residential heating costs (per million B.t.u.). 1980-1990.
U.S. News Jul 30, 1990 p. 39

Look who's cooking with gas. Known reserves of oil and natural gas. [By world regions.]
U.S. News Jul 31, 1989 p. 38

NAVY - UNITED STATES

The Lehman Navy. Ships purchased by the Navy since 1981.
U.S. News Jul 11, 1988 p. 19

NETHERLANDS

The sump of Europe. About half the pollution in the Netherlands comes from other countries. Percentage of acid rain coming from [various countries]. Map.
U.S. News Sep 11, 1989 p. 68

NEW ENGLAND - ECONOMIC CONDITIONS

New England. Mid-Atlantic. Employment. Population. Median home price. Annual rate of change from previous year. 1989-1990.
U.S. News Nov 13, 1989 p. 64

NEW YORK (CITY)

New York. Population. Unemployment rate. Poverty rate. Bond rating. Outstanding debt per capita. Budget surplus or deficit.
U.S. News Oct 29, 1990 p. 81

Percent change in employment. Percentage of Manhattan offices vacant. New York City tax revenues. Service cuts and tax hikes.
Business Week Jun 18, 1990 pp. 182–183

New York's overall job growth stalls. Total New York City employment. 1984-1989.
Business Week Nov 20, 1989 p. 108

The financial sector shrinks. Total New York City employment in finance, insurance, and real estate. 1984-1989.
Business Week Nov 20, 1989 p. 108

Tax revenues lag behind inflation. Corporation tax. Sales tax. Inflation. 1987-1990.
Business Week Nov 20, 1989 p. 109

We'll take Manhattan. Goods cost more in Tokyo than in New York City. Product. Maker. Tokyo. New York.
U.S. News Nov 20, 1989 p. 13

The Big Apple's troubled racial landscape. Attacks against blacks. Attacks against whites. 1981-1987.
Black Enterprise Mar 1988 p. 20

NEW YORK (STATE)

Bicoastal pain. This recession has hit New York and California harder than the rest of the country. Civilian unemployment rate. 1989, 1990, 1991.
U.S. News Dec 23, 1991 p. 52

Big spenders. Per capita spending, 1988. Education. Welfare. Highways. Health. Capital improvements. New York. California.
U.S. News Dec 23, 1991 p. 53

NEWS MEDIA

The tuned-out generation [by age group]. % who "read a newspaper yesterday." % who "watched TV news yesterday." % who followed these news events "very closely."
Time Jul 9, 1990 p. 64

Where we get our news. TV is now America's number one source of news. Primary news sources: 1959-1988. (Percent of Americans who named each medium as a top source). Television. Newspapers. Radio. Magazines.
Scholastic Update Sep 8, 1989 p. 6

NEWSPAPER PUBLISHERS AND PUBLISHING

The bad news. Newspaper advertising revenues. 1987-1990.
Newsweek May 27, 1991 p. 39

Final delivery. [List of] recently deceased major newspapers.
U.S. News Feb 4, 1991 p. 71

Still read all over. TV and radio have not pushed newspapers out of circulation just yet. Nearly two thirds of adults still read one newspaper a day. Daily newspapers. Total circulation.
U.S. News Aug 28, 1989 p. 102

His-and-her newspapers. Percentages of men and women who usually read selected newpaper items.
U.S. News Aug 15, 1988 p. 78

NEWSPAPERS, TABLOID

Callahan's newsstand. Magazines [list of]. Circulation.
Business Week May 15, 1989 p. 139

NICARAGUA - POLITICS AND GOVERNMENT

How the Sandinista defeat affects Nicaragua's neighbors. Guatemala. El Salvador. Cuba. Honduras. Costa Rica. Panama. Map.
Time Mar 12, 1990 p. 14

NICARAGUA AND THE SOVIET UNION

Comrades in arms. Aid from the Soviet bloc. Economic aid. Military aid. To Cuba. To Nicaragua.
Newsweek Dec 5, 1988 p. 30

NICARAGUA AND THE UNITED STATES

What the U.S. has spent on Nicaragua. 1979-1989.
Time Mar 12, 1990 p. 12

NOISE

Creating the sounds of silence: Fighting noise with anti-noise. Noise levels of common sounds. Permanent damage begins after 8-hour exposure. Diagram.
U.S. News Sep 9, 1991 p. 60

NONPROFIT ORGANIZATIONS

How much it takes to give. Of the major nonprofit organizations that released figures for fund-raising and administrative costs last year, these spent the least as a percentage of total expense.
U.S. News Dec 18, 1989 p. 82

NORTH AMERICA

The Columbian Exchange. New world portrait. New world to old [animals and plants brought]. Old world portrait. Old world to new. Diagram.
U.S. News Jul 8, 1991 pp. 29–31

NORTH ATLANTIC TREATY ORGANIZATION - MILITARY STRENGTH

NATO's leaner fighting machine. Central front reductions. Current forces. Anticipated forces.
Time Jun 10, 1991 p. 38

Setting limits. Overall treaty limits. NATO. Warsaw Pact.
Time Oct 15, 1990 p. 62

NATO nuclear weapons in Western Europe, 1990. U.S. nuclear warheads deployed by country. Nuclear delivery systems deployed by country. NATO nuclear forces in Germany. U.S. and NATO nuclear airbases in Europe.
Bull Atomic Sci Oct 1990 pp. 48–49

Current strength. NATO. Warsaw Pact. Troops. Tanks. Artillery. Aircraft.
Time May 14, 1990 p. 27

Force levels. Troops. Tanks. Artillery. Aircraft. NATO. Warsaw Pact.
Time Dec 11, 1989 p. 38

Counting down. Existing forces. NATO's count of its own forces. NATO's count of Warsaw Pact. Warsaw Pact's count of NATO. Warsaw Pact's count of its own forces.
Time Jun 12, 1989 pp. 30–31

The balance of power in Europe. Current NATO forces. Current Warsaw Pact forces. NATO's proposal. Warsaw Pact's proposal.
U.S. News Jun 12, 1989 p. 27

A mean bean count. Mean of estimates [military power]. Warsaw Pact. NATO. Warsaw Pact advantage.
Bull Atomic Sci Mar 1989 p. 31

In-place and immediate reinforcing divisions in central Europe. NATO country. Warsaw Pact country. Total divisions. Total armor division equivalents.
Bull Atomic Sci Mar 1989 p. 32

NATO vs. Warsaw Pact in the central front. Troops. Tanks. Map.
Time Dec 19, 1988 p. 25

So who's counting [conventional forces in Europe]. Troops. Tanks. Combat aircraft. Warsaw Pact. NATO.
Bull Atomic Sci Sep 1988 p. 17

NORTHWESTERN STATES - ECONOMIC CONDITIONS

Alaska, Northwest. California, Southwest. Employment. Population. Median home price. Annual rate of change from previous year. 1989-1990.
U.S. News Nov 13, 1989 p. 56

NUCLEAR FUSION

See FUSION

NUCLEAR POWER PLANTS

Power puzzle. Net generation of electricity by U.S. utilities. Nuclear. Coal. Natural gas. Petroleum. Renewal sources.
Time Apr 29, 1991 pp. 56–57

Nuclear power. Fossil fuel power. Hydroelectric power. Solar power. Wind power. Pros. Cons. Comments.
Time Apr 29, 1991 pp. 56–57

Who's gone nuclear. Percent of electricity derived from nuclear power. [By country.]
Time Apr 29, 1991 p. 61

When reactors get old. Symptoms of aging in nuclear power plants. Diagram.
Time Mar 18, 1991 p. 87

Tough choices. Electricity generated from: Nuclear power. Coal. Natural gas. Hydropower. Petroleum. Geothermal and other. Sweden. U.S. 1987.
U.S. News Jul 18, 1988 p. 43

See Also ELECTRIC POWER

NUCLEAR REACTORS

A new age in reactor design. Westinghouse AP600 design. Diagram.
U.S. News May 29, 1989 p. 52

A ceramic shield. General Atomics' high-temperature gas-cooled reactor design. Diagram.
U.S. News May 29, 1989 p. 53

NUCLEAR SHIPS

Nuclear weapons at sea, 1990. Weapons. Nuclear-capable ships and submarines. Nuclear reactors on naval vessels. U.S. Soviet. U.K. France. China. Total.
Bull Atomic Sci Sep 1990 p. 49

Nuclear weapons at sea, 1989. Weapons. Nuclear-capable ships and submarines. Nuclear reactors on naval vessels. U.S. Soviet. U.K. France. China. Total.
Bull Atomic Sci Sep 1989 p. 48

Seawolf attack sub: What it can do. Diagram.
U.S. News Apr 24, 1989 p. 29

Nuclear weapons at sea, 1988. Nuclear-capable ships and submarines. Nuclear reactors on naval vessels. U.S. Soviet. U.K. France. China. Total.
Bull Atomic Sci Sep 1988 p. 64

NUCLEAR TEST BAN

See ARMS CONTROL

NUCLEAR WEAPONS

Scramble for the bomb. Declared nuclear powers. Working on obtaining nuclear weapons. Undeclared nuclear powers. Stopped working on nuclear weapons. [By country].
Time Dec 16, 1991 p. 47

Nuclear pursuits [various historical data]. United States. Soviet Union. Britain. France. China.
Bull Atomic Sci May 1991 p. 49

Nuclear weapons at sea, 1990. Weapons. Nuclear-capable ships and submarines. Nuclear reactors on naval vessels. U.S. Soviet. U.K. France. China. Total.
Bull Atomic Sci Sep 1990 p. 49

Countries that have not signed the Non-Proliferation Treaty. Nuclear parties. Non-nuclear parties. Non-parties. Map.
Bull Atomic Sci Jul 1990 pp. 24–25

The numbers game. START was billed as cutting long-range nuclear forces by 50 percent. But loopholes exempt many warheads. Current. After START. Missile warheads. Bomber. Sea-launched cruise. Total.
U.S. News Jun 11, 1990 p. 31

The charge and the bomb. Capacitors: A vital part in making nuclear weapons. Diagram.
Newsweek Apr 9, 1990 p. 28

How it works. An H-Bomb explodes in two stages. Diagram.
Time Jan 15, 1990 p. 79

First A-bomb. First H-bomb. United States. Soviet Union. Britain. France. China.
Bull Atomic Sci Jan 1990 p. 29

U.S., Soviet nuclear weapons stockpile, 1945-1989: Number of weapons.
Bull Atomic Sci Nov 1989 p. 53

Nuclear weapons at sea, 1989. Weapons. Nuclear-capable ships and submarines. Nuclear reactors on naval vessels. U.S. Soviet. U.K. France. China. Total.
Bull Atomic Sci Sep 1989 p. 48

Where the nuclear weapons are: The three legs of triad. Bombers. Land-based missiles. Submarine-launched ballistic missiles. U.S. U.S.S.R.
U.S. News Jun 12, 1989 pp. 32–33

Nuclear pursuits. United States. Soviet Union. Britain. France. China.
Bull Atomic Sci May 1989 p. 57

In a general nuclear war, B-2 Stealth bombers have the crucial mission of destroying mobile Soviet SS-24 and SS-25 missiles. [How this mission will be accomplished.] Diagram.
U.S. News Nov 28, 1988 pp. 20–21

U.S. and Soviet nuclear-capable aircraft (1988). Country and type. Year introduced. Number. Comments.
Bull Atomic Sci Nov 1988 p. 48

Nuclear network. Plutonium. Uranium. Tritium. Bomb components. Nuclear waste. Map of U.S.
Time Oct 31, 1988 pp. 62–63

U.S. and Soviet nuclear weapons under development, 1988. United States. Soviet Union.
Bull Atomic Sci Oct 1988 p. 56

Nuclear weapons at sea, 1988. Nuclear-capable ships and submarines. Nuclear reactors on naval vessels. U.S. Soviet. U.K. France. China. Total.
Bull Atomic Sci Sep 1988 p. 64

Members of the club. Established nuclear powers. Nations that now produce nuclear weapons and can deploy them. Nations known to actively seek nuclear weapons. Map.
Newsweek Jul 11, 1988 p. 42

U.S. and Soviet strategic nuclear forces, 1972-1987. ICBMs. SLBMs. Bombers. Totals.
Bull Atomic Sci May 1988 p. 56

Nuclear pursuits. Nuclear warheads in the U.S. arsenal today. Countries that host U.S. nuclear weapons. [Various statistics.]
Bull Atomic Sci Apr 1988 p. 60

How tritium works in a boosted fission explosion. Diagram.
Bull Atomic Sci Jan 1988 p. 41

U.S. and Soviet strategic nuclear forces, end of 1987. Type. Name. Launchers. Year deployed. Warheads x yield (megatons). Warheads. Total megatons.
Bull Atomic Sci Jan 1988 p. 56

NUCLEAR WEAPONS - ACCIDENTS

Recipe for disaster. The warhead [W-79 artillery shell]. Normal explosion. Accidental explosion.
Time Jun 4, 1990 p. 53

How a nuclear-weapons plant poisoned a community. The Feed Materials Production Center at Fernald, Ohio, has been contaminating the air, water and ground around it almost from the day its doors opened 37 years ago. Diagram.
U.S. News Oct 23, 1989 pp. 22–23

Lost at sea 1945-89: 50 weapons, 11 reactors. [Chronology and description of events.]
Bull Atomic Sci Jul 1989 p. 22

NUCLEAR WEAPONS - CHINA

Chinese nuclear forces, 1990. Delivery vehicle. Year deployed. Number. Range. Warhead, yield. Total warheads in stockpile.
Bull Atomic Sci Nov 1990 p. 49

Known Chinese nuclear tests, 1964-1988. Date. Time. Yield. Description.
Bull Atomic Sci Oct 1989 p. 48

The Chinese nuclear force. Strategic. Nonstrategic. Total warheads. Total megatons. Weapons under development.
Bull Atomic Sci Jun 1989 p. 34

NUCLEAR WEAPONS - COMMONWEALTH OF INDEPENDENT STATES

How many fingers on the button? Number of warheads [by republic]. [Types of] nuclear sites. Map of Soviet Union.
Newsweek Dec 16, 1991 p. 33

Four new members of the club. Ukraine. Belorussia. Kazakhstan. Russia. ICBMs. Heavy bombers. Total warheads.
Time Dec 16, 1991 p. 41

NUCLEAR WEAPONS - EUROPE

NATO nuclear weapons in Western Europe, 1990. U.S. nuclear warheads deployed by country. Nuclear delivery systems deployed by country. NATO nuclear forces in Germany. U.S. and NATO nuclear airbases in Europe.
Bull Atomic Sci Oct 1990 pp. 48–49

The last nuclear weapons in Europe. NATO. Number. Range. Warsaw Pact. Number. Range. 1987.
Newsweek May 8, 1989 p. 16

Nuclear weapons: Playing the numbers game. NATO. Warsaw Pact.
U.S. News Jan 23, 1989 p. 25

Short-range nuclear weapons on Battlefield Europe. Map.
U.S. News Jan 23, 1989 p. 25

NUCLEAR WEAPONS - FRANCE

French nuclear forces, 1990. Delivery vehicle. Year deployed. Number. Range. Warhead x yield. Type. Total warheads in stockpile.
Bull Atomic Sci Dec 1990 p. 57

NUCLEAR WEAPONS - GREAT BRITAIN

British nuclear forces, 1990. Delivery vehicle. Year deployed. Number. Range. Warhead, yield. Total warheads in stockpile.
Bull Atomic Sci Nov 1990 p. 49

NUCLEAR WEAPONS - INDIA

Nuclear enterprise on the subcontinent. Pakistan. India. Map.
Bull Atomic Sci Jun 1989 p. 23

NUCLEAR WEAPONS - IRAQ

Iraq's lethal plants. Iraqi nuclear facilities. Map.
Newsweek Oct 7, 1991 p. 33

The evidence: How Saddam planned to build a bomb. Evidence uncovered by U.N. inspectors. Plutonium. Uranium. The bomb. Equipment. Delivery. Chart.
Newsweek Oct 7, 1991 pp. 34–35

Iraq's atomic future. If it chooses, there are two main routes Iraq can use to produce plutonium or highly enriched uranium for an atomic bomb.
U.S. News Jun 4, 1990 p. 50

Can Baghdad make the bomb? [What is] needed to make a nuclear weapon. Does Iraq have it? Chart.
Time Apr 9, 1990 p. 45

NUCLEAR WEAPONS - PAKISTAN

Nuclear enterprise on the subcontinent. Pakistan. India. Map.
Bull Atomic Sci Jun 1989 p. 23

NUCLEAR WEAPONS - SOUTH AMERICA

Bomb potential for South America. Uranium mining area. Map.
Bull Atomic Sci May 1989 p. 16

NUCLEAR WEAPONS - SOVIET UNION

Estimated Soviet nuclear stockpile, July 1989. Category/type. Weapon system. Launchers. Warheads.
Bull Atomic Sci Jul 1991 p. 56

Soviet strategic nuclear forces, end of 1990. Type: ICBMs, SLBMs, bomber/weapons. Name. Launchers. Year deployed. Warheads x yield (megaton). Total warheads. Total megatons.
Bull Atomic Sci Mar 1991 p. 49

Estimated Soviet nuclear stockpile (July 1990). Category/type. Weapon system. Launchers. Warheads.
Bull Atomic Sci Jul 1990 p. 49

Soviet strategic nuclear forces, end of 1989. Type: ICBMs, SLBMs, bomber/weapons. Name. Launchers. Year deployed. Warheads x yield. Total warheads. Total megatons.
Bull Atomic Sci Mar 1990 p. 49

Soviet strategic nuclear forces, end of 1988. Type: ICBMs, SLBMs, bombers/weapons. Name. Launchers. Year deployed. Warhead x yield (megatons). Total warheads. Total megatons.
Bull Atomic Sci Mar 1989 p. 52

Estimated Soviet nuclear stockpile, July 1988. Category/type. Weapon system. Launchers. Warheads.
Bull Atomic Sci Jul 1988 p. 56

NUCLEAR WEAPONS - TESTING

Known nuclear tests worldwide, 1945 to December 31, 1990. Year. United States. Soviet Union. United Kingdom. France. China. Total.
Bull Atomic Sci Apr 1991 p. 49

Known nuclear tests worldwide, 1945 to December 31, 1989. Year. U.S. Soviet Union United Kingdom France China Total.
Bull Atomic Sci Apr 1990 p. 57

Secret U.S. nuclear tests, 1963-1988. Number of tests.
Bull Atomic Sci Apr 1989 p. 36

Known nuclear tests worldwide, 1945 to December 31, 1988. Year. U.S. U.S.S.R. United Kingdom France. China. Total.
Bull Atomic Sci Apr 1989 p. 48

Known nuclear tests by year, 1945 to December, 1987. United States. Soviet Union. Britain. France. China. Total.
Bull Atomic Sci Mar 1988 p. 56

Nuclear test explosions 1945 through 1987. United States. Soviet Union. Britain. France. China. [Number and year.]
Bull Atomic Sci Mar 1988 p. 56

NUCLEAR WEAPONS - UNITED STATES

U.S. nuclear weapons stockpile (June 1991). Warhead/weapon. First produced. Yield (kilotons). User. Number. Status.
Bull Atomic Sci Jun 1991 p. 49

U.S. strategic nuclear forces, end of 1990. Type: ICBMs, SLBMs, bombers. Name. Launchers/SSBNs. Year deployed. Warheads x yield (megaton). Total warheads. Total megatons.
Bull Atomic Sci Jan 1991 p. 48

U.S. nuclear weapons stockpile (June 1990). Warhead/weapon. First produced. Yield (kilotons). User. Number (warheads). Status.
Bull Atomic Sci May 1990 p. 48

U.S. strategic nuclear forces, end of 1989. Type: ICBM's, SLBMs, bomber/weapons. Name. Launchers/SSBNs. Year deployed. Warheads x yield. Total warheads. Total megatons.
Bull Atomic Sci Jan 1990 p. 49

U.S. nuclear weapons stockpile (June 1989). Warhead/weapon. First produced. Yield. User. Number. Status.
Bull Atomic Sci Jun 1989 p. 49

U.S. strategic nuclear forces, end of 1988. Type: ICBMs, SLBMs, bombers/weapons. Name. Launchers. Year deployed. Warheads x yield (megatons). Total warheads. Total megatons.
Bull Atomic Sci Jan 1989 p. 68

Ronald Reagan's military budget dreams. Defense Department five-year budget requests. Energy Department five-year budget requests. Nuclear weapons. Fiscal years 1982-1989.
Bull Atomic Sci Dec 1988 p. 52

U.S. nuclear weapons stockpile (June 1988). Warhead/weapon. First produced. Yield. User. Number. Status.
Bull Atomic Sci Jun 1988 p. 56

Department of Energy warhead production complex. Research and development. Materials production. Component production. Testing. Map.
Bull Atomic Sci Jan 1988 p. 13

NURSING HOMES

Who will look after mama? Need. Nursing homes. Cost. At home. Caregivers. Insurance.
Newsweek Mar 12, 1990 p. 73

Picking up the tab. Payments to nursing homes. Out of pocket. Medicaid. Federal programs. Medicare. Private insurance plans.
U.S. News Jan 23, 1989 p. 58

NUTRITION

Bread-and-butter issues. 1958: The basic four [food groups]. The proposed pyramid scheme [1991].
Time Jul 15, 1991 p. 57

Boob-tube diet. The menu on prime-time TV. Number of times these foods were displayed or referred to in a 14-hour sampling of shows.
U.S. News Sep 24, 1990 p. 99

Eating right the rest of the year. Breakfast. Lunch. Dinner. [Typical foods]. Saturated fat (grams). Cholesterol (milligrams). Calories.
U.S. News Nov 27, 1989 p. 89

Fiber facts. [Foods] richer in water-soluble fiber and more likely to reduce heart disease. [Foods] richer in water-insoluble fiber and more likely to reduce colon cancer.
U.S. News Nov 27, 1989 p. 91

Giving in to temptation. Percentages of respondents who say they are making *no* effort to cut down on: Caffeine. Red meat. Preservatives, colorings and artificial flavors. Fried foods. Sugar. Cholesterol.
U.S. News Nov 13, 1989 p. 89

Are Americans eating a healthier diet? Yes. Palm and coconut oil. Whole milk. Beef. Butter. Change in per-capita sales. 1977-87.
Business Week Oct 9, 1989 p. 116

Pass the vegetables, please. The percentage of people in various regions who said they eat a lot of the following foods.
U.S. News Aug 28, 1989 p. 102

Picking a healthy picnic. Nutritional values compare as follows: Fast-food picnic. Homemade picnic.
U.S. News Aug 7, 1989 p. 62

Today's menu. Better menu. Fat. Cholesterol. Carbohydrates. Protein.
Time Mar 13, 1989 p. 51

The changing American diet. [Recommended calories, protein, and carbohydrates in] 1977. Recent. Goal.
U.S. News Aug 8, 1988 p. 59

Diet-health trends. Shoppers who "pay attention" to ingredient lists. Shoppers who use lists to "avoid" or limit.
FDA Consumer May 1988 pp. 7–8

NUTS

What's in a nut. Fat type and content of some popular nuts.
U.S. News Jun 5, 1989 p. 69

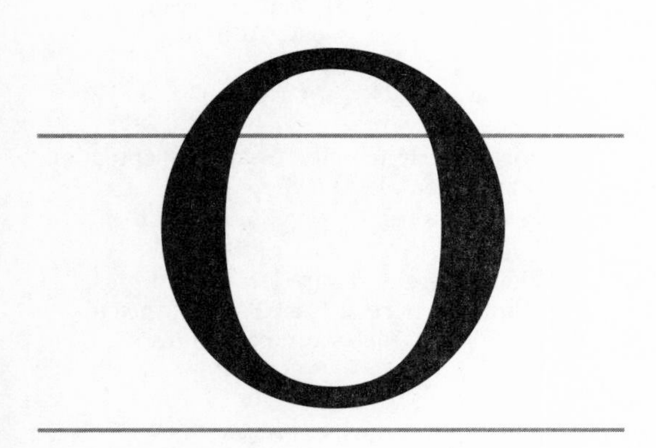

OCCUPATIONS

See JOBS

OCEANOGRAPHY

Probing the deep. To understand the ocean processes that affect climate and mineral deposits, researchers need to sample the seas on an unprecedented scale. A variety of new tools is helping.
U.S. News Aug 21, 1989 p. 50

OFFICE BUILDINGS

The office glut gets worse. Downtown office vacancy rates. 1987-1991.
Business Week Jun 17, 1991 p. 86

The office glut goes on. Home sales slide, but prices keep rising. Sales of existing one-family homes. Median sale price of existing one-family homes. 1985-1989.
Business Week Dec 25, 1989 p. 140

It all comes down to real estate. Comparative look at what you'll have to spend for office space in 10 major urban areas. [City]. Office-space rental, per square foot. 1987-1989.
Newsweek Oct 2, 1989 p. 8

Tight quarters. Annual rent per square foot in prestigious downtown areas. Employees per 1,000 square feet. Tokyo. New York. Washington. Chicago. Los Angeles.
U.S. News Jan 30, 1989 p. 59

Office vacancies remain high. Values lag behind inflation. Home prices keep rising, although far from evenly. Median sale price of existing one-family homes. Northeast. Midwest. 1983-1988.
Business Week Dec 26, 1988 p. 172

OFFICE EMPLOYEES

The compact workplace. Here's the average work space per worker in downtown areas of the following cities. Average square feet per office worker. Average rental cost per square foot.
U.S. News Sep 18, 1989 p. 77

OFFICE EQUIPMENT AND SUPPLIES

A small boom. Personal computers. Fax machines. Cellular phones. Home copiers.
Newsweek Apr 24, 1989 p. 59

OIL INDUSTRY

See PETROLEUM INDUSTRY

OIL PRICES

Price spike. Retail gasoline. Retail home-heating oil (cents per gallon). 1989.
U.S. News Jan 22, 1990 p. 49

See Also PETROLEUM PRICES

OIL SPILLS

[Oil spills in the Persian Gulf.] Iraq. Kuwait. Saudi Arabia. Oil spills as of Feb. 2. Projected up to Feb. 14-15. Map.
Time Feb 11, 1991 p. 41

Iraq's crude weapon. Iraqi troops torched oilfields and released more than 6 million barrels of crude. Map.
Newsweek Feb 4, 1991 p. 37

As soon as one leak stops. Oil spills in the United States. Foreign oil spills. Date. Size of oil spill in gallons.
Newsweek Feb 19, 1990 p. 78

The spill. Where the oil went. The spread. Maps.
Newsweek Sep 18, 1989 p. 53

Drop by drop: A box score [damage and costs].
Newsweek Sep 18, 1989 p. 55

Legacy of a spill: The damage done...and not done. First 48 hours. By 14 days. After 14 days.
U.S. News Sep 18, 1989 p. 64

Portrait of a state. How the population has followed Alaska's fortunes. Damage from the Exxon Valdez. Map.
Time Apr 17, 1989 p. 59

'A disaster of enormous potential.' In one week, the oil had spread over 900 square miles, coating thousands of marine mammals, birds, fish, and threatening hatcheries and parklands. Map.
Newsweek Apr 10, 1989 p. 55

Oil and troubled waters. Largest spills. Offshore-well explosions. Map.
U.S. News Apr 10, 1989 p. 48

OIL WELL DRILLING

See PETROLEUM PROSPECTING

OILS AND FATS

Healthier frying. Proportions of fatty acids in cooking oils and fats [by type of oil or fat].
U.S. News Oct 31, 1988 p. 82

OLDER AMERICANS

Employment status of civilian labor force by age, sex and race. Employed—age 45-64. Unemployed—age 45-64. Black males. White males. Black females. White females. 1989.
Black Enterprise Dec 1990 p. 41

Time off to help out. Hours of work missed in one year [to care for elderly relative].
Newsweek Jul 16, 1990 p. 51

Who will care for us? Aging America. As a percent of total population. Children under age 17. Adults over age 65. 1980-2040.
Newsweek Jul 16, 1990 p. 53

Choosing death. Elderly suicides. People 65 years or older. National average. 1977-1987.
Newsweek Jun 18, 1990 p. 46

The elderly are really poorer than others. Percent living in poverty. Rate if same poverty criteria were used for elderly as for nonelderly.
Business Week May 21, 1990 p. 32

Who will look after mama? Need. Nursing homes. Cost. At home. Caregivers. Insurance.
Newsweek Mar 12, 1990 p. 73

More elderly people. Share of population 65 and over. 1920-1987.
Business Week Sep 25, 1989 p. 99

Where the elderly are left behind. Counties where at least 17.5% of the residents in 1980 were 65 and over.
Business Week Sep 25, 1989 p. 99

A senior citizens' geography. U.S. population 65 years of age and over. Map.
Newsweek Jul 17, 1989 p. 6

Their money's worth. Percentage of consumers age 55 and older who told pollsters they are charged a fair price by [various businesses and services].
U.S. News May 15, 1989 p. 73

The aging of America. People 50 or over. Percent of U.S. populaton. 1988-2025.
Business Week Apr 3, 1989 p. 66

Seniors: Big spenders. Purchases by consumers over 50 [by category of expenditure].
Business Week Apr 3, 1989 p. 67

The aging of America. Percent of the population 65 or older. 1940. 1990. 2040. 2090.
FDA Consumer Oct 1988 p. 22

Common ailments in older Americans. Percent of American men and women aged 65 or older who suffer from one or more of six common ailments.
FDA Consumer Oct 1988 p. 23

Older people are an untapped resource. Population 55 and older. 1980-1995.
Business Week Sep 19, 1988 p. 118

A baby-boomer future. People who will reach age 65, year by year. 1990-2050.
U.S. News Aug 15, 1988 p. 65

[Social Security]. Population. Life expectancy. Federal spending. 1950, 1985, 2020.
Time Feb 22, 1988 p. 70

OLDER AMERICANS - HOME CARE

Who are the caregivers? Those who provide unpaid care for frail old people. Wives. Daughters. Other women. Husbands. Sons. Other men.
Ms. Mar 1988 p. 67

OLDER WOMEN - ECONOMIC CONDITIONS

The "Golden Years." Percent of women 65 and over below poverty level. All women. Hispanic. Black. White. Separated. Widowed. Divorced.
Ms. Oct 1988 p. 68

OLYMPIC GAMES

Olympic money. 1996 Olympics in Atlanta. Expenditures. Revenues. Estimated profit. Jobs created.
Newsweek Oct 1, 1990 p. 8

OPIUM

The road to America: The global smuggling maze. Suspected opium production. Map.
U.S. News Aug 14, 1989 p. 32

ORANGE JUICE INDUSTRY - MARKET SHARE

Tropicana pours it on. Market share, by volume, of total U.S. orange juice market. 1985, 1990.
Business Week May 13, 1991 p. 48

Way behind with Citrus Hill. Share of U.S. orange juice market. Minute Maid. Tropicana. Citrus Hill.
Business Week Jan 23, 1989 p. 38

ORGAN TRANSPLANTS

See TRANSPLANTATION OF ORGANS, TISSUES, ETC.

ORGANIZATION OF PETROLEUM EXPORTING COUNTRIES

OPEC's new windfalls. Revenues will be quickly swallowed up by the staggering costs of the gulf crisis and foreign debts. Libya. Nigeria. United Arab Emirates. Venezuela. Iran. Saudi Arabia.
U.S. News Jan 14, 1991 pp. 42–43

OPEC production may be nearing its limit. Maximum sustainable capacity. Annual production. 1985-1991.
Business Week Dec 31, 1990 p. 45

Where the oil is. Daily average output in millions of barrels. Estimated reserves in billions of barrels. (OPEC). Other oil-producing nations [by country].
Scholastic Update Oct 5, 1990 p. 20

OPEC is running out of wiggle room. OPEC's spare capacity as a percent of global demand. 1979-1990.
Business Week Apr 2, 1990 p. 32

OPEC is running out of slack. Sustainable capacity. Production. 1979-1990.
Business Week Feb 12, 1990 p. 36

ORGANIZED LABOR

See LABOR UNIONS

OZONE

An atmospheric Catch-22. There are two crises in the skies. [Results of] more CFCs. [Results of] less CFCs. Diagram.
Newsweek Nov 4, 1991 p. 49

Danger: Sunlight. [Consequences of] reduction in ozone.
Newsweek Apr 15, 1991 p. 64

Saving the ozone layer. Atmospheric chlorine. Without cooperation from key developing nations. With cooperation from developing nations. 1985-2065.
Business Week Jul 2, 1990 p. 58

Cutting down on ozone-eating chlorine. Cumulative reductions in parts per billion. Montreal Protocol only. [Montreal Protocol plus other methods]. Phasing out CFCs. Phasing out carbon tetrachloride. Phasing out methyl chloroform. 1980-2100.
Business Week Jun 12, 1989 p. 56

A tattered umbrella. The ozone layer over Antarctica. Diagram.
Newsweek Jul 11, 1988 p. 21

The dirty dozen. The average number of days per year that ozone levels exceeded federal health standards in 12 [U.S.] cities.
U.S. News May 16, 1988 p. 10

Ozone's thinning shield. Percentage change in ozone levels at these latitudes from 1969 through 1986. Diagram.
U.S. News Mar 28, 1988 p. 10

Wayward cities. Some 60 areas failed to meet the standard for ozone, exceeding it by up to 148 days last year.
Newsweek Jan 4, 1988 p. 62

See Also GREENHOUSE EFFECT

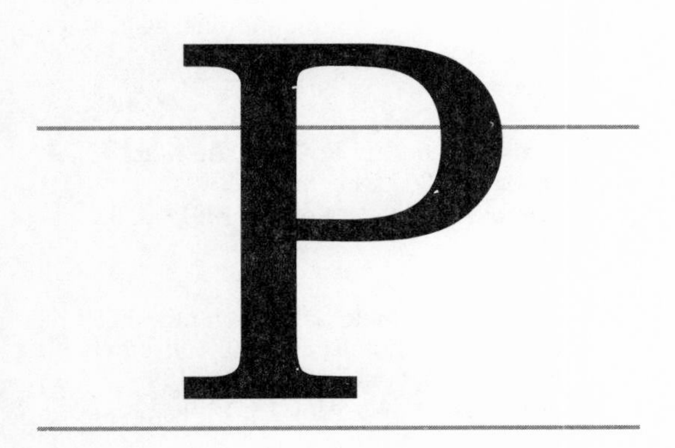

PACIFIC RIM COUNTRIES

Pacific overtures. Free-for-all. Economic muscle in the Pacific Rim. Economic growth rate. Per capita income. Economic strength. [By country.]
U.S. News Nov 20, 1989 p. 65

Growth. Increase in per capita GNP. Average annual percent change, 1973-86. U.S. Japan. China. Hong Kong. Singapore. S. Korea. Per capita GNP, 1986.
Newsweek Feb 22, 1988 p. 44

Trade. Increase in exports. 1980-86. Market share. 1986. U.S. Japan. China. Hong Kong. Singapore. S. Korea. Taiwan.
Newsweek Feb 22, 1988 p. 45

Defense. Military spending as a percentage of GNP, average, 1982-86. U.S. Japan. China. Hong Kong. Singapore. S. Korea. Taiwan.
Newsweek Feb 22, 1988 p. 54

PACS

See POLITICAL ACTION COMMITTEES

PAIN

The pain pathway. The impulses transmitting the pain sensation. Diagram.
FDA Consumer Jul 1989 p. 30

PAKISTAN

Pakistan. Population. GDP. Current account balance. Religion. Adult literacy rate.
U.S. News Aug 29, 1988 p. 76

PALESTINIAN ARABS

Palestinians comprise [percentage of labor force and various jobs in Israel].
Time Nov 26, 1990 p. 47

Facts on the ground. Estimated population of the West Bank and Gaza. Jewish. Palestinian. 1967-1987. Settlements founded between 1967 and mid-1977. Settlements as of April, 1987. Map.
U.S. News Feb 22, 1988 p. 53

PAN AMERICAN FLIGHT 103 DISASTER, 1988

Deadly cargo. Both the wreckage and the bodies of victims provide clues. Anatomy of a bomb. Diagrams.
Newsweek Jan 9, 1989 p. 29

PANAMA

The flow goes on. Average monthly cocaine seizures in Panama. 1989-1991.
Time Aug 26, 1991 p. 18

Americans in Panama. Top five users [of the Panama Canal]. Goods through the canal in millions of tons.
Time May 22, 1989 p. 44

PANAMA-AMERICAN INVASION, 1989-1990

U.S. battle deaths. Killed in action. Vietnam. Lebanon. Grenada. Panama. Persian Gulf.
Newsweek Mar 11, 1991 p. 28

How Operation Just Cause "decapitated" Panama's Defense Forces, then bogged down in scattered, and surprisingly tough, street fighting. Friday, 12/15 - Wednesday [12/20]. Map.
Time Jan 1, 1990 p. 24

PAPER

See FOREST PRODUCTS INDUSTRY

PARENTAL LEAVE

The company we keep. Number of paid weeks and percent of normal pay. Canada. France. W. Germany. Italy. Japan. Sweden.
Black Enterprise Oct 1988 p. 33

PARENTING

See CHILD REARING

PATENTS

Foreigners gain in U.S. patents. Total number of patents granted. Japan. Other. U.S. 1978-1990.
Business Week Dec 2, 1991 p. 114

The top 10 in patents [by company]. Number of patents granted in fiscal year 1990.
Business Week Dec 2, 1991 p. 115

Patent laws: The U.S. vs. Japan [differences].
Business Week Aug 19, 1991 p. 48

Top 10 companies receiving patents. 1978, 1988.
Scholastic Update Sep 7, 1990 p. 21

Bright ideas. Percentage of total U.S. patents granted. U.S. Japan. West Germany. 1975, 1989.
U.S. News Jul 16, 1990 p. 24

An imbalance of patents. U.S. patents issued to residents of foreign countries in 1987. Companies granted U.S. patents in 1987.
U.S. News May 16, 1988 p. 79

PATRIOT (MISSILE)

How the Patriot works. Diagrams.
U.S. News Feb 4, 1991 p. 48

PEACE CORPS

Peace Corps enrollment in 1961. In 1991. Number of people taught English by the Peace Corps. Volunteers over the years. [Other selected statistics.]
U.S. News Aug 5, 1991 p. 8

The Peace Corps starts climbing back. Thousands of volunteers. 1961-1993.
Business Week Aug 22, 1988 p. 63

PEARL HARBOR, ATTACK ON, 1941

Pearl Harbor. December 7, 1941. First wave. Second wave. Warships sunk or severely damaged. Map.
Time Dec 2, 1991 p. 35

Bloody Sunday: The lines of attack. Major damage to fleet. Sunk and unsalvageable. Hit but salvageable.
Newsweek Nov 25, 1991 pp. 34–35

Across the ocean: Route of the raid. November 26 - December 7, 1941.
Newsweek Nov 25, 1991 p. 37

PENSIONS

Seeing red. Uncle Sam's potential bill from underfunded company retirement plans. Pension shortfall. 1980-1990.
Newsweek Nov 25, 1991 p. 50

Where's my pension? Percent of work force in employer-financed pension plans. All workers. Men under 35. 1979, 1983, 1988.
Business Week Jul 30, 1990 p. 46

Pension fund assets are increasing mightily. Pension fund assets. Trillions of dollars. 1980-1989.
Business Week Nov 6, 1989 pp. 154–155

Washington slows the rush to tap [pension funds]. Amounts captured by companies through closing pension plans. 1980-1989.
Business Week Nov 6, 1989 pp. 154–155

The pension cup is overflowing. Pension plans with assets equal to 100% or more of the present value of accrued benefits. 1980-1988.
Business Week Jul 3, 1989 p. 31

The pension burden gets heavier. France. W. Germany. Italy. Britain. U.S. Japan. Average annual cost per working-age adult. 1950-2040.
Business Week Mar 13, 1989 p. 55

Pension headaches. Private pensions. % of population, 65 and over, receiving pensions. Average amount received per year. 1976, 1987.
Ms. Dec 1988 p. 85

Where do the pension funds invest and how big is the pension nest egg. 1981-1987.
Black Enterprise Apr 1988 p. 52

PENTAGON PROCUREMENT SCANDAL

Scandal watch. Procurement fraud investigations have increased sharply. Fiscal years. Cases under investigation. Procurement fraud. Bribery.
Newsweek Jun 27, 1988 p. 21

PERSIAN GULF WAR, 1991

Wearing down the enemy. Damage assessment [before ground campaign]. Tanks. Armored personnel carriers. Artillery. Planes. Prisoners of war.
Newsweek Mar 4, 1991 p. 41

[Kuwait and Kuwait City]. Refineries. Oil fields. Pipelines. Roads. Area of smoke. Map.
Time Mar 4, 1991 p. 35

Desert Storm. Week 6. Scud launches. Personnel and equipment. Iraq. Allies. Scorching the oil fields. Military engagements. Map.
U.S. News Mar 4, 1991 pp. 28–29

How badly hurt is Saddam's war machine? [Damage assessment of] armored personnel carriers. Artillery. Tanks. Planes. Prisoners of war. Bridges.
Newsweek Feb 25, 1991 p. 18

The Gulf War: Week five. Map.
Newsweek Feb 25, 1991 p. 28

War action, Feb. 11-17. Diplomacy and bombs. Maps.
Time Feb 25, 1991 p. 20

Desert Storm. Week 5. Scud launches. Bombing the bunkers. Weather extremes. Map.
U.S. News Feb 25, 1991 pp. 24–25

The Gulf War: Week four. U.S. and its allies stepped up the bombardment of Iraq and Kuwait. Border skirmishes continued. Map.
Newsweek Feb 18, 1991 p. 30

Links in the chain of command. How orders get from George Bush to the troops in the field.
Newsweek Feb 18, 1991 p. 42

Desert Storm. Week 4. Fuel-air explosives. Scud launches. Bridges. Oil slicks. Ground troops.
U.S. News Feb 18, 1991 pp. 24–25

The wages of war. Companies that pay [employees on reserve duty in the Gulf]. Full salary. Difference only. How long.
U.S. News Feb 18, 1991 p. 67

War action. Jan. 28 - Feb.3. Desert fighting. Naval action. Targets in Iraq. Map and summary.
Time Feb 11, 1991 pp. 24–25

Patriot missiles fired in gulf war. Cost per missile. Daily cost to supply U.S. gulf troops with food. Average pieces of mail delivered. [Other selected statistics.]
U.S. News Feb 11, 1991 p. 9

The Gulf War: Week two. Aerial battering of Iraq and Kuwait. Troop distribution at the center of Desert Storm. Active air base. Other air base. Scud launching area. Oil fire. Map.
Newsweek Feb 4, 1991 p. 32

War action, Jan. 21-27. The Tornado IDS. Targets in Iraq. Saddam's air-force bunkers. Naval action. U.S. compared with Gulf area. Maps and summary.
Time Feb 4, 1991 pp. 22–23

War action, Jan. 16-20. Targets in Iraq. Allied ships. Allied combat aircraft. Maps and summary.
Time Jan 28, 1991 pp. 20–21

PERSIAN GULF WAR, 1991 - AERIAL OPERATIONS

Updating the 'Big Ugly Fellow' [B52]. Major modifications. 1959-1984. Where the B-52s are flying from. Diagram.
Newsweek Feb 18, 1991 p. 47

Pinpointing the target. Present. Vietnam. World War II. Sorties needed to destroy target. Bombs needed to destroy target.
U.S. News Feb 11, 1991 p. 27

Preparing the battlefield. AH-64s. A-10s. Harriers.
U.S. News Feb 11, 1991 p. 29

Rescues: High risk, high reward. [Steps in a rescue]. Aircraft down. Team flies in. The pickup. Diagram.
Newsweek Feb 4, 1991 p. 55

Controlling the skies. Targets: Main operating bases. Other bases. Air defense centers. Weapons plants. Map.
Newsweek Jan 28, 1991 pp. 18–19

Plane with electronic jamming. Clearing a corridor. Missiles. Jamming pod. Chaff dispenser. Infrared flares. Diagram.
Newsweek Jan 28, 1991 p. 21

Bombs away. The air campaign against Iraq has sent more than a dozen types of combat aircraft from the air forces of seven nations. Maps.
U.S. News Jan 28, 1991 pp. 22–23

The high-technology war. Controlling the air battle. Defeating enemy air defenses. Specialized weapons. Diagram.
U.S. News Jan 28, 1991 pp. 30–31

'All the tools in the toolbox.' Allied forces. Iraqi forces.
U.S. News Jan 28, 1991 pp. 33–38

F/A-18A. A-10 Thunderbolt. Tornado GR-1. EF-111A. B-52G. Service. Mission. Combat range. Armament.
Newsweek Jan 21, 1991 pp. 34–35

PERSIAN GULF WAR, 1991 - CAMPAIGNS AND BATTLES

How 43 days of war ended seven months of Iraqi occupation. Aug. 3-Jan. 15. Jan. 16-Feb. 23. Feb. 23. Feb. 25-28. Maps.
Newsweek Mar 11, 1991 pp. 38–39

The 100-hour war, Feb. 24-28. Allied movements. Iraqi targets. Maps.
Time Mar 11, 1991 pp. 17–18

Schwarzkopf's 'Hail Mary.' Jan. 17-Feb. 23. Map.
U.S. News Mar 11, 1991 p. 16

Ground action. Feb. 24-26. 12 hours. 48 hours. Map.
U.S. News Mar 11, 1991 p. 17

The final day. Feb. 27. 80 hours. 100 hours. Map.
U.S. News Mar 11, 1991 p. 17

The great 'cavalry charge.' 24th mechanized infantry division. Map.
U.S. News Mar 11, 1991 p. 36

Week six: Going in on the ground. Allied attacks. Amphibious assault. Map.
Newsweek Mar 4, 1991 p. 22

War action, Feb. 18-24. Map and summary.
Time Mar 4, 1991 p. 24

War action, Feb. 4-10. Military powwow. Naval action. Targets in Iraq. Maps.
Time Feb 18, 1991 p. 22

The Gulf War: Week three. First major ground battles of the war [air war continues]. Map.
Newsweek Feb 11, 1991 p. 22

PERSIAN GULF WAR, 1991 - CASUALITIES

U.S. battle deaths. Killed in action. Vietnam. Lebanon. Grenada. Panama. Persian Gulf.
Newsweek Mar 11, 1991 p. 28

PERSIAN GULF WAR, 1991 - COST

The Gulf War: Final payments. Six biggest contributors [excluding U.S.]. Country. Totals. Pledged. Payments. Percent paid.
Newsweek Jun 17, 1991 p. 6

Weapons of war. Some of the main weapons in the U.S. and Iraqi arsenals. What do they do? How much do they cost?
Scholastic Update Mar 22, 1991 pp. 4–7

Kuwait's tab for the war. Total. Kuwait's financial reserves.
Business Week Mar 11, 1991 p. 34

If Allied pledges of support are made good, U.S. war costs will be more than covered. U.S. cost. Allied pledges. Where the pledges come from.
Time Mar 11, 1991 p. 50

Fighting numbers. The human cost. The war bill. The weapons. The troops. War miscellany.
U.S. News Mar 11, 1991 p. 74

Patriot missiles fired in gulf war. Cost per missile. Daily cost to supply U.S. gulf troops with food. Average pieces of mail delivered. [Other selected statistics.]
U.S. News Feb 11, 1991 p. 9

Still cheap, relatively. Gulf War (estimate). World War II. Vietnam War. World War I. Korean War. Costs in 1991 dollars.
Newsweek Feb 4, 1991 pp. 63–65

[Cost of a] Patriot. Tomahawk. F-15E.
Newsweek Feb 4, 1991 p. 65

The price of war. U.S. daily cost of running the war. Flight time. "Smart bomb." Patriot missile. Tomahawk cruise missile.
Time Feb 4, 1991 p. 22

Running up the tab. Short war. Long war.
Time Feb 4, 1991 p. 57

Where the money is going. Combat aircraft. Missiles. Ground forces. Personnel expenses.
U.S. News Feb 4, 1991 pp. 46–47

The allies ante up. Pledged to U.S. Contributed to date [by country].
Business Week Jan 21, 1991 p. 34

OPEC's new windfalls. Revenues will be quickly swallowed up by the staggering costs of the gulf crisis and foreign debts. Libya. Nigeria. United Arab Emirates. Venezuela. Iran. Saudi Arabia.
U.S. News Jan 14, 1991 pp. 42–43

Costly operation. Projected costs for Desert Shield through end of 1990. Defense Cooperation Account for U.S. Desert Shield [by country]. Pledged for 1990. Cash received. Material assistance.
Time Dec 24, 1990 p. 38

The checks are in the mail. Promised aid to Operation Desert Shield. Pledged. Paid. Kuwait. Saudi Arabia. Germany. South Korea. Japan.
U.S. News Nov 26, 1990 p. 28

Order of battle. Costs of Desert Shield through September 1991. Item. Total cost (in millions).
Newsweek Sep 17, 1990 p. 31

Paying the price. The cost of containing Saddam Hussein. Estimated costs of lost remittances, higher oil prices and reduced trade [by country]. Contributors [by country].
U.S. News Sep 17, 1990 p. 33

The price of peace. Cost of deploying U.S. military forces to the Middle East by September 30 [1990]. Army. Navy. Air Force. Some of the increased costs.
U.S. News Aug 27, 1990 p. 23

PERSIAN GULF WAR, 1991 - DEFENSES

What ever happened to the Republican Guard. Chemical weapons. Air defenses. The front line.
Time Mar 11, 1991 p. 19

Is this where the Guards are hiding? Diagram.
Time Feb 11, 1991 p. 25

Pinpointing the target. Present. Vietnam. World War II. Sorties needed to destroy target. Bombs needed to destroy target.
U.S. News Feb 11, 1991 p. 27

Hunkering in the bunkers. [Iraqi] underground fortifications. [Allies'] B-52G. Diagram.
Newsweek Feb 4, 1991 p. 42

Finding the enemy. Firefinder radar. Tank laser range finder. Diagrams.
U.S. News Feb 4, 1991 pp. 40–41

Lines in the sand. Ground-troop deployments. Iraqi forces in theater. U.S. and allied forces in theater. Map.
U.S. News Jan 28, 1991 p. 24

The high-technology war. Controlling the air battle. Defeating enemy air defenses. Specialized weapons. Diagram.
U.S. News Jan 28, 1991 pp. 30–31

Attacking Iraqi defenses. Iraqi installations. U.S. military equipment.
U.S. News Jan 14, 1991 pp. 26–27

Deploying—and supplying—the troops. Iraqi defenses and allied forces will rely heavily on a network of roads to get supplies to the front. Map.
U.S. News Jan 14, 1991 p. 28

PERSIAN GULF WAR, 1991 - ENVIRONMENTAL ASPECTS

Snuffin' and cappin' in the desert. 520 wells still ablaze in Kuwait. [Three-step plan to extinguish them]. Diagram.
Newsweek Mar 25, 1991 p. 30

[Oil spills in the Persian Gulf.] Iraq. Kuwait. Saudi Arabia. Oil spills as of Feb. 2. Projected up to Feb. 14-15. Map.
Time Feb 11, 1991 p. 41

Iraq's crude weapon. Iraqi troops torched oilfields and released more than 6 million barrels of crude. Map.
Newsweek Feb 4, 1991 p. 37

PERSIAN GULF WAR, 1991 - EQUIPMENT AND SUPPLIES

Anatomy of an arms deal. Sale of cluster bombs to Iraq involved some odd routes. Map.
Newsweek Apr 8, 1991 p. 28

U.S. arsenal: Planned obsolescence. U.S. weapons used in the Gulf conflict. Weapon system. Last year of procurement. Prewar inventory. Years until replacement on line.
Newsweek Mar 18, 1991 p. 43

Keeping the 24th [Mechanized Infantry Division] rolling. Fuel. Water. Food. Ammo. Medical. Misc. Trucks needed to carry these supplies.
U.S. News Mar 11, 1991 p. 41

When it absolutely, positively has to be there. Daily mail deliveries. Meals served. Trucks. Ships.
Business Week Mar 4, 1991 pp. 42–43

Who built the arsenal? [War materials supplied to Iraq by] France. Soviet Union. Italy. West Germany. United States.
Newsweek Feb 4, 1991 p. 57

PERSIAN GULF WAR, 1991 - JOURNALISTS

Media workout. To travel with U.S. troops in the Persian Gulf, newshounds must pass this U.S. military fitness test. Age. Men. Women. Sit-ups (in 2 minutes). Push-ups (in 2 minutes). 1 1/2-mile run.
U.S. News Jan 21, 1991 p. 84

PERSIAN GULF WAR, 1991 - MILITARY STRENGTH

Forces in the Gulf. Iraq. U.S. Allies. Troops. Tanks. Armored personnel carriers. Artillery. Combat aircraft. Helicopters. Warships.
Scholastic Update Mar 22, 1991 p. 6

Personnel and equipment. Iraq. Allies. Troops. Tanks. Artillery. APCs. Helicopters. Aircraft. Ships. Before war. Losses.
U.S. News Mar 11, 1991 p. 16

U.S. firepower in the Gulf. Deployed. Total available. Airplanes. Tanks. Marines. Army. Aircraft carriers.
Time Mar 4, 1991 pp. 38–39

The allies' firepower. Air. Land. Sea.
Newsweek Feb 18, 1991 p. 42

Saddam's war machine. Fighter jets, bombers, tanks, high-tech weaponry, chemical and biological weapons.
Newsweek Feb 18, 1991 p. 42

Ground assault. Iraqi forces in theater. U.S. and allied forces in theater. Map.
U.S. News Feb 4, 1991 p. 34

Lines in the sand. Ground-troop deployments. Iraqi forces in theater. U.S. and allied forces in theater. Map.
U.S. News Jan 28, 1991 p. 24

'All the tools in the toolbox.' Allied forces. Iraqi forces.
U.S. News Jan 28, 1991 pp. 33–38

Might vs. might. United States. Allies. Iraq. Troops. Tanks. Armored personnel carriers. Artillery. Helicopters.
U.S. News Jan 14, 1991 p. 29

PERSIAN GULF WAR, 1991 - PERSONALITIES

Top dollar. How much would key players in Gulf war fetch on the lecture circuit?
Newsweek Mar 18, 1991 p. 6

PERSIAN GULF WAR, 1991 - SOLDIERS

Women at war. Operation Desert Storm. Total number of females. Number who died. Number of POWs.
Newsweek Aug 5, 1991 p. 25

Blacks in uniform. Percent of blacks in: Forces. Desert Storm. Army. Navy. Marine Corps. Air Force.
Newsweek Mar 11, 1991 p. 54

By the numbers. [Organization of] the Marines and the U.S. Army.
Newsweek Feb 18, 1991 p. 41

Arms and the men. An American Marine and his equipment. Diagram.
Newsweek Feb 18, 1991 p. 42

PERSIAN GULF WAR, 1991 - TECHNOLOGY

Clearing a pathway. Aerial sorties. M-9 Ace. Mine-clearing line charges. Ditches. Fascine bridges. Mines.
Newsweek Feb 11, 1991 p. 26

Exotic weapons—and their limitations. Reconnaissance. Tactical weapons.
Business Week Feb 4, 1991 pp. 38–39

Finding the enemy. Firefinder radar. Tank laser range finder. Diagrams.
U.S. News Feb 4, 1991 pp. 40–41

High-tech payoff. Costly arms face their first combat use. Purpose. Uses. Distinction. Cost.
Time Jan 28, 1991 pp. 30–31

The high-technology war. Controlling the air battle. Defeating enemy air defenses. Specialized weapons. Diagram.
U.S. News Jan 28, 1991 pp. 30–31

The games radars play [electronic warfare]. "Deception" jammers. "Noise" jammers. Decoys. Chaff. Diagram.
U.S. News Sep 10, 1990 p. 44

PERSIAN GULF WAR, 1991 - WEAPONS

U.S. firepower in the Gulf. Deployed. Total available. Airplanes. Tanks. Marines. Army. Aircraft carriers.
Time Mar 4, 1991 pp. 38–39

'Discount A-bomb.' [Four-step diagram of how the CBU cluster bomb works.]
Newsweek Feb 25, 1991 p. 27

The path to a target. [Four-step diagram of how the Tomahawk cruise missile works.]
Newsweek Feb 18, 1991 pp. 40–41

Where Saddam's best weapons come from. [By country of origin.]
Time Feb 11, 1991 p. 35

Exotic weapons—and their limitations. Reconnaissance. Tactical weapons.
Business Week Feb 4, 1991 pp. 38–39

Tomahawk. The missile's route is planned days or months in advance. Diagram.
Time Feb 4, 1991 p. 46

The weapons of the ground war. Allied forces.
U.S. News Feb 4, 1991 p. 42

How the Patriot works. Diagrams.
U.S. News Feb 4, 1991 p. 48

Patriot vs. Scud [how Patriot missiles track and destroy Scud missiles].
Newsweek Jan 28, 1991 p. 20

The Tomahawk. Diagram.
Newsweek Jan 28, 1991 p. 22

'Smart' bombs at work. GBU-15(V)2 glide bomb. Weight. Length. Guidance.
Newsweek Jan 28, 1991 p. 23

High-tech payoff. Costly arms face their first combat use. Purpose. Uses. Distinction. Cost.
Time Jan 28, 1991 pp. 30–31

PERSONAL COMPUTERS

See COMPUTERS, PERSONAL

PESTICIDES

Heartier bugs. Number of pesticide-resistant species. Insects. Plant pathogens.
U.S. News Apr 23, 1990 p. 62

PESTICIDES - HEALTH ASPECTS

Suspect sprays. Pesticides. Some of the crops affected. Potential hazards.
Time Mar 27, 1989 p. 29

Minimizing the risks from nature's bounty. Food. Predicted cancers per 1 million people. What to do.
Newsweek Jan 30, 1989 p. 75

FDA's monitoring for pesticide residues. Number of food samples tested. Import. Domestic. 1982-1987.
FDA Consumer Oct 1988 p. 11

PETROCHEMICALS

The pervasive petrochemical. Petrochemicals are squeezed out of each barrel of oil. Here are some of the products.
U.S. News Oct 8, 1990 p. 59

PETROLEUM - SOVIET UNION

Proposed oil deals. Despite problems, a bonanza looms. Map.
Business Week May 20, 1991 p. 60

Running on empty? The Soviet Union is the world's largest petroleum producer. [Map showing location of oil fields and gas fields in Soviet Union.]
U.S. News Feb 5, 1990 p. 46

PETROLEUM CONSUMPTION

Barrels of oil used per person in a year in leading industrial countries.
Scholastic Update Oct 5, 1990 p. 21

Gas pains. Total energy consumption. Amount coming from petroleum. Residential. Transportation. Industrial. Electric utilities.
U.S. News Sep 10, 1990 p. 68

Transportation soaks up the most oil. Total U.S. oil consumption. Industrial. Residential & commercial. Electric generation. Transportation. Share of transportation's 1988 oil consumption.
Business Week Aug 27, 1990 p. 29

The world's economic engine gets better mileage. B.t.u.'s used to generate each dollar of real gross domestic product. Canada. United States. Britain. West Germany. France. Japan. Percentage change (1978-88).
U.S. News Aug 27, 1990 p. 37

Industrial democracies are less reliant on oil. Consumption per unit of output in 23 OECD economies. 1980-1990.
Business Week Aug 20, 1990 p. 29

Energy eaters. Per capita energy consumption (in oil equivalent, barrels per year): U.S. Sweden. Soviet Union. West Germany. Japan. China. India. Kenya. 1970, 1989.
U.S. News Apr 23, 1990 p. 70

PETROLEUM INDUSTRY

The consequences of $20 oil [oil production, oil demand, cost of products]. 1990-1991.
Business Week Mar 11, 1991 p. 37

The U.S. tanker shortage. U.S.-flagged. World supply.
Business Week Jan 28, 1991 p. 35

Another gusher for the oil industry. Total net income of 13 major domestic oil companies. 1986-1991.
Business Week Jan 14, 1991 p. 109

The struggle to control oil [chronology 1859-1990].
U.S. News Oct 8, 1990 p. 60

This time, an oil shock might not be so shocking. Average world production. Average world consumption. Petroleum inventories. Government strategic oil reserves.
Business Week Aug 20, 1990 p. 27

Oil, money and power. Reserves of crude oil. Percentage of world crude-oil production. Primary pipelines.
U.S. News Aug 13, 1990 pp. 22–23

Petropower shift. Ranking of the world's top oil companies. 1980, 1988.
Time Jul 16, 1990 p. 43

The leap in oil imports. Energy-related petroleum products. 1989-1990.
Business Week Apr 2, 1990 p. 27

Average monthly prices, West Texas intermediate crude. Worldwide exploration and production expenditures. Average daily rates for jackup rigs.
Business Week Mar 12, 1990 pp. 80–81

Springing a leak. Crude oil production in millions of bbl. a day. Total. Non-OPEC. OPEC. 1973-1987.
Time May 9, 1988 p. 67

Turning the corner on prices and prospecting. Price per barrel of imported oil. Drilling rigs in operation in U.S. 1980-1988.
U.S. News Mar 28, 1988 p. 46

PETROLEUM INDUSTRY - SOVIET UNION

Rusty rigs. Soviet oil production. 1980-1991.
U.S. News Sep 16, 1991 p. 47

Soviet oil exports plunge. 1986-1991.
Business Week Feb 11, 1991 p. 46

PETROLEUM PRICES

[Percent] increase in the price of crude oil. % increase in the price of gasoline. Week before the invasion of Kuwait. Week ending Oct. 26.
Time Nov 5, 1990 p. 52

Gas prices: Well off their peak. Price per gallon of gasoline, in 1990 dollars. 1970-1990.
Business Week Aug 27, 1990 p. 28

Who's gouging whom? Gas stations blame the oil companies for higher prices; the companies point to the spot market. A guide to help you decide. Chart.
Newsweek Aug 20, 1990 pp. 34–35

Oil's roller-coaster price ride. Price per barrel of crude oil. 1970-1990.
U.S. News Aug 13, 1990 p. 25

Oil in millions of bbl. per day. U.S. production. U.S. imports. Monthly closing price per bbl. of West Texas Intermediate. 1986-1989.
Time Apr 3, 1989 p. 40

Real oil prices are near a 15-year low. Constant 1982 dollars. Current dollars. 1968-1989.
Business Week Mar 13, 1989 p. 26

Paying through the hose. Average refiner's price per barrel of crude oil. 1981-1988.
U.S. News Nov 28, 1988 p. 57

As oil supplies build. 1987, 1988. Oil prices lose ground. Near contract futures settlement prices. Mar.-Sept. 1988.
Business Week Sep 26, 1988 p. 45

Oil's slippery slope. Price per barrel of domestic crude oil. 1986-1988.
U.S. News Sep 19, 1988 p. 8

Oil prices are sliding. West Texas intermediate crude, price per barrel. 1985-1988.
Business Week Aug 1, 1988 p. 50

Turning the corner on prices and prospecting. Price per barrel of imported oil. Drilling rigs in operation in U.S. 1980-1988.
U.S. News Mar 28, 1988 p. 46

PETROLEUM PRODUCTION

OPEC production may be nearing its limit. Maximum sustainable capacity. Annual production. 1985-1991.
Business Week Dec 31, 1990 p. 45

Where the oil is. Daily average output in millions of barrels. Estimated reserves in billions of barrels. (OPEC). Other oil-producing nations [by country].
Scholastic Update Oct 5, 1990 p. 20

Declining market share. Saudi oil production as share of OPEC output. 1978-1989.
Business Week Sep 17, 1990 p. 30

If Iraq can't sell the oil it controls, others could take up the slack. Country. Production. Exports.
Business Week Aug 20, 1990 p. 26

Oil, money and power. Reserves of crude oil. Percentage of world crude-oil production. Primary pipelines.
U.S. News Aug 13, 1990 pp. 22–23

America's oil patch. Percentage of U.S. domestic crude-oil production in 1989 from: Alaska. Offshore. Onshore.
U.S. News Jul 9, 1990 p. 35

OPEC is running out of slack. Sustainable capacity. Production. 1979-1990.
Business Week Feb 12, 1990 p. 36

Outside OPEC, oil production will stagnate. Free world, five biggest non-OPEC oil producers. Millions of barrels per day. 1985-1992.
Business Week Sep 25, 1989 p. 34

Oil in millions of bbl. per day. U.S. production. U.S. imports. Monthly closing price per bbl. of West Texas Intermediate. 1986-1989.
Time Apr 3, 1989 p. 40

Springing a leak. Crude oil production in millions of bbl. a day. Total. Non-OPEC. OPEC. 1973-1987.
Time May 9, 1988 p. 67

PETROLEUM PROSPECTING

Drilling deeper under the sea. Fixed platform. Tension leg platform. Diagram.
U.S. News Jun 10, 1991 p. 53

Drilling overtime. United States. Saudi Arabia. Production per well. Producing wells. Proven oil reserves.
U.S. News Oct 8, 1990 pp. 56–57

Drilling down. Average number of oil and gas rigs in operation in the U.S. 1970-1990.
Time Aug 27, 1990 p. 47

A stepped-up search for new supplies. Exploration and development investment outside the U.S. by the 14 largest oil companies. 1984-1989.
Business Week Jul 30, 1990 p. 26

Drilling takes a new turn. Horizontal well. Vertical wells. Diagram.
Business Week Jan 22, 1990 p. 57

The hunt for oil nudges back up. Rotary drilling rigs operating in the U.S. 1985-1990.
Business Week Jan 8, 1990 p. 108

A new slant on drilling. A technique known as horizontal drilling is enabling petroleum companies to pump more oil and gas from fractured geological formations. Diagram.
U.S. News Dec 18, 1989 p. 35

Leaner but meaner. After years of decline, U.S. oil companies are beginning to invest in new drilling equipment. Rotary rigs in operation. 1980-1989.
U.S. News Dec 18, 1989 p. 35

Oil exploration comes back. Domestic oil and gas drilling. 1983-1988.
Business Week Jan 11, 1988 p. 121

PETROLEUM REFINERIES

The fine art of refining [four steps]. Diagram.
U.S. News Oct 8, 1990 pp. 58–59

Scraping the bottom of the barrel. Percentage of oil refined into [by products].
U.S. News Oct 8, 1990 p. 58

PETROLEUM SUPPLY

Sources of U.S. oil, by country and region. 1960-1988.
Scholastic Update Apr 19, 1991 p. 5

Before and after the invasion. U.S. oil imports.
Time Feb 25, 1991 p. 64

Sources of oil used in the United States. Average number of barrels a day in millions.
Scholastic Update Oct 5, 1990 p. 21

Foreign oil as a percentage of all oil used in the U.S. 1970-1990.
Scholastic Update Oct 5, 1990 p. 21

America's hidden oil. Inside the U.S. Strategic Petroleum Reserve. Diagram.
U.S. News Sep 10, 1990 p. 69

This time, an oil shock might not be so shocking. Average world production. Average world consumption. Petroleum inventories. Government strategic oil reserves.
Business Week Aug 20, 1990 p. 27

Tapping out. Oil reserves [by region].
Newsweek Aug 20, 1990 p. 40

Petroleum supply. Reserves in millions of bbl. Current production in millions of bbl. per day. Possible increases [various countries]. And demand. 1990 average in millions of bbl. per day. U.S. OECD Europe. Japan.
Time Aug 20, 1990 p. 43

Built on sand. Source [country]. Oil imports as a % of total U.S. consumption, 1989.
Newsweek Aug 13, 1990 p. 29

Drip, drip. Imports of crude oil from Arab OPEC countries in thousands of bbl. per day. U.S. Japan. 1973-1989.
Time Aug 13, 1990 p. 22

More gas use could curb rising oil imports. U.S. oil (barrels per day). U.S. production. Net imports. 1980-1990.
U.S. News Jul 30, 1990 p. 39

OPEC is running out of wiggle room. OPEC's spare capacity as a percent of global demand. 1979-1990.
Business Week Apr 2, 1990 p. 32

Look who's cooking with gas. Known reserves of oil and natural gas. [By world regions].
U.S. News Jul 31, 1989 p. 38

Oil imports surge as output falls. Domestic oil production. Oil imports. 1982-1989.
Business Week Apr 3, 1989 p. 24

As oil supplies build. 1987, 1988. Oil prices lose ground. Near contract futures settlement prices. Mar.-Sept. 1988.
Business Week Sep 26, 1988 p. 45

PETROLEUM TANKERS

Cruising with crude. Running speed. Capacity. Typical crew.
U.S. News Oct 8, 1990 p. 58

PHILADELPHIA (PENNSYLVANIA)

Philadelphia. Population. Unemployment rate. Poverty rate. Bond rating. Outstanding debt per capita. Budget surplus or deficit.
U.S. News Oct 29, 1990 p. 81

PHOTOGRAPHY

Shooting spree. Number of pictures taken per person in: United States. Japan. West Germany. France. Britain.
U.S. News Nov 27, 1989 p. 91

PHYSICAL FITNESS

Fit for life. Physically fit person. Sedentary person. Activity level. Age.
Newsweek May 13, 1991 p. 61

The body's peak performance. Endurance. Strength. Coordination. Twenties. Thirties. Forties. Fifties.
Time Apr 22, 1991 p. 80

The price of fun. Americans parted with more than $13 billion for sports clothing last year. Here is how much the average participant spent. [By activity.]
U.S. News Oct 9, 1989 p. 81

On a roll. Americans in pursuit of fitness and recreation spent $677 million on these balls last year. Here's where the money went.
U.S. News Aug 21, 1989 p. 66

PHYSICIANS

Failing doctors. Science exam most medical students must pass. Failure rate. 1984-1990.
U.S. News Nov 12, 1990 p. 93

Dollars for docs. Average income [by specialty].
U.S. News Dec 4, 1989 p. 11

Where J.D.'s meet M.D.'s. States with the most lawyers. States with the fewest lawyers. States with the most doctors. States with the fewest doctors.
U.S. News Feb 13, 1989 p. 84

PHYSICS

The Nobel experiment. Quarks, the basic building blocks of matter. Diagram.
Newsweek Oct 29, 1990 p. 68

PILGRIMS (NEW ENGLAND COLONISTS)

The real first families. The *Mayflower* carried just over 100 passengers. Here are the surnames of those known to have made the trip.
U.S. News Nov 27, 1989 p. 91

POACHING

From the bloody hands of poachers into the stashes of smugglers, ivory moves across Africa under the noses of often corrupt officals. 1979, 1989. Map.
Time Oct 16, 1989 p. 68

POISONOUS PLANTS

Poison plants. Those most commonly cited. Chart.
U.S. News May 16, 1988 p. 79

POLAND - ECONOMIC CONDITIONS

Adding up the damage. Bulgaria. Czechoslovakia. Hungary. Poland. Romania. Real economic growth. Hard-currency balance. Oil imports.
U.S. News Nov 5, 1990 pp. 55–56

Getting less for their money. Average prices and percent increase. Wages. Refrigerator. Gasoline. Rent. Pork. Bread. Men's shoes.
Time Jun 11, 1990 p. 31

Troubled economies. Yugoslavia. Hungary. Poland. Change in real wages. Inflation. Gross domestic product per capita. 1984-1988.
Scholastic Update Oct 20, 1989 pp. 16–17

Cost of living. The Polish economy suffers from rampant inflation. Increase in consumer-price index since 1980. [By selected Eastern European country.]
Newsweek Sep 5, 1988 p. 40

POLICE

Cops and crime. Cops under fire. [Crime rates and number of police for] Washington D.C. Dallas. San Francisco. The 10 cities with the largest police forces.
U.S. News Dec 3, 1990 p. 34

Cops' lives. Cops under fire. Officers working police agencies. Police killed in the line of duty. 1961-1989. Assaults [on police]. Percent female, black, Hispanic. [Alcoholism]. Divorce rate. Public spending.
U.S. News Dec 3, 1990 p. 44

The police pay spread. Entry-level base pay for 1989.
U.S. News Jan 8, 1990 p. 32

The short arm of the law. Change in homicide rate and police strength in largest U.S. cities since 1978.
U.S. News May 1, 1989 p. 76

POLICE BRUTALITY

Federal investigations of police abuse 1984-1989. State. Number of investigations.
Black Enterprise Jul 1991 p. 13

POLITICAL ACTION COMMITTEES

Cash and carry. Senate campaign costs. House campaign costs. [Number of] political action committees. Biggest PAC givers. 1974-1988.
U.S. News Oct 22, 1990 p. 31

The biggest PACs [Political Action Committees]. Contributions [from various groups] from January 1989 to June 1990. Group. $ in millions.
Scholastic Update Oct 19, 1990 p. 11

POLITICAL PARTIES

Party strength. Total registered voters. Democrats. Republicans. Independents and minority parties.
Scholastic Update Jan 29, 1988 p. 2

POLLUTION

Getting tough with polluters [including fines]. Ashland Oil. Texaco. Ocean Spray Cranberries. Exxon.
Time Mar 12, 1990 p. 54

As soon as one leak stops. Oil spills in the United States. Foreign oil spills. Date. Size of oil spill in gallons.
Newsweek Feb 19, 1990 p. 78

America's most hazardous waste sites (number per state). Map.
Scholastic Update Jan 12, 1990 p. 29

Uncle Sam against the world. The U.S. has 5% of the earth's population, but [has the following % of oil use, nitrogen oxide and carbon dioxide emissions, toxic waste generated].
Time Dec 18, 1989 p. 63

Stiff slaps. Fines for environmental offenses. 1983-1989.
Newsweek Sep 25, 1989 p. 35

The sump of Europe. About half the pollution in the Netherlands comes from other countries. Percentage of acid rain coming from [various countries]. Map.
U.S. News Sep 11, 1989 p. 68

Threats to the environment. Name. Description. Causes. Effects. Chart.
Scholastic Update Apr 21, 1989 p. 3

A guide to some of the scariest things on earth. What environmentalists consider some of the worst problems in the past year [by geographic region]. Map.
Scholastic Update Apr 21, 1989 pp. 4–5

Portrait of a state. How the population has followed Alaska's fortunes. Damage from the Exxon Valdez. Map.
Time Apr 17, 1989 p. 59

Toxins from the tap. Five types of contamination in the U.S. water supply: Substance. Source. Risk.
Time Mar 27, 1989 p. 38

Nowhere to hide. Air pollution. Tropical forests [annual loss]. Desertification. Map.
Newsweek Jul 11, 1988 p. 24

POLLUTION, AIR

A change of atmosphere. Major provisions of the Clean Air Act. Acid rain. Auto exhaust. Toxic emissions. Ozone depletion.
Time Nov 5, 1990 p. 33

Fueling pollution. These gas guzzlers emit the most carbon dioxide. Carbon dioxide emitted per 100,000 miles.
U.S. News Jul 2, 1990 p. 62

Where it hurts [effects of smog on the human body]. Nose, throat, lungs. Smog index. Diagram.
U.S. News Jun 12, 1989 pp. 50–51

Dirt, coast to coast. Average number of days per year in violation of ozone standard. Counties that emit more than 20 million lbs. of toxic air pollutants per year. Map.
U.S. News Jun 12, 1989 p. 52

Transformations in the air. Smog. Acid rain. Major sources of air pollutants. Diagram.
U.S. News Jun 12, 1989 p. 53

An environmental scorecard. Air. Where acid rain falls [by pH number]. Map.
Scholastic Update Apr 21, 1989 p. 16

Pollution progress. Fresher air. Pollutants emitted annually (metric tons). Lead. Ozone pollutants. Nitrogen oxides. Carbon monoxide. Sulfur dioxide. 1978-1987.
U.S. News Apr 17, 1989 p. 29

The toxic 10. Of 2.4 billion pounds of toxics emitted in 1987, more than half came from 10 states. States. Toxic pollutants released.
Newsweek Apr 3, 1989 p. 25

The global carbon budget. Carbon added to atmosphere (in metric tons per year). Carbon removed from atmosphere (in metric tons per year).
U.S. News Oct 31, 1988 p. 58

Hold your breath. Dirty air can do more than make you cough: it can hurt the heart and increase the risk of cancer. What makes air dirty.
Newsweek Aug 29, 1988 p. 47

The dirty dozen. The average number of days per year that ozone levels exceeded federal health standards in 12 [U.S.] cities.
U.S. News May 16, 1988 p. 10

Wayward cities. Some 60 areas failed to meet the standard for ozone, exceeding it by up to 148 days last year.
Newsweek Jan 4, 1988 p. 62

POLLUTION, AIR - EUROPE

Cloud over a continent. Sulfur emissions in thousands of tons [by country].
Time May 28, 1990 p. 42

POLLUTION, MARINE

How Massachusetts plans to clean up the Harbor. Diagram and map.
U.S. News Sep 24, 1990 pp. 58–59

Trashy shores. Here are the pounds of debris collected per mile in [various states].
U.S. News Jan 15, 1990 p. 67

The spill. Where the oil went. The spread. Maps.
Newsweek Sep 18, 1989 p. 53

Drop by drop: A box score [damage and costs].
Newsweek Sep 18, 1989 p. 55

'A disaster of enormous potential.' In one week, the oil had spread over 900 square miles, coating thousands of marine mammals, birds, fish, and threatening hatcheries and parklands. Map.
Newsweek Apr 10, 1989 p. 55

Oil and troubled waters. Largest spills. Offshore-well explosions. Map.
U.S. News Apr 10, 1989 p. 48

Where the fish no longer jump. A look at America's best-known bays [selected harbors] shows that every one is in trouble.
Newsweek Aug 1, 1988 p. 44

A witch's brew of pollutants. Toxic, bacterial and nutrient contamination now affects virtually the entire coastline of the continental United States. Pollutant. Source. Effects.
Newsweek Aug 1, 1988 p. 45

Threats to the ocean. The major long-term hazard is chronic land-based pollution. Diagram.
Time Aug 1, 1988 pp. 46–47

America's troubled coasts. Toxic chemicals. PCB or pesticide contamination. Areas of oxygen depletion. Areas closed to commercial shellfish harvest. Map.
Time Aug 1, 1988 p. 50

See Also OIL SPILLS

POLLUTION, RADIOACTIVE

Neither bird nor plane. The uncertain fate of Cosmos 1900 underscores the growing threat from radioactive space debris. Total launches 1957-1988. Total nuclear launches 1961-1988.
Newsweek Aug 29, 1988 p. 53

POLLUTION, WATER

An environmental scorecard. Water. More sewage treatment. 1972, 1986. Ground water is threatened. Number of states reporting each pollutant as the primary threat to its ground water in 1986.
Scholastic Update Apr 21, 1989 p. 16

POPULATION - AFRICA

A troubled continent. The population explodes. Africa's total population. 1950-2010.
Scholastic Update Mar 23, 1990 p. 13

POPULATION - EUROPE

As Europe's population ages [by country]. 1950-2040.
Business Week Mar 13, 1989 p. 54

POPULATION - INDIA

India's complex society. % of total population. The government job market [quotas].
Time Oct 15, 1990 p. 63

POPULATION - ISRAEL

Palestinians comprise [percentage of labor force and various jobs in Israel].
Time Nov 26, 1990 p. 47

POPULATION - MEXICO

Labor is plentiful. Mexico's population. Millions. 1960-2010.
Business Week Nov 12, 1990 p. 105

Mexico's total population and the percentage who live in urban and rural areas. 1900-1988.
Scholastic Update Nov 18, 1988 p. 9

POPULATION - SOVIET UNION

Demographics. Percentage of the U.S.S.R. population that is Russian, Central Asian, other non-Russian. 1970-2050.
U.S. News Apr 3, 1989 p. 44

POPULATION - UNITED STATES

A sharp slowing in the growth of households. Annual rise in number of U.S. households. 1980-1990.
Business Week Mar 25, 1991 p. 18

U.S. population by race and ethnicity, 1970-2030. Non-Hispanic white. Black. Hispanic origin. Other races.
Scholastic Update Jan 11, 1991 p. 4

Population increase, 1980-1989. White. Black. Asian and Pacific Islander. American Indian, Eskimo, and Aleut. Hispanic origin.
Scholastic Update Jan 11, 1991 p. 4

Rate of population increase per 1,000. Population. White. Black. Other races. Hispanic origin. Natural increase. Net immigration.
Scholastic Update Jan 11, 1991 p. 4

Comparing state stats. Percent population change. Metropolitan areas. Population under 18. Population age 65 and over. [By state.]
Scholastic Update Jan 11, 1991 pp. 26–27

More people live in the suburbs. Share of metropolitan population outside core city. 1970, 1988.
Business Week Nov 26, 1990 p. 81

Who will care for us? Aging America. As a percent of total population. Children under age 17. Adults over age 65. 1980-2040.
Newsweek Jul 16, 1990 p. 53

By 2056, whites may be a minority group. Non-Hispanic whites as a % of total population. 1920-2056.
Time Apr 9, 1990 p. 30

[Percent] increase in each population group (1980-1988). Asians and others. Hispanics. Blacks. Whites.
Time Apr 9, 1990 p. 30

U.S. population, ages 14-24, by sex. 1980-2050.
Black Enterprise Feb 1990 p. 65

Characteristics of persons ages 14-24. Male. Female.
Black Enterprise Feb 1990 p. 65

[Two] centuries of the census. Our population has grown. Population (in millions). 1790-1980.
Scholastic Update Jan 12, 1990 p. 6

[Two] centuries of census. We've moved west. U.S. center of population. 1790-1980. Map.
Scholastic Update Jan 12, 1990 p. 6

Comparing state stats. Percent population change. Metropolitan areas. Population under 18. Population age 65 and over. [By state.]
Scholastic Update Jan 12, 1990 pp. 30–31

Where the kids are. The median age in the U.S. was just over 32 last year, but some parts of the country have a younger population. The metropolitan areas with the lowest median age. [Selected cities.]
U.S. News Oct 2, 1989 p. 70

More elderly people. Share of population 65 and over. 1920-1987.
Business Week Sep 25, 1989 p. 99

A rising median age. Median age of population. 1920-1987.
Business Week Sep 25, 1989 p. 99

The aging of America. People 50 or over. Percent of U.S. populaton. 1988-2025.
Business Week Apr 3, 1989 p. 66

The spread of city living. State residents who live outside metropolitan areas. [State]. Nonmetro-population. Percentage change since 1950.
U.S. News Mar 13, 1989 p. 73

A population dip. How the nation's profile will shift. White. Black. Other. Over 65. Under 35. 1990-2080.
Newsweek Feb 13, 1989 p. 7

The new melting pot. Percentage of U.S. population. White. Hispanic. Black. Asian and other. 1990, 2080.
U.S. News Feb 13, 1989 p. 31

Going gray. Median age of U.S. population. 1970-2030.
U.S. News Feb 13, 1989 p. 31

Americans: Who are we? By age. Percent of U.S. population in each age group. Male. Female. Chart.
Scholastic Update Dec 2, 1988 p. 5

Americans: Who are we? Race. Percent of population. White. Black. Hispanic. Native American. Other.
Scholastic Update Dec 2, 1988 p. 5

U.S. Almanac: Comparing state stats. Percent population change. Population in metropolitian areas. Divorces. Marriages. [By state.]
Scholastic Update Dec 2, 1988 pp. 18–19

The aging of America. Percent of the population 65 or older. 1940. 1990. 2040. 2090.
FDA Consumer Oct 1988 p. 22

Minorities: Fast growth and too many dropouts. Population growth, 16 and older. 1986-2000. Whites. Blacks. Hispanics.
Business Week Sep 19, 1988 p. 114

America's going gray. Number of Americans age: 20-64. 19 and younger. 65 and older. 1960-2050.
U.S. News Aug 29, 1988 p. 60

Population centers past and present. Center of U.S. population. 1970-1980. Metropolitan area. Rank. Population. Map.
Scholastic Update Jan 15, 1988 p. 12

U.S. in focus: Comparing the states. Percent population change. Metropolitan areas. Divorces. Marriages. [By state.]
Scholastic Update Jan 15, 1988 pp. 16–17

POPULATION - WORLD

People power. Percentage of the total world population. U.S. Japan. United Germany. European Community.
U.S. News Jul 16, 1990 p. 23

The lure of city lights. Here's how today's 50 biggest cities will change in population by the year 2000.
U.S. News Feb 19, 1990 p. 70

While the world's population has climbed steadily in the past ten years, real U.S. spending on family planning has declined. 1979-1989.
Time Dec 18, 1989 p. 62

Declining fertility rates. Average number of children born to women in less developed countries during their lifetimes. 1965-1990. 1990-2025 [projected].
U.S. News Apr 17, 1989 p. 32

[As Europe's population] begins to shrink. EC. U.S. Japan. 1950-2040.
Business Week Mar 13, 1989 p. 55

World population growth. From 1750, projected to the year 2100. Less developed regions. Africa, Latin America, and Asia minus Japan. More developed regions. Europe, U.S.S.R., Japan, North America, Australia, and New Zealand.
Scholastic Update Sep 23, 1988 pp. 36–37

PORTER, COLE

Songs Porter wrote. Number of the 100 most recorded songs between 1890 and 1954 written by Porter. Gershwin. Berlin. Gross revenue generated by the recording industry in 1940. In 1990.
U.S. News May 27, 1991 p. 14

PORTLAND (OREGON)

Portland. Population. Median family income. Median home. Unemployment.
Newsweek Feb 6, 1989 p. 44

POSTAL RATES

Please, Mr. Postman, not another price hike. Cost to send a 1-ounce first class letter. 1963-1991.
Business Week Mar 5, 1990 p. 28

Postal rates. 1st class. Postal volume. Billions of pieces. 1980-1989.
Time Dec 25, 1989 p. 31

Instant delivery. Cost to have mail delivered overnight [by carrier/company].
U.S. News May 23, 1988 p. 76

Third-class mail keeps growing despite big hikes in postage. Percentage of total pieces of mail. (Typical prices in cents.) 1984-1988.
Business Week Mar 21, 1988 p. 53

Bigger licks. U.S. domestic postal rates for first class letters. 1958-1985.
U.S. News Jan 25, 1988 p. 34

POSTAL SERVICE

Mail call. Third-class mail as a percentage of total. Pieces of mail per person each year. 1945-1988.
U.S. News Jun 19, 1989 p. 74

The check is in the mail. Cities that get the most mail. Pieces of mail per person in 1988. Total mail.
U.S. News Mar 20, 1989 p. 90

POTASSIUM (MINERAL SUPPLEMENTS)

Debunking bananas. Bananas are commonly considered a rich source of potassium. Other foods actually offer more [by various foods].
U.S. News Apr 16, 1990 p. 62

POVERTY

Overcrowding has come down but more 'couch people' go to shelters. Black. Hispanic. White. 1978, 1983, 1989.
U.S. News Dec 23, 1991 p. 31

The poverty rate remains stubbornly high. Share of the population living in poverty. 1977-1991.
Business Week Nov 18, 1991 p. 85

Deep-rooted poverty. Percentage of persons below the poverty line 1990. White. Black. Black children.
U.S. News Nov 4, 1991 p. 63

In poverty: 34 million Americans. Median family income. Inflation-adjusted dollars. Current dollars. 1980-1990.
U.S. News Oct 7, 1991 p. 12

The poverty line. White. Black. Hispanic. Americans living below it.
Time Sep 30, 1991 p. 30

More children and young families live in poverty. Share of U.S. population living in poverty. 1973, 1989.
Business Week Aug 19, 1991 p. 83

Poverty rates for black children: 1969-1989.
Black Enterprise May 1991 p. 36

People visiting soup kitchens. Number of children in the United States who are hungry or at risk of becoming hungry.
U.S. News Apr 22, 1991 p. 8

Comparing state stats. Per capita income. Poverty rate. Job growth. Federal taxes paid per person. [By state.]
Scholastic Update Jan 11, 1991 pp. 26–27

Who they are [the "hyper-poor"]. How many. What race. Where they live. Household type. Welfare use.
U.S. News Oct 15, 1990 p. 40

The elderly are really poorer than others. Percent living in poverty. Rate if same poverty criteria were used for elderly as for nonelderly.
Business Week May 21, 1990 p. 32

Below the poverty level: White males. Black males. 16-34. 1988.
Black Enterprise May 1990 p. 41

Social problems are getting worse. Percentage increases. High school dropouts. 1970-88. Violent crime. 1979-88. Poverty rate. 1975-85. Drug abuse cases. 1980-87.
Scholastic Update Feb 23, 1990 p. 3

Comparing state stats. Per capita income. Poverty rate. Job growth. Federal taxes paid per person. [By state.]
Scholastic Update Jan 12, 1990 pp. 30–31

Poverty rate of black families: 1977 to 1987. Poverty rate of black children under 18 years old: 1977 to 1987. Poverty thresholds, by size of family unit: 1987.
Black Enterprise Nov 1989 p. 49

Persons below poverty line. 1939, 1969, 1984. Black. White.
Black Enterprise Oct 1989 p. 26

The need for housing aid will climb. Poverty-level households. 1974-2003.
Business Week Jul 10, 1989 p. 75

Rich vs. poor: The gap widens. Percent change in household income from 1979 to 1987. Married couples with children. Single mothers with children. Heads of household under 25. Heads of household over 65. Blacks.
Business Week Apr 17, 1989 pp. 78–79

Income. Living in poverty. Unemployment. Median income. Blacks. Whites. 1966, 1987.
Scholastic Update Apr 7, 1989 p. 21

Four causes of homelessness. More people live in poverty. Number of Americans below the poverty line. 1969-1986.
Scholastic Update Feb 10, 1989 p. 12

America's poor. People below the poverty level. People receiving AFDC. 1977-1986.
U.S. News May 23, 1988 p. 27

Poverty in America. Population above and below the poverty line. Black. White. Other. 1966, 1976, 1986.
Newsweek Mar 7, 1988 p. 21

Ghetto woes. The underclass. Growth, in millions. 1970, 1980.
Newsweek Mar 7, 1988 p. 43

The poor aren't who you think they are. Poor adults who work. [Who] receive welfare. 1975-1986.
U.S. News Jan 11, 1988 p. 20

Percent of persons below the poverty level, by race and Hispanic origin: 1986, 1985, and 1983. White. Black. Hispanic.
Black Enterprise Jan 1988 p. 54

PREGNANCY

Couples who have difficulty conceiving a child. In vitro fertilizations. Average paid for human sperm. Fertilized human embryos in frozen storage.
U.S. News Dec 30, 1991 p. 21

Prenatal tests. When/weeks. Cost. Fetal risk.
U.S. News Oct 21, 1991 p. 20

Or pay later. The cost of prenatal care for a healthy pregnant woman. The cost of medical care for an extremely premature baby.
Time Oct 8, 1990 p. 45

See Also TEENAGE PREGNANCY

PRESIDENTIAL CAMPAIGNS, 1984

Campaign costs. Primary spending in 1984. Democrats. Republicans.
Scholastic Update Jan 29, 1988 p. 2

PRESIDENTIAL CAMPAIGNS, 1988

The big picture. Michael Dukakis. Jesse Jackson. George Bush. Total number of delegates. Amount of money spent per delegate received. Amount of money raised. Amount of money spent.
Black Enterprise Jul 1988 p. 48

[Campaign spending]. Republicans. Democrats. Contributions in 1987. Cash on hand as of 1-1-88.
Time Feb 15, 1988 p. 19

PRESIDENTIAL ELECTION, 1984

Primary numbers. Small vote, big impact. Total population. Voting age population. Registered voters. Voters in general election. Voters in primary elections.
Scholastic Update Jan 29, 1988 p. 2

PRESIDENTIAL ELECTION, 1988

[Suburban voters] traditionally vote Republican. Share voted Republican, 1988 presidential election. City. Suburb.
Business Week Nov 26, 1990 p. 81

The election. Electoral votes. Bush. Dukakis. Map of U.S.
Time Nov 21, 1988 p. 33

Where the votes are. Campaign strategy and the Electoral College [number of electoral votes by state]. Region-by-region look at the electoral prospects of the Dukakis and Bush campaigns. Map.
Scholastic Update Oct 7, 1988 pp. 20–21

The Democrats' electoral game plan. [Electoral votes by state]. Map.
Newsweek Jul 25, 1988 p. 18

Spotlight on Super Tuesday. States holding primaries or caucuses on Super Tuesday. States holding primaries or caucuses prior to Super Tuesday. Democrats. Republicans. [Number of delegates.] Map.
Scholastic Update Jan 29, 1988 p. 2

1988 election countdown calendar. [Caucuses, primaries, conventions - by month.]
Scholastic Update Jan 29, 1988 pp. 14–15

PRESIDENTIAL ELECTIONS

Will rising taxes sink Bush next year? Change in personal tax rate since previous presidential election. Party in White House loses. Party in White House wins. 1956-1988.
Business Week Dec 9, 1991 p. 24

Dow Jones picks the President. Percentage change in Dow Jones industrial average. Year incumbent party won. Year incumbent party lost. 1900-1984.
U.S. News Sep 19, 1988 p. 72

Voter turnout by gender and age, presidental elections. Women 18-44. Men 18-44. 1972-1984.
Ms. Apr 1988 p. 77

Voter turnout by gender and race, presidential elections. Black women. Black men. 1972-1984.
Ms. Apr 1988 p. 77

PRESIDENTS - UNITED STATES

Bush's days at Kennebunkport in 1989. In 1990. Vacation days of Eisenhower. Reagan. [Other selected statistics.]
U.S. News Jul 29, 1991 p. 6

First use of "state of the union." Longest annual message ever given. Bills passed during George Bush's tenure. President who vetoed the most. [Other selected statistics.]
U.S. News Jan 28, 1991 p. 10

Line of succession. In the event that the President were incapacitated in a nuclear attack, federal law specifies that these officials would succeed him in the following order.
U.S. News Aug 7, 1989 p. 29

Picture-perfect chiefs. These Presidents are portrayed most frequently in movies.
U.S. News Jul 10, 1989 p. 62

Two hundred years of leadership. President's birthplace, political party, age when inaugurated, term of office, and notable accomplishment.
Scholastic Update Jan 13, 1989 pp. 10–11

Washington's John Hancock. Value of a presidental signature, as most commonly signed.
U.S. News Oct 24, 1988 p. 81

White House soldiers. Presidents who served in the military. President. War.
U.S. News Sep 12, 1988 p. 71

Paying the President. What past Presidents would earn if their salaries were adjusted for inflation in 1988 dollars.
U.S. News Aug 8, 1988 p. 65

PRICE INDEXES

What's up, what's down. Changes in consumer prices (Dec., 1986 - Dec., 1987). 25 worst. 25 best.
U.S. News Feb 1, 1988 p. 73

PRISONS AND PRISONERS

Prisons. 1971 versus 1991. Prison and jail inmates. Incarceration rate. Inmates on death row. Reported serious crime. 1990 average prison operating cost.
U.S. News Sep 16, 1991 p. 16

The U.S. population behind bars is soaring. Inmates. State and federal prisons. Local jails. 1950-1990.
Business Week Mar 18, 1991 p. 20

Filling prisons. Women behind bars. In both federal and state prisons. 1980-1989.
Newsweek Jun 4, 1990 p. 37

A bleak indictment of the inner city. Based on the lastest data, scholars believe that the percentage of the state-prison population that is black will reach 50 percent by the year 2000. 1930-2000.
U.S. News Mar 12, 1990 p. 14

Safe and secure but lacking amenities. Item. General population at other U.S. penitentiaries. At Marion.
Newsweek Jan 15, 1990 p. 69

The jailhouse crunch. Corrections population. Out of custody. In custody. 1983-1989.
U.S. News Nov 20, 1989 p. 76

Courts and prisons are swamped. Number of prison inmates, state and federal. 1984-1988.
Scholastic Update Nov 17, 1989 p. 9

Most released criminals commit new crimes. Percent of prisoners released in 1983 who were rearrested, reconvicted, or returned to jail within 3 years. All prisoners. Males prisoners. Female prisoners. Prisoners age 17 or younger.
Scholastic Update Nov 17, 1989 p. 9

Average sentence length, in years. Murder. Rape. Robbery. Arson. Fraud.
Time Nov 6, 1989 p. 62

The states pay the price. While violent crime climbs, so do prison costs. Percent increase 1988 to 1989 in state spending on: Penal institutions. Medicaid. Education. Welfare. Overall spending.
Time Aug 21, 1989 p. 25

Jammed jail cells. U.S. prison population. Federal inmates. State inmates. 1978-1988.
U.S. News Jul 24, 1989 p. 66

Spending on Medicaid and prisons is up. Outlays as a percentage of total state spending. Medicaid. Corrections. 1987-1989.
Business Week Jun 5, 1989 p. 88

Inmates. Number of prisoners in state and federal prisons, in thousands. Crime rate. Number of crimes committed per 1,000 population. 1980-1988.
Time May 29, 1989 p. 29

America's swelling prison population. 1978-1988.
Business Week May 8, 1989 p. 81

Crime and punishment. Average time served in state prisons (in months). Blacks. Whites.
U.S. News Aug 22, 1988 p. 54

PRIVACY

States that have passed employee privacy protection laws. Prohibits employer discrimination for: Off-the-job smoking. Any legal off-the-job activity. Map.
Business Week Aug 26, 1991 p. 70

PRO-DEMOCRACY MOVEMENT

Freedom House's latest analysis of political liberty around the world. Free. Partly Free. Not Free. Map.
U.S. News Jan 8, 1990 p. 16

PRODUCTIVITY, INDUSTRIAL

Feeding the bottom line. Japan's productivity has pulled ahead. Output per hour in manufacturing. Overhead as a percent of manufacturing cost. U.S. Japan. Germany. 1986-1990.
Business Week Oct 25, 1991 p. 66

The productivity gap. Manufacturing productivity. Service productivity. 1980-1991.
U.S. News Jul 22, 1991 p. 44

A spotlight on 1990 industry profits. Aftertax profits. Industries with the biggest dollar change. Winners and losers [by industry] in the year's profit race.
Business Week Mar 18, 1991 p. 54

Productivity index, nonfarm business. Change in manufacturing output. Manufacturing productivity. 1980-1990.
U.S. News Feb 18, 1991 p. 12

How the competition stacks up. Labor. Japan. West Germany. U.S. Productivity index. 1980-1990. Unit labor costs. 1980-1990. Scientists and engineers engaged in R&D. 1980-1987.
U.S. News Jul 16, 1990 pp. 24–25

Productivity leaders [various countries]. Real output per worker. Rank. 1950-1988.
U.S. News Jan 8, 1990 p. 58

Four keys to success as the expansion matures. Monthly ratio of inventories to sales in manufacturing. Exports. Index of nonfarm productivity. Employment cost index for private industry workers. 1984-1989.
Business Week Jan 9, 1989 pp. 64–65

Tight capacity, higher spending. Utilization rate. Percent. Projected capital outlays 1988 vs. 1987. Percent increase. [By industry.]
Business Week Oct 3, 1988 p. 90

Yearly output per factory worker in thousands of dollars. U.S. W. Germany. Japan. 1972-1987.
Scholastic Update Sep 9, 1988 p. 2

How the U.S. ranks in the productivity race. Output per hour in manufacturing. 1979-87, 1987. U.S. Japan. France. West Germany. Britain. Percent increase, annual rate.
Business Week Jul 25, 1988 p. 14

Capacity utilization. The average is only edging up but for some industries, rates are soaring. Industry. 1987, 1988.
Business Week May 2, 1988 p. 31

PROFIT

The leaders in 1989. The top 25 in sales. The top 25 in earnings.
Business Week Mar 19, 1990 pp. 64–65

PROMOTIONS

Measuring the glass ceiling. Why white males monopolize the top of the corporate ladder. Men. Women. Minorities. Employees. Managers. Top execs.
U.S. News Aug 19, 1991 p. 14

PROPERTY

She's gotta have it. Who owns what? Single women heads of households vs. single male heads of households.
Ms. Nov 1988 p. 86

PROPERTY TAXES

Costly counties. The 10 counties charging the highest per capita property taxes in the nation. County. Property-tax revenue per person. Property tax as share of total county revenue.
U.S. News Aug 6, 1990 p. 39

PROTECTIONISM

See FREE TRADE AND PROTECTIONISM

PROTEINS

How to get your protein. Amount needed [of selected foods] for 20 grams of protein. Cost of 20 grams of protein. 1987.
U.S. News Mar 21, 1988 p. 73

PROTESTANT CHURCH

Catholics give less to the church. Share of income contributed by U.S. Catholics. Protestants. 1960-1989.
Business Week Jun 10, 1991 p. 60

Growing crowd of believers. Number of Protestants in Latin America. 1930s, 1960s, 1980s.
Time Jan 21, 1991 p. 69

PSYCHIATRISTS

Help with mental health. How the states rank in serving the seriously mentally ill. 1988 ranking. 1986 ranking. Psychiatrists/psychologists per 100,000 population.
U.S. News Dec 5, 1988 p. 93

PSYCHOLOGISTS

Help with mental health. How the states rank in serving the seriously mentally ill. 1988 ranking. 1986 ranking. Psychiatrists/psychologists per 100,000 population.
U.S. News Dec 5, 1988 p. 93

The therapy business. Percentage of psychologists who break confidentiality. Limit treatment. Discuss clients. Terminate therapy. Are sexually attracted to a client.
U.S. News Mar 21, 1988 p. 73

PUBLIC LANDS

See NATIONAL PARKS AND RESERVES

PUBLIC SCHOOLS - CHICAGO (ILLINOIS)

Big and troubled. The problems of the inner city are reflected in Chicago's classrooms. Student population. By race. Students limited in English proficiency. Students living below poverty level. Dropout rate.
Newsweek Jul 4, 1988 p. 60

PUBLIC SCHOOLS - INTEGRATION

Federal school desegregation lawsuits, since 1963. States with ten or more districts that are under permanent injunction to remain desegregated.
Black Enterprise Sep 1988 p. 25

PUBLIC SCHOOLS - NEW YORK (CITY)

Comparative report card. New York City. Public schools. Catholic schools. Students. Student-to-teacher ratio. Percentage graduated on time. Special-education classes. Spending. Average teacher salary. Administrators.
Time May 27, 1991 p. 49

PUBLIC SCHOOLS - SPORTS

The sporting life. Top ten sports for high school girls. 1971. 1986-87. Percent increase.
Ms. Aug 1988 p. 83

PUBLIC SCHOOLS - VIOLENCE

Blackboard jungle. Percentage of teens who say that in the last year they [were involved in various violent acts]. Boys. Girls.
U.S. News Feb 13, 1989 p. 84

PUBLIC WELFARE

[The young] are hurt most by welfare cuts. Monthly benefit. 1970-1990.
Business Week Aug 19, 1991 p. 85

Four causes of homelessness. Federal aid hasn't kept up. Percent rise in federal budget for social welfare programs. 1960-1985.
Scholastic Update Feb 10, 1989 p. 12

Significant changes in the welfare system [new welfare bill].
Black Enterprise Jan 1989 p. 21

Families on the dole. The number of families on public assistance. 1970-1988.
U.S. News Oct 10, 1988 p. 11

Americans receiving AFDC [aid to families with dependent children] benefits. Black. White. 1977-1986.
Black Enterprise Mar 1988 p. 35

The poor aren't who you think they are. Poor adults who work. [Who] receive welfare. 1975-1986.
U.S. News Jan 11, 1988 p. 20

PUBLIC WORKS

A sharp slowdown in infrastructure spending. Capital outlays for public works. Percent of gross national product. 1968-1988.
Business Week Aug 7, 1989 p. 18

PUBLISHERS AND PUBLISHING

See BOOK PUBLISHERS AND PUBLISHING

PUERTO RICO

Puerto Ricans favoring statehood. 1989 per capita income. Percentage below U.S. poverty line. U.S. adults who favor making Washington, D.C. a state.
U.S. News Feb 18, 1991 p. 10

PULITZER PRIZE

Pulitzers awarded. Top journalism winners. Most awards. [Other selected statistics.]
U.S. News Sep 16, 1991 p. 11

QUALITY OF LIFE

How America compares [to seven industrial countries]. Health care costs. Public spending on health. Infant mortality. National maternity-leave policy. Public spending on education. Labor force participation.
Business Week Dec 17, 1990 pp. 88–89

The good life [in California].
Newsweek Jul 31, 1989 p. 24

The tough life [in California].
Newsweek Jul 31, 1989 p. 27

Quality of life [in the Soviet Union]. Automobile registrations per 1,000 people. Miles of paved roads. Percentage of families with telephones in cities and towns, in rural areas.
U.S. News Apr 3, 1989 p. 44

The top 10: Having it all. Population. Median family income. Median home. Unemployment.
Newsweek Feb 6, 1989 pp. 42–44

Putting kids first. How adults rank the importance of different aspects of their lives. U.S. Japan. Europe.
U.S. News Aug 1, 1988 p. 62

Comparing two worlds. U.S. U.S.S.R. Geography. The people. Quality of life. The economy. Military might. Map of Soviet Republics.
Scholastic Update Mar 25, 1988 pp. 8–9

QUARKS

The Nobel experiment. Quarks, the basic building blocks of matter. Diagram.
Newsweek Oct 29, 1990 p. 68

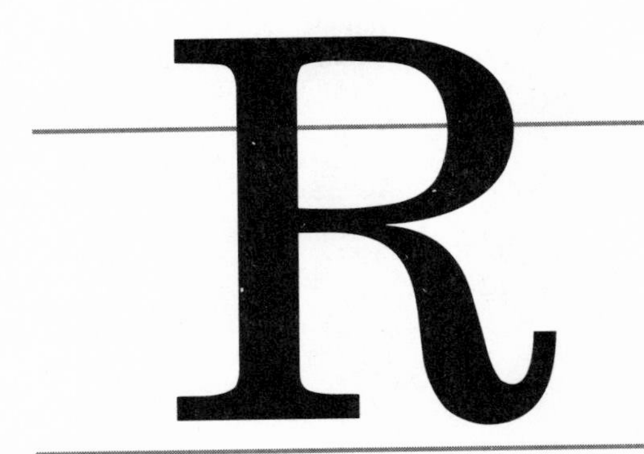

RACE RELATIONS - UNITED STATES

Growing tolerance? Reported discriminatory acts against: Asians. Hispanics. Blacks. Jews. 1980-1987.
U.S. News Aug 29, 1988 p. 103

The Big Apple's troubled racial landscape. Attacks against blacks. Attacks against whites. 1981-1987.
Black Enterprise Mar 1988 p. 20

RADAR DEFENSE NETWORKS

How Aegis works. Radar (Aegis Command and Control System) puts up an electronic shield. Firepower.
Time Jul 18, 1988 pp. 14–15

RADIATION

Sources of radiation exposure. Estimated annual radiation exposure, of an average American.
Bull Atomic Sci Sep 1990 p. 13

Changing radiation standards. U.S. recommended maximum permissible whole-body doses. Occupational exposure. General public exposure. 1934-1990.
Bull Atomic Sci Sep 1990 p. 14

RADIO INDUSTRY

Money in the air. Average sale price for radio and TV station. 1960s, 1970s, 1980s.
Business Week Jul 23, 1990 p. 52

RADON

Danger from below. Radon makes its way from basement to attic. One way to get it out. Diagram.
Newsweek Sep 26, 1988 p. 69

The ten worst states. Highest percentage of houses with radon levels.
Time Sep 26, 1988 p. 69

Where it [radon] gets in. When to take steps. Exposure. Lung-cancer deaths per 1,000 people exposed. Comparable exposure level. Comparable risk. Action recommended. How to get it out. Diagram.
U.S. News Sep 26, 1988 pp. 62–63

Areas of potentially high radon levels. Map.
U.S. News Jan 18, 1988 p. 13

RAILROADS

Gross railroad industry revenues. Freight carried by trains. Miles of railroad track abandoned since 1980. Average cars per freight train. Locomotives in service.
U.S. News Apr 29, 1991 p. 13

The rising cost of railroad injuries. Year. Railroad employees. Injuries. Payouts. 1982-1988.
Business Week Nov 6, 1989 p. 93

How the big railroads stack up financially. Cash. Long-term debt.
Business Week Mar 14, 1988 p. 38

RAIN AND RAINFALL

What's stopping the rain? A lingering high-pressure area and split air currents prevent precipitation. June, 1988.
U.S. News Jul 4, 1988 p. 46

RAIN FORESTS

The world's tropical forest. Tropical rain forests, located along the earth's equator, have declined by 15 to 20 percent since preagricultural time. South America. Africa. Asia.
U.S. News Jun 4, 1990 p. 64

Nowhere to hide. Air pollution. Tropical forests [annual loss]. Desertification. Map.
Newsweek Jul 11, 1988 p. 24

RAPE

Rape: Relationship of victim to offender. Kansas, 1988. Relationship. Number. % of total.
Ms. Sep 1990 p. 37

The portrait of a nightmare [rape statistics].
Newsweek Jul 23, 1990 p. 48

READING

Adults in the United States who used a library in 1990. Most popular reference books. Adults who didn't read a single book in 1990. Average time spent per day reading a book. Watching TV.
U.S. News Apr 15, 1991 p. 12

Why Johnny must read. Average time spent reading for the job. Occupation. Avg. time. Type of material.
U.S. News Jun 26, 1989 p. 46

REAL ESTATE - JAPAN

The jump in land prices. Index of land prices for Japan's six largest cities. Commercial. Residential. 1985-1990.
Business Week Nov 26, 1990 p. 72

REAL ESTATE, COMMERCIAL

The office glut gets worse. Downtown office vacancy rates. 1987-1991.
Business Week Jun 17, 1991 p. 86

Prime locations. Price of commercial land per square foot in high-rent districts. Tokyo. New York. Washington. Chicago. Los Angeles.
U.S. News Jan 30, 1989 p. 58

REAL ESTATE INDUSTRY

Sales of existing one-family homes. Median sale price of existing one-family homes. [Office] vacancy rate. 1986-1991.
Business Week Dec 31, 1990 p. 139

Sagging profits. Real-estate values helped erode commercial-bank profits. Net income of commercial banks. 1985-1989.
U.S. News Jul 2, 1990 p. 38

Bottom lines. Banks with the highest total in deadbeat real-estate loans. Bank. Problem real-estate loans [dollar amount]. Percentage of all real-estate loans.
U.S. News Jul 2, 1990 pp. 38–39

Less lending. Banks are reducing their real-estate exposure. Growth in real-estate loans. Commercial real-estate loans. Construction and land development loans. 1985-1990.
U.S. News Jul 2, 1990 p. 39

Lending by commercial banks. Percentage change in growth of real-estate loans. 1988-1990.
U.S. News Apr 23, 1990 p. 52

Big gains on real estate are over. Annual returns including rents and appreciation. Increase in median price of existing single-family home. 1984-1989.
Business Week Jan 22, 1990 p. 61

See Also HOUSING

RECESSION (FINANCIAL)

A painful recession, but others have been worse. Unemployment rate. Prime rate. Consumer price index. [Recession years:] 1970. 1974. 1980. 1982. Today.
Business Week Dec 30, 1991 p. 60

Percentage change in real GNP. 1987-1991. GNP decline. Business investment. Residential construction. Government spending.
U.S. News May 6, 1991 p. 19

A tougher recession than the last one. State & local operating surplus or deficit. Federal aid as a percent of state & local receipts. 1981-1991.
Business Week Apr 22, 1991 p. 26

Economic slump to rival '81-82 debacle. Dollar drops. Value of U.S. dollar. Deficit deepens. U.S. budget deficit. 1980-1991.
U.S. News Jan 7, 1991 p. 46

Why this recession is different. Ratio of private debt to GNP. Office vacancy rate. FDIC-insured banks closed or assisted. Total private employment. Service industries. Manufacturing. 1981-1990.
Business Week Dec 24, 1990 pp. 58–59

Signs on the road to recession. Deteriorating job picture. Income and spending pinch. Financial distress. Poor expections. Recent level. '81-'82 recession level.
Business Week Nov 19, 1990 p. 50

Recession's signs and portents. Stock Market. S&P 500 index. 1989, 1990.
U.S. News Sep 24, 1990 p. 80

Recession's signs and portents. Consumer attitude. Index of consumer expectations. 1989, 1990.
U.S. News Sep 24, 1990 p. 80

Recession's signs and portents. Factory orders. Monthly increase in unfilled orders. 1989, 1990.
U.S. News Sep 24, 1990 p. 81

Recession's signs and portents. Unemployment claims. New claims for unemployment benefits. 1989, 1990.
U.S. News Sep 24, 1990 p. 81

A recessions'-eye view of the Dow. Dow Jones industrial average. 1945-1990. [Chronology including eight recessions.]
U.S. News Sep 24, 1990 pp. 82–83

Recessions: The recent record. Starting date. Months.
Business Week Aug 13, 1990 p. 32

Remembrances of recessions past. 1969-1970. 1973-1975.
U.S. News Aug 14, 1989 p. 42

Rain or shine? [Chronology of recessions and the stock market.] 1929-1988. Chart.
Time Apr 4, 1988 pp. 54–55

Growth cycles. Periods of recession. Oct. 1949 - Mar. 1988.
Time Mar 21, 1988 p. 53

RECOMBINANT DNA

See GENETIC ENGINEERING

RECREATION

See LEISURE

RECYCLING (WASTE, ETC.)

Paper chase. Old newspapers recovered, in short tons. 1982-1995.
U.S. News Aug 26, 1991 p. 48

Not so green. Here is the percentage of households that [practice various conservation methods].
U.S. News Feb 4, 1991 p. 71

Worldwide waste. Percent of all garbage recycled in— Switzerland. U.S. Sweden. West Germany. Netherlands. Canada.
U.S. News Apr 23, 1990 p. 64

Do as I say, not as I do. People recently surveyed said businesses have a "definite" responsibility to reduce their air and water pollution. But here is how often the respondents themselves said they take the following steps [various kinds of recycling].
U.S. News Oct 16, 1989 p. 125

An environmental scorecard. Land. Material recovered—and recycled—from municipal solid waste each year. 1960-1984. Total garbage generated by cities, counties and towns each year.
Scholastic Update Apr 21, 1989 p. 16

REDUCING

See DIETING

REFUGEES

The world's refugees—where they come from. The nations and regions that have generated the greatest number of the world's homeless.
Scholastic Update Oct 18, 1991 pp. 2–3

Whom the U.S. lets in. Refugees the U.S. has admitted each year since 1980, and the region from which they have come. No. of refugees. 1980-1991.
Scholastic Update Oct 18, 1991 p. 12

The forgotten millions. More than 17 million refugees [by country]. Map.
U.S. News Apr 29, 1991 p. 28

The world's wanderers. Of the more than 15 million refugees in the world, only a few resettle in Western countries like the U.S. Most live in stateless limbo. Refugees. Where they have fled. Map.
U.S. News Oct 23, 1989 p. 36

Displaced. The heaviest refugee flows take place in Africa and Asia. Map.
Newsweek Oct 9, 1989 p. 44

World refugees in millions. Where they came from. 1984-1987.
Time Jul 3, 1989 p. 26

Huddled masses. Applications for political asylum by Hispanics and Haitians now in the U.S. Country. Cases pending. As of Nov. 1, 1988.
Business Week Feb 6, 1989 p. 52

REFUGEES, IRAQI

Iraq's huddled masses. 2 million displaced Kurds and Shiites. Map.
U.S. News Apr 29, 1991 p. 26

REFUGEES, KURDISH

[Kurdish] refugee centers in northern Iraq. Staging areas. Map.
Time Apr 29, 1991 p. 41

Who are the Kurds? A people apart. Years of defeat. Where they live. Oil.
Time Apr 15, 1991 p. 27

The Kurdish exodus. Refugees. American protection. Map.
U.S. News Apr 15, 1991 p. 30

REFUSE AND REFUSE DISPOSAL

Worldwide waste. Percent of all garbage recycled in— Switzerland. U.S. Sweden. West Germany. Netherlands. Canada.
U.S. News Apr 23, 1990 p. 64

Going up. Garbage anthropologists say that the average American household is a round-the-clock trash factory.
Newsweek Nov 27, 1989 p. 67

How much time do we have? Years to landfill capacity [by state].
Newsweek Nov 27, 1989 p. 68

No more open pits. The days of outright dumping are gone. Waste-management companies now employ a range of sophisticated techniques.
Newsweek Oct 3, 1988 p. 39

REGULATORY AGENCIES

Uncle Sam hires a few more watchdogs. Work force levels for federal regulators. Thousands of employees. 1980-1990.
Business Week Jun 26, 1989 p. 58

RELIGION

Growing crowd of believers. Number of Protestants in Latin America. 1930s, 1960s, 1980s.
Time Jan 21, 1991 p. 69

Membership in U.S. churches or synagogues. Churches with the most members. Membership declines since 1965. Membership increases since 1965. Per capita church contributions.
U.S. News Nov 19, 1990 p. 16

Faith and doubt. Breakdown of Soviet population by affiliation.
Time Oct 15, 1990 p. 71

Untithed. Religious denomination. % of income members give to charity.
Newsweek Jan 22, 1990 p. 8

Mainline vital signs. U.S. Christians 1987. Foreign missionaries. Membership in millions. Sunday-school enrollment in millions. 1965, 1987.
Time May 22, 1989 p. 96

Religion. Majority and minority religions in various parts of the world. Map.
Scholastic Update Dec 16, 1988 pp. 10–11

Americans: Who are we? Religion. Religious affiliation, by percent of population. Chart.
Scholastic Update Dec 2, 1988 p. 5

The importance of prayer. Number of Americans who regularly go to a place of workship. 1960, 1987.
Scholastic Update Feb 26, 1988 p. 2

REMARRIAGE

Second chances. Marriage rates per 1,000 divorced men and women. 1986.
U.S. News Jan 29, 1990 p. 50

RENT AND RENTERS

Cheap pads. Out of 100 [selected Western cities] metropolitan areas, here's where rents were lowest.
U.S. News Oct 8, 1990 p. 83

Median income ratios of owners and renters to all households: 1950-1985. Owners. All households. Renters.
Black Enterprise Jan 1990 p. 45

REPUBLICAN PARTY

Bush's box score. Bush campaigned for 63 candidates since May. Here's how they fared.
Time Nov 19, 1990 p. 31

Westward ho, Republicans! Population change in percent. Congressional seats. Gained. Lost. No change. 1980-1990.
Newsweek Sep 10, 1990 p. 33

Which party is better for the economy? How the economy fared under recent presidents [1960-1988]. Democrats. Republicans. Real GNP growth. Unemployment rate. Inflation rate. Stock prices. Prime rate.
U.S. News Oct 17, 1988 p. 82

RESEARCH - FEDERAL GRANTS

University research: The squeeze is on. Funding from federal government, state and local government, corporate. 1980-1991. Who foots the bill.
Business Week May 20, 1991 p. 125

Focus on funds. Top 10 recipients of federal research-and-development expenditures and their indirect cost rates.
Time Mar 18, 1991 p. 75

Building empires of science. Major research universities [leading recipients of federal research funds] with their overhead charges as a percentage of funding.
U.S. News Mar 4, 1991 p. 52

How the federal government spends its R&D money. Federal R&D outlays. Basic. Applied. Development. Defense. Nondefense. 1980, 1985, 1989.
Business Week Jun 15, 1990 p. 47

Government R&D spending, fiscal 1981-1988. Nondefense R&D. Defense R&D.
Bull Atomic Sci Jan 1989 p. 47

RESEARCH AND DEVELOPMENT

America's R&D gap. Spending on non-defense research and development as a percent of GNP. Japan. West Germany. U.S. 1971-1989.
Time Dec 23, 1991 p. 55

Corporate funding for R&D: The new lineup. Business contributions to research universities [ranked by university]. 1989, 1982.
Business Week May 20, 1991 p. 124

R&D spending as a share of GNP. Defense and space. Federal non-defense. Private. 1953-1988.
Bull Atomic Sci Jan 1991 p. 39

How the competition stacks up. Labor. Japan. West Germany. U.S. Productivity index. 1980-1990. Unit labor costs. 1980-1990. Scientists and engineers engaged in R&D. 1980-1987.
U.S. News Jul 16, 1990 pp. 24–25

How the competition stacks up. Investment. Japan. West Germany. U.S. Investment in plant and equipment. 1980-1990. Nondefense R&D expenditures. 1980-1988. Public investment. 1967-1985.
U.S. News Jul 16, 1990 p. 25

How the big three compare. Emphasis on nonmilitary R&D. Scientists and engineers per 10,000 workers. Source of R&D funds. Cost of capital for research. Japan. West Germany. U.S.
Business Week Jun 15, 1990 p. 35

Unequal opportunities are holding down R&D. Nondefense research-and-development spending as a share of gross national product. Japan. W. Germany. U.S. 1980-1987.
U.S. News Dec 25, 1989 p. 41

The research race. Japan. U.S. 1984, 1988.
Time Dec 4, 1989 p. 68

Takeovers haven't squelched R&D spending. Total value of M&A transactions. R&D expenses per employe. 1975-1988.
U.S. News Dec 4, 1989 p. 59

Technological edge. R&D spending as a percentage of GNP. R&D—funding sources. U.S. Japan. West Germany. Patent gap. 1970-1987.
U.S. News Jul 10, 1989 p. 45

RESOLUTION TRUST CORPORATION

The RTC's caseload continues to mount along with its inventory of unsold assets. Number of thrifts seized. Mortgages. Securities. Real estate. Other loans. Other assets.
Business Week Dec 25, 1990 p. 67

RESTAURANTS

Midprice eateries still set the pace. Average same-store sales growth. Fast-food hamburger chains. Midprice restaurant chains. 1986-1991.
Business Week Jan 14, 1991 p. 92

New-age cooks. Restaurants are responding to their customers' interest in healthier eating [various cooking methods]. 1984, 1989.
U.S. News Feb 26, 1990 p. 63

An abundance of restaurants. U.S. restaurants, excluding those in hotels and convenience stores. 1985-1990.
Business Week Jan 8, 1990 p. 90

Pigging out at the mall. Food. Sales share. Annual sales.
Newsweek Sep 25, 1989 p. 6

Dining out, still very in. Of 500 restaurants surveyed last year, these were the top moneymakers: Dining seats. Average dinner check per person.
U.S. News Aug 7, 1989 p. 62

Menu for modern times. Residents of the largest metropolitan areas in the U.S. rack up the following annual bar and restaurant tabs.
U.S. News May 15, 1989 p. 73

Family restaurants lag in growth. Family-style. Fast food. Sales. 1985-1988.
Business Week Sep 5, 1988 p. 63

McDonald's: Staying ahead. Systemwide restaurant revenues. McDonald's. Burger King. Wendy's. 1983-1987.
Business Week May 9, 1988 p. 92

Nibbling away at the grocer's pie. Food eaten away from home [at restaurants, fast-food outlets, cafeterias, bars and taverns] as a share of total U.S. food sales. 1983-1988.
Business Week Jan 11, 1988 p. 86

RETAIL TRADE

Burgers and books. Retailers that flourished. 1980-1990.
U.S. News Jul 8, 1991 p. 57

Percentage in change in retail sales. 1990-1991. Sales increase by stores. Most notable sales increases.
U.S. News May 27, 1991 p. 16

Percentage change in retail sales. Decline in retail sales from January 1990 to January 1991. Auto sales. Gas station sales. 1989, 1990.
U.S. News Feb 25, 1991 p. 14

Why merchants are miserable. Store space keeps growing. Square feet of U.S. retail space per capita. But sales are not keeping pace. Retail sales per square foot in constant 1990 dollars. 1972-1990.
Business Week Nov 26, 1990 p. 134

Why merchants are miserable. Retailers' debt keeps climbing. Total long and short-term debt of top 30 U.S. retail companies. 1980-1990.
Business Week Nov 26, 1990 p. 135

Why merchants are miserable. Bankruptcies are on the rise. Number of retailers filing for bankruptcy protection. 1985-1990.
Business Week Nov 26, 1990 p. 135

Shopper's world. Retail stores per 1,000 people. Britain. West Germany. France. U.S. Japan.
U.S. News Apr 24, 1989 p. 48

Clip file. Supermarket coupons redeemed. 1983-1987. Average value of a coupon for health and beauty items [and for] grocery items.
U.S. News Dec 5, 1988 p. 93

The nation's largest convenience store chains. Company. Number of stores.
Business Week Jun 13, 1988 p. 86

RETIREMENT

Who is eligible for early retirement. Percent of companies. Eligibility.
Black Enterprise Dec 1990 p. 68

Popular early retirement packages. Percent of companies. Incentives.
Black Enterprise Dec 1990 p. 70

Where's my pension? Percent of work force in employer-financed pension plans. All workers. Men under 35. 1979, 1983, 1988.
Business Week Jul 30, 1990 p. 46

Retirement. To estimate how much of your savings you could spend each month were you to retire today. Interest rate. Life expectancy in years.
U.S. News Jul 30, 1990 p. 74

Permissible reasons for 401(k) hardship withdrawals.
Black Enterprise Jun 1990 p. 73

Quitting the race. A greater proportion of men are retiring earlier, but women below 65 are hanging in longer. 1948-1989.
U.S. News May 14, 1990 p. 49

Capital needed to replace 60% of your current income for 20 years. Number of years to retirement. Today's income level.
Black Enterprise Apr 1990 p. 70

Senior savings. Looking for a retirement home at a good price? City. Average condo price.
Newsweek Mar 5, 1990 p. 8

Will your money last? Percent of original capital, withdrawn annually. Will last this many years, if invested at these interest rates.
Newsweek Oct 9, 1989 p. 74

Where retirees are settling. Counties where 15% or more of new arrivals from 1970 to 1980 were 60 and over.
Business Week Sep 25, 1989 p. 99

What you can expect from the government. Estimated monthly Social Security benefits at retirement.
U.S. News Aug 14, 1989 p. 61

Better start now. How much you'll have to save each year until age 65 to generate various levels of annual retirement income.
U.S. News Aug 15, 1988 p. 68

Fate of the retirement kitty. Number of Social Security beneficiaries. Social Security funds. 1985-2048.
U.S. News Jun 13, 1988 p. 71

RICH
See WEALTH

ROADS
The rocky road to everywhere. More than 5,000 miles of America's interstates are badly in need of repair. Here's where you can expect the bumpiest ride [by state]. Repairs needed. Percentage of interstate. Miles.
U.S. News Nov 6, 1989 p. 108

ROBOTS, INDUSTRIAL
Unequal opportunities. Capital investment. Industrial robots in use at the end of 1988. Britain. France. Italy. West Germany. U.S. Japan.
U.S. News Dec 25, 1989 p. 41

ROCK CONCERTS
Rock on the road. Most successful touring rock groups of 1988. Group. Total revenues. Total attendance.
U.S. News Feb 27, 1989 p. 75

ROMANIA - ECONOMIC CONDITIONS
Adding up the damage. Bulgaria. Czechoslovakia. Hungary. Poland. Romania. Real economic growth. Hard-currency balance. Oil imports.
U.S. News Nov 5, 1990 p. 55

RUMANIA
See ROMANIA

RUSSIA
See SOVIET UNION

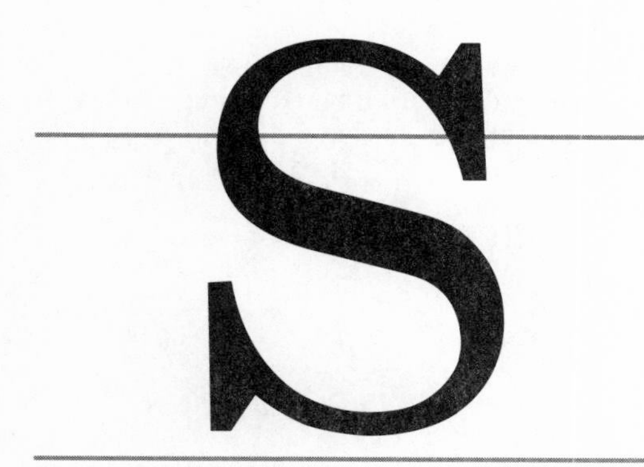

SALARIES

See WAGES AND SALARIES

SALES PERSONNEL

A long way from "Avon calling." Industries with the most saleswomen. Industries with fewest saleswomen.
U.S. News Feb 6, 1989 p. 42

SALT IN THE BODY

A little dab'll do ya. Amount of sodium and calories in various condiments (per tablespoon).
U.S. News Aug 22, 1988 p. 65

SAUDI ARABIA

Kuwait. Iraq. Saudi Arabia. Population 1988. Area. GNP. Oil production. Oil reserves. Troops. Tanks. Aircraft. Ships. [Map of oil pipelines.]
Time Aug 13, 1990 p. 17

SAUDI ARABIA - ECONOMIC CONDITIONS

Saudi Arabia's export revenues are up but the kingdom is running a big tab. Revenues. New commitments. Item cost.
Business Week Feb 4, 1991 p. 44

The Saudi's new oil billions. Annual revenues. Billions of U.S. dollars. 1986-1991.
Business Week Nov 19, 1990 p. 66

SAVING AND SAVINGS - UNITED STATES

Heavy lifting. How America's debt burden threatens the economic recovery. Debt-to-GNP ratio. 1952-1989. Total U.S. debt. 1980-1989. U.S. net national savings as a percentage of GNP. 1953-1989.
U.S. News May 6, 1991 pp. 52–54

Squeezing savings. Personal saving rate. 1989, 1990.
U.S. News Oct 22, 1990 p. 59

The savings slump. Personal savings rate. 1970-1990.
U.S. News Oct 8, 1990 p. 68

In the bank. Percentage of personal disposable income saved. 1960-1990.
U.S. News Jul 30, 1990 p. 48

Savings. Disposable income saved. Families headed by a person [various ages]. Percent saved. 1988.
U.S. News Jul 30, 1990 p. 71

Consumers are saving more. Savings as a percent of aftertax income. 1988-1990.
Business Week Apr 16, 1990 p. 20

Private savings still weak. Personal savings. Business and local government pension funds. Share of GNP. Budget surplus/deficit as share of GNP. 1950-1989.
Business Week May 29, 1989 pp. 88–89

Not what it used to be. Net national saving as a percent of GNP. 1960s-1980s.
Black Enterprise Mar 1989 p. 24

The savings rate is in the cellar. Savings as a percent of disposable personal income. 1966-1988.
Business Week Nov 21, 1988 p. 24

The savings rate: How big a rebound in 1988? Ratio of savings to disposable personal income. 1980-1988.
Business Week Mar 14, 1988 p. 24

Personal savings. Japan's frugality is one reason that it has so much money to invest. [Savings] as a percentage of disposable personal income. Japan. U.S. 1980-86.
Newsweek Feb 22, 1988 p. 46

SAVING AND SAVINGS - WORLD

Japanese household savings as a percentage of disposable income. 1983-1988.
U.S. News Sep 17, 1990 p. 43

Building a bigger piggy bank. Savings rate. United Kingdom. United States. Switzerland. West Germany. Japan.
Newsweek Jan 8, 1990 p. 45

Personal savings. Japan's frugality is one reason that it has so much money to invest. [Savings] as a percentage of disposable personal income. Japan. U.S. 1980-86.
Newsweek Feb 22, 1988 p. 46

SAVINGS AND LOAN ASSOCIATIONS

Banks. Is big trouble brewing? Loan losses for federally insured banks, after recoveries. 1976-1990. Noncurrent loans as a percent of all real estate loans. Change in real estate loans. Map.
Business Week Jul 16, 1990 pp. 146–147

Asset distribution among minority thrifts. Institutions. Assets. Women. Blacks. Hispanics. Asians.
Black Enterprise Jun 1990 p. 226

Shaky foundation. S&L profits and losses. 1978-1988. Value of foreclosed S&L real-estate loans. 1982-1988. Commercial real-estate loans made by S&L's. 1982-1988.
U.S. News Dec 12, 1988 p. 69

Mounting problems. Number of troubled savings and loans. 1978-1987.
U.S. News Jun 27, 1988 p. 60

SAVINGS AND LOAN ASSOCIATIONS - FAILURES

Scorecard for the S&L mess.
U.S. News Jul 23, 1990 p. 18

Where did the money go? Paying for all the bankrupt S&L's. The breakdown of the money trail.
U.S. News Jul 23, 1990 p. 20

Where has all the money gone? Assets of failed thrifts. Type of asset. Book value. Fair value. Loss.
U.S. News Apr 9, 1990 p. 38

Shaky S&L's. A dying breed. Bank failures resolved by: Mergers & acquisitions. Liquidations. Changes in management. 1978-1987.
U.S. News Aug 29, 1988 p. 84

SAVINGS AND LOAN ASSOCIATIONS - FEDERAL AID

The S&L debacle haunts U.S. Total budget deficit with S&L cleanup. Deficit excluding S&L cleanup. 1989, 1990, 1991.
U.S. News Jul 8, 1991 p. 8

The RTC's caseload continues to mount along with its inventory of unsold assets. Number of thrifts seized. Mortgages. Securities. Real estate. Other loans. Other assets.
Business Week Dec 25, 1990 p. 67

Guessing game. Estimates of the total cost of the savings and loan bailout. 1986-1990.
Time Aug 13, 1990 p. 51

What will the bailout cost? Treasury's optimistic projection. Goldman Sachs. GAO's pessimistic projection.
U.S. News Jul 9, 1990 p. 39

Huh? How much? The government's estimate of the cost of the S&L bailout has increased at least three times. March 1985. Spring 1986. August 1989. May 1990.
Newsweek Jun 4, 1990 p. 60

S&L scorecard. Total S&Ls. Taken over. Resolved. Unhealthy. Close to the brink.
Newsweek May 21, 1990 p. 22

A bailout that needs bailing out. The government's bailout agency is still having trouble selling and closing S&Ls and peddling foreclosed real estate. Where the properties are. Map.
Newsweek May 21, 1990 p. 22

Meet Uncle Sam, thrift owner. Thrift assets under RTC management.
Business Week Apr 9, 1990 p. 21

The burgeoning S&L bureaucracy. The thrift crisis has spawned a labyrinthine bureaucracy to oversee the cleanup. Chart.
U.S. News Apr 9, 1990 p. 37

The nation's biggest bailout states. The states with the highest percentage of these thrifts. 1989.
Black Enterprise Apr 1990 p. 33

Paying the tab. Taxpayers will foot almost two-thirds of the cost of the huge federal S&L bailout. Cost to industry. Cost to taxpayer.
Newsweek Aug 21, 1989 p. 39

Costly life supports. Cost of merging insolvent thrifts as a percentage of liquidation costs. 1985-1988.
U.S. News Jan 23, 1989 p. 46

Cracks in the system. Costs to the FDIC from bank failures. Bank failures. Costs to the FSLIC from savings and loan failures. Savings and loan liquidations, mergers and acquisitions. 1981-1987.
Time Aug 29, 1988 pp. 54–55

Shaky S&L's. Spendthrifts. Cost of savings and loan failures resolved by: Mergers & acquisitions. Liquidations. 1980-1988.
U.S. News Aug 29, 1988 p. 84

Shaky S&L's. Budget-bursting burials. Net outlays by FSLIC (negative numbers indicate additions to the federal budget deficit). 1980-1989.
U.S. News Aug 29, 1988 p. 84

To the rescue. Thrift institutions insured by FSLIC. Total. Insolvent. Potentially insolvent. FSLIC actions. Assisted mergers and acquisitions. Liquidations.
Newsweek Jun 20, 1988 p. 40

That sinking feeling. The FSLIC's total reserve has gone into deficit. 1982-1987.
Time Jun 20, 1988 p. 49

SAVINGS AND LOAN ASSOCIATIONS - FRAUD

S&L fraud: A scorecard. Pending investigations. Indictments. Convictions. Acquittals. Restitution ordered.
Business Week Sep 10, 1990 p. 84

SCANDINAVIA - ECONOMIC CONDITIONS

To market, to market. Nordic nations. Trade balance.
U.S. News Jul 2, 1990 p. 42

SCHOLASTIC APTITUDE TEST

SAT slide. A sampling of scores for students planning to major in [various subjects].
U.S. News Sep 16, 1991 p. 69

Money and SAT scores. Family income. SAT averages. Verbal. Math.
U.S. News Sep 9, 1991 p. 10

The nation's average combined SAT score. The best. The worst. 1990 average scores. 10-year change. [Various states.]
Newsweek Sep 10, 1990 p. 33

Setting hurdles. Average SAT scores at [various universities]. NCAA cutoff.
U.S. News Jan 30, 1989 p. 69

Key inconsistencies. SAT [scores] compared to performance. SAT scores (national average). Verbal. Math. Men. Women. 1978-1988. Percentage of Phi Beta Kappa electees who are female.
Ms. Jan 1989 p. 138

Testing the college-bound. Scholastic Aptitude Test scores and proportion of high-school graduates who attend college. [By state.] 1988 average SAT scores. Verbal, math. Percentage attending college.
U.S. News Oct 3, 1988 p. 71

The gates of academe. SAT scores. Percentage tested. 1977-1987.
Newsweek May 2, 1988 p. 59

SCHOOLS

See PUBLIC SCHOOLS

SCIENCE - SOVIET UNION

Soviet strength in science compared with U.S. strength.
Bull Atomic Sci Oct 1988 p. 30

SCIENTISTS

Percentage of U.S. scientists and engineers employed in defense industries. Percentage of 1989 engineering Ph.D.'s from U.S. colleges granted to foreigners. Percentage of college freshmen who plan to be scientists.
U.S. News Feb 4, 1991 p. 14

Engineering women. Scientists. Engineers. Number of women in field. Percentage in field. 1978, 1988.
U.S. News Jul 16, 1990 p. 66

Turning away from high tech. Science and engineering degrees awarded to women as a share of all such degrees in 1986 [by discipline and type of degree]. Baccalaureates. Masters. Doctorates.
Business Week Aug 28, 1989 p. 89

SECURITIES EXCHANGE COMMISSION

The SEC's workload keeps growing. Filings fully reviewed. Unread or partially reviewed. 1983-1987. It has barely added to its staff. Staff positions. 1983-1987.
Business Week Oct 10, 1988 p. 121

SECURITY GUARD SERVICES

Crime breeds business. Security-guard services' revenues. 1967-1995.
U.S. News Jul 11, 1988 p. 36

SELF-EMPLOYED

More people are working for themselves. Self-employed nonfarm workers as percent of civilian employment. 1989-1991.
Business Week Jul 22, 1991 p. 14

SELF-PERCEPTION

Fanciful fat. About 24 percent of men and 27 percent of women are truly overweight. Here is the percentage who said they are [various weights]. Male. Female.
U.S. News Feb 19, 1990 p. 70

SELLING

Better times for middlemen. Wholesalers' gross profits. 1986-1991.
Business Week Jan 14, 1991 p. 86

SEMICONDUCTOR INDUSTRY

Japan topped Silicon Valley by outinvesting the U.S. Share of world chip production. Average capital spending as share of chip revenues. 1979-1989.
Business Week Feb 5, 1990 p. 55

The U.S. loses more ground in chips. Share of worldwide semiconductor sales. 1985-1990.
Business Week Jan 8, 1990 p. 100

Japanese producers [of memory chips] are earning record profits. Sales. Increase. Net profits. Increase [by company].
Business Week Jul 4, 1988 p. 108

As the chip shortage keeps prices high, makers of small computers feel the pinch. Average selling prices for DRAM chips in the U.S. Estimated memory chip costs as percentage of total computer costs.
Business Week Jun 27, 1988 p. 29

Another pickup in chip sales. U.S. sales. 1983-1988.
Business Week Jan 11, 1988 p. 114

SENSES

How the brain really works its wonders. The things memories are made of. Diagram.
U.S. News Jun 27, 1988 pp. 48–49

SERVICE INDUSTRIES

The highs and lows of America's service trade. U.S. trade balance in selected services (1988). Winners. Losers.
U.S. News Nov 6, 1989 p. 66

SEX DISCRIMINATION IN EDUCATION

Key inconsistencies. SAT [scores] compared to performance. SAT scores (national average). Verbal. Math. Men. Women. 1978-1988. Percentage of Phi Beta Kappa electees who are female.
Ms. Jan 1989 p. 138

SEXUAL HARASSMENT

A disturbing pattern. Sexual harassment complaints filed with the EEOC. 1981, 1990.
Newsweek Oct 21, 1991 p. 36

Percent of women experiencing [various forms of sexual harassment]. Federal employees, 1987 survey.
Newsweek Oct 21, 1991 p. 36

SEXUALLY TRANSMITTED DISEASES

The other dangers of close encounters. Number of new cases reported yearly to the Centers for Disease Control. Syphilis. Chlamydia. Genital herpes. 1985-1988.
Time Jan 30, 1989 p. 62

SHAKESPEARE, WILLIAM, 1564-1616

The role's the thing. Shakespeare's longest roles. Role (play). Number of lines.
U.S. News Jul 2, 1990 p. 62

SHOPLIFTING

What shoplifters like. Types of goods most frequently stolen from supermarkets. Age of shoplifter.
U.S. News Mar 13, 1989 p. 73

SHOPPING

Shop till you drop. Consumer spending [Japan]. Adjusted for inflation, in trillions of yen. 1980-1990.
Newsweek Aug 13, 1990 p. 47

More shopping from home. Mail orders as a share of total general merchandise sales. 1985-1990.
Business Week Jan 8, 1990 p. 85

Why shopping takes forever. Average number of items in a supermarket. 1978-1987. 1987's new products.
U.S. News Jun 20, 1988 p. 78

SHOPPING CENTERS AND MALLS

Why merchants are miserable. As consumers shop less. Average shopping mall use per month. Trips to mall. Stores visited. Hours spent at mall. 1980-1990.
Business Week Nov 26, 1990 p. 135

Pigging out at the mall. Food. Sales share. Annual sales.
Newsweek Sep 25, 1989 p. 6

Attention, shoppers. [Here are] the shopping centers in the U.S. with the most stores.
U.S. News Jul 31, 1989 p. 66

SINGLE PARENT FAMILIES

A mixed picture. Rate of black children living with mother only. 1960-1988.
U.S. News Jan 22, 1990 p. 28

Single file. Who are the single mothers? Divorced. Separated. Widows. Never married. "Single mothers by choice."
Ms. Mar 1989 p. 84

She's gotta have it. Who owns what? Single women heads of households vs. single male heads of households.
Ms. Nov 1988 p. 86

See Also UNMARRIED MOTHERS

SINGLE PEOPLE

The singles count. Unmarried men and women [by age]. Males. Females.
U.S. News Sep 19, 1988 p. 72

SKIING

Slope numbers. Ski trips [by geographic area in U.S.] Daily visits in millions [by region]. 1986-87. 1988-89.
Newsweek Mar 12, 1990 p. 6

The price of skiing climbs and the danger drops. Average lift-ticket price. Injuries per 1,000 skier visits.
U.S. News Jan 11, 1988 p. 63

SLEEP

[Sleep patterns of] the trucker. The mother. The executive. The doctor. The student.
Time Dec 17, 1990 pp. 78–85

How much is enough? How much rest a person needs. Newborns. Children. Teenagers. Adulthood. Old age.
Time Dec 17, 1990 p. 80

The biology of sleep. Non-REM sleep. REM (rapid eye movement) sleep. The biological clock. Homeostasis.
Time Dec 17, 1990 p. 83

SMALL BUSINESSES

Job machine. Percentage of new jobs created by: Firms with fewer than 100 employees. Firms with more than 10,000 employees. 1986-1990.
U.S. News Jun 3, 1991 p. 52

Capital punishment. Venture capital invested in small businesses. 1980-1990.
U.S. News Jun 3, 1991 p. 62

Demand for SBA [Small Business Administration] loan guarantees. In billions of dollars. 1976-1989.
Black Enterprise Nov 1989 p. 22

Start-ups on a shoestring. Percentages of firms started with the following capital. Capital. Percentage of businesses.
U.S. News Oct 23, 1989 p. 78

A shrinking gender gap. Nonfarm self-employed individuals. Women. Men. 1963-1988.
U.S. News Oct 23, 1989 p. 78

Starts and stops. The number of small businesses that opened and closed their doors in 1987, by U.S. region.
U.S. News Oct 23, 1989 p. 80

Getting the right price. The following will give you a rough idea of the value of various kinds of businesses. Business. Rule of thumb.
U.S. News Jun 26, 1989 p. 64

The Yellow Pages index. 15 fastest-growing [business listings]. 15 businesses with biggest drops [in Yellow Page listings]. 1987.
U.S. News May 2, 1988 p. 77

SMOKING

The hunt for killer DNA. Smokers with lung cancer are more likely to have a deadly enzyme in their blood. Diagram.
Newsweek Oct 21, 1991 pp. 56–57

Teens and tobacco. Percentages of ninth to 12th graders who smoke. Male. Female. White. Black. Hispanic.
U.S. News Oct 14, 1991 p. 103

Puffing up. % who gained weight. Quitters. Men. Women.
Time Mar 25, 1991 p. 55

Substance abuse: The toll on blacks. Smoking-attributable deaths. Cirrhosis of the liver caused by alcohol abuse. Deaths caused by drug abuse.
Black Enterprise Jul 1990 p. 39

Stubbing out. % of 20-to-24-year-olds who smoke. Male. Female. 1965-1990.
Time Mar 5, 1990 p. 41

Leaders of the pack. Share of the U.S. cigarette market held [by company]. 1985-1989.
U.S. News Mar 5, 1990 p. 58

A smoking gun for smokers. Four of 5 people who die of lung disease each year are smokers. The number of deaths from lung disease by state. Rate of deaths per 100,000. Deaths.
U.S. News Nov 13, 1989 p. 89

A consumer's guide to highs and lows. Alcohol. Cocaine. Nicotine. How it works. How it feels. How it hurts. How to get help.
Newsweek Feb 20, 1989 p. 56

Who lights up? The demographics. Proportion of smokers by education and race. 1980-1987. Smokers by occupation.
Newsweek Feb 13, 1989 p. 20

Smoking, drinking and oral cancer. The risk of getting oral cancer for smokers and drinkers. Nondrinkers. Nonsmokers. Moderate smokers and drinkers. Heavy smokers and drinkers.
FDA Consumer Feb 1989 p. 40

Kicking the habit. Percent of each group that smokes. All. Men. Women. Whites. Black. High school dropouts. College graduates. 1965, 1987.
Time Jan 23, 1989 p. 54

Smoke signals. Per person consumption of cigarettes. 1925-1980.
U.S. News Jan 23, 1989 p. 9

No guarantees. Success rates of stop-smoking methods, ranked by percentage of smokers who quit for at least a year.
U.S. News Aug 1, 1988 p. 60

Where there's smoke. [Effects of nicotine on the body].
Time May 30, 1988 p. 56

Blowing smoke. Percentage of Americans who smoke. Men. Women. Total. 1955-1986.
U.S. News Feb 8, 1988 p. 20

Get married, stop smoking. Estimated time lost [life expectancy] because of [various factors].
U.S. News Jan 25, 1988 p. 77

See Also TOBACCO INDUSTRY

SMOKING - COSTS

Smoking and drinking: Taxes vs. the social cost. Taxes per pack or ounce. Social costs per pack or ounce (medical care, property loss, deaths, etc., not borne by smoker or drinker).
Business Week Jun 5, 1989 p. 27

The burning costs of smoking [due to premature death, disability, medical treatment and lost productivity]. Cost per person. Map.
U.S. News Apr 17, 1989 p. 26

SMOKING - LAWS AND REGULATIONS

States that have passed employee privacy protection laws. Prohibits employer discrimination for: Off-the-job smoking. Any legal off-the-job activity. Map.
Business Week Aug 26, 1991 p. 70

Snuffing out office smokers. Percentage of companies that restrict smoking [in various office areas].
U.S. News Apr 10, 1989 p. 72

A guide to lighting up. States with tough restrictions on smoking. States with limited anti-smoking laws. States with no curbs on smoking. Map.
Newsweek Apr 18, 1988 p. 25

SNACK FOODS

When good snacks go bad. Packaging and preservatives determine how long snack goods stay on the supermarket shelves. Here's how long different snacks can be sold before they're pulled.
U.S. News Jun 11, 1990 p. 74

SOCCER

I Mondiali: The greatest draw on earth. The World Cup's total cumulative worldwide TV audience. 1974-1990.
Business Week Jun 18, 1990 p. 169

SOCIAL PROGRAMS

How social investment in children pays off. $1 invested in [a social program] saves [this amount in remedial care].
Business Week Sep 19, 1988 p. 123

SOCIAL SECURITY

What you can expect from the government. Estimated monthly Social Security benefits at retirement. Worker's current age. Worker's earnings in 1989.
U.S. News Jul 30, 1990 pp. 62–63

How higher taxes stack up. The effects of raising Social Security benefit taxes on several elderly taxpayers in 1991. Moderate-income seniors. High-income seniors.
U.S. News Jul 23, 1990 pp. 46–47

Floating on a sea of green. Beating the deficit. Social-Security surplus. National debt. Payouts in 1989. To retirees. To hospital insurance. To disability insurance.
Newsweek May 7, 1990 p. 55

The graying of the numbers. Life expectancy has increased the demands on Social Security. [Various statistics.]
Newsweek May 7, 1990 p. 56

Federal outlays for retirees are climbing while other spending is falling. Domestic outlays. Defense. 1980, 1990, 1995.
Business Week Apr 2, 1990 p. 33

How Social Security masks the deficit. Federal budget deficit. Excluding Social Security surplus. 1985-1994.
Business Week Feb 5, 1990 p. 26

Headed for long-term trouble. Social Security trust fund assuming no payroll tax cut. 1989-2030.
Business Week Feb 5, 1990 p. 26

The Social Security take is gaining on income taxes. Individual income taxes. Social Security taxes. Deficit without Social Security. Social Security surplus. Unified budget deficit. 1980-1989.
Business Week Jan 29, 1990 p. 67

Maximum Social Security tax. 1980-1990.
Time Jan 22, 1990 p. 26

What you can expect from the government. Estimated monthly Social Security benefits at retirement.
U.S. News Aug 14, 1989 p. 61

The surplus story: The future of the Social Security system. Future scenarios that are deemed most probable by experts. Today. 1993-2005. 2010-2030. 2035-2050.
U.S. News Dec 26, 1988 pp. 27–28

The surplus scenario. Social Security trust funds. End-of-year assets. Income. Outgo. 1988-2048.
U.S. News Dec 26, 1988 p. 30

The surplus scenario. How the surplus masks the deficit. Reported budget deficit. Projected Social Security surplus. Real projected budget deficit. 1989-1994.
U.S. News Dec 26, 1988 p. 30

The rising Social Security Trust Fund balance. 1988-2040.
Black Enterprise Nov 1988 p. 41

Shaky Social Security. Social Security funds. Social Security payouts. 1988-2048.
U.S. News Aug 15, 1988 p. 67

The Social Security surplus is growing and helping to finance the deficit. Trust fund. Deficit without Social Security surplus. Unified budget deficit. 1986-1993.
Business Week Jul 18, 1988 p. 93

Now you see it, now you don't. Social Security Trust Fund assets in trillions of dollars. 1980-2050.
Time Jul 4, 1988 p. 58

Fate of the retirement kitty. Number of Social Security beneficiaries. Social Security funds. 1985-2048.
U.S. News Jun 13, 1988 p. 71

[Social Security]. Population. Life expectancy. Federal spending. 1950, 1985, 2020.
Time Feb 22, 1988 p. 70

SOCIAL VALUES

Putting kids first. How adults rank the importance of different aspects of their lives. U.S. Japan. Europe.
U.S. News Aug 1, 1988 p. 62

SOFT DRINK INDUSTRY - MARKET SHARE

Pouring it on. Share of the domestic market. Coca-Cola. Pepsi-Cola. 1985-1991.
U.S. News Dec 9, 1991 p. 64

Cola competition. Market share. Coke. Coca-Cola Classic. Pepsi. 1985-1989.
Newsweek Mar 19, 1990 p. 38

Soda wars. Top 10 soft drinks and their market shares. Average yearly per-person consumption of soft drinks and coffee (in gallons) in U.S.
U.S. News May 23, 1988 p. 76

In soft drinks, the big get bigger. Brand. Market share. 1986, 1987.
Business Week Feb 8, 1988 p. 94

SOUTH AFRICA

U.S. corporate divestment. Total companies with direct investment in South Africa: U.S. Non-U.S. 1984-1988.
Black Enterprise Mar 1989 p. 41

South Africa at a glance. Area. Population. Per capita gross national product. Natural resources. Industries. Black homelands. Map.
Scholastic Update Feb 12, 1988 p. 4

SOUTH AFRICA - ETHNIC GROUPS

South Africa at a glance. Whites. Blacks. Colored. Asians. Background. Population. Where they live. Average monthly income. Life expectancy. Infant mortality. Education.
Scholastic Update Feb 12, 1988 p. 4

SOUTH AFRICA - RACE RELATIONS

Proposed Afrikanerland. Homelands. [Proposed home for whites]. Map.
Time Dec 10, 1990 p. 66

Scrambling for seats. If negotiations do begin, it will not be easy to settle who sits at the table. Some possible players.
Time Feb 5, 1990 p. 28

SOUTH AMERICA

The Columbian Exchange. New world portrait. New world to old [animals and plants brought]. Old world portrait. Old world to new. Diagram.
U.S. News Jul 8, 1991 pp. 29–31

SOUTH AMERICA - ANTIQUITIES

Parallel worlds. Andes region. Other regions. 13,000 B.C. - 1532 A.D. Map. Chart.
U.S. News Apr 2, 1990 pp. 48–49

SOUTHEAST ASIA

Indochina. A history of foreign domination. Thumbnail histories. Vietnam. Cambodia. Laos. Map.
Scholastic Update Apr 6, 1990 pp. 10–11

SOUTHERN STATES - ECONOMIC CONDITIONS

The Southeast. South Central. Employment. Population. Median home price. Annual rate of change from previous year. 1989-1990.
U.S. News Nov 13, 1989 p. 62

SOVIET UNION

From the Arctic to Asia. Russian Republic. U.S.S.R. Area. Population. Production of: Grain. Hard coal. Steel. Map.
Newsweek Jun 24, 1991 p. 29

Drop in Soviet GDP. Average monthly wage in U.S.S.R. Official ruble exchange rate. Percentage who say they are dissatisfied with their life. Percentage of marriages that end in divorce: U.S.S.R. U.S.A.
U.S. News Nov 12, 1990 p. 12

Boris's bailiwick: Mother Russia. Population. Energy. Industry. Agriculture. Security. Map.
Newsweek Jun 11, 1990 p. 22

[What] the Soviet Union relies on Lithuania for. [What] Lithuania relies on the Soviet Union for.
Time Apr 30, 1990 p. 45

The 15 republics of the Soviet Union. Population. [Year] established as a separate republic of the Soviet Union. Map. [Other selected statistics.]
Time Mar 12, 1990 pp. 29–31

Medicine & health. Infant deaths per 1,000 births. Official life expectancy at birth. Kidney transplants. Medical supplies. CAT scanners. Rural hospitals.
U.S. News Apr 3, 1989 p. 43

Quality of life. Automobile registrations per 1,000 people. Miles of paved roads. Percentage of families with telephones in cities and towns, in rural areas.
U.S. News Apr 3, 1989 p. 44

SOVIET UNION - BREAKUP OF THE UNION

What in common? The new marriage of Slavic and Central Asian republics may be rife with tensions. Founders of the Commonwealth. Agreed to join. Interested. Not interested. Map.
Newsweek Dec 23, 1991 p. 28

Re-forming itself. Founding members of the Commonwealth. Agreed to join. Considering the pact. Baltic states. Map.
Time Dec 23, 1991 p. 20

Chaos or commonwealth? Old animosities, economic chaos and the presence of some 27,000 nuclear weapons will make the transition from the Soviet Union to a Commonwealth of Independent States a difficult one. [Other selected statistics.]
U.S. News Dec 23, 1991 p. 38

The union of ...? [Republics and their latest moves to independence.] Map.
Scholastic Update Dec 6, 1991 pp. 6–7

State of the union. Republics. Population. Per capita GNP. [Date] established as a separate republic of the Soviet Union. [Description of recent political events]. Map.
Time Sep 9, 1991 pp. 30–31

Coming apart. All of the republics but Russia are dependent on one another for basic necessities, and their defenses also are integrated. [Various statistics.]
U.S. News Sep 9, 1991 p. 24

What they voted on: "Do you consider it necessary to preserve the Union of Soviet Socialist Republics?" Six republic governments that refused to participate. Republics voting on referendum as written. Republics with a second ballot question. Map.
Time Mar 25, 1991 p. 30

[Independence or not.] Noncommunist republics seeking independence. Communist republics wishing to retain the present federation. Seeking compromise. Average monthly net wages. Map.
Time Feb 25, 1991 p. 57

A disintegrating empire. Ethnic makeup of the Soviet Republics. Russian. Predominant nationality. Other. Percentage earning less than 75 rubles per month.
U.S. News Nov 19, 1990 p. 36

Breaking away. Russian Republic. Ethnic regions asserting political or economic autonomy. Other Soviet republics. Map.
Newsweek Nov 12, 1990 p. 41

The Soviet Empire: The center cannot hold. The Soviet Union beset by ethnic, economic and religious strife. Estonia. Latvia. Lithuania. Ukraine. Moldavia. Georgia. Azerbaijan. Map.
Newsweek Jun 4, 1990 p. 20

Trouble spots. Strikes, ethnic unrest, mediocre grain harvest. Environmental problem areas. Map.
U.S. News Nov 20, 1989 p. 28

SOVIET UNION - ECONOMIC CONDITIONS

Soviet oil output is falling steadily. Crude oil. Millions of barrels per day. 1988-1992.
Business Week Sep 9, 1991 p. 36

Shattered economy. Percent change in Soviet GNP. 1983-1991.
U.S. News Sep 9, 1991 p. 36

Decaying industry. Percent change in Soviet industrial production. 1983-1991.
U.S. News Sep 9, 1991 p. 37

Soaring prices. Soviet retail price index.
U.S. News Sep 9, 1991 p. 39

The new cost of Soviet living. [Selected] item. Old price. New price. Black market.
Newsweek Apr 15, 1991 p. 39

Soviet oil exports plunge. 1986-1991.
Business Week Feb 11, 1991 p. 46

Moscow's money trap. Trade collapses as foreign debt soars. 1987-1991.
Business Week Feb 4, 1991 p. 66

Moscow's lopsided balance sheet. Soviet deposits in Western banks. Western bank loans to Soviets. 1986-1990.
Business Week Dec 10, 1990 p. 44

The shattered Soviet system. Percent change in gross national product. Production drops. 1985-1990.
Scholastic Update Dec 7, 1990 p. 11

Percent change in the rate of inflation. Prices keep going up. 1985-1990.
Scholastic Update Dec 7, 1990 p. 11

Exchange rate for one Soviet ruble. (Versus one U.S. dollar). Official rate. Black market rate. The ruble plunges. 1985-1990.
Scholastic Update Dec 7, 1990 p. 11

Yes, we have no potatoes. Many potatoes remain unharvested because of poor planning and lack of incentive. Diagram.
Time Dec 3, 1990 p. 75

A disintegrating empire. Ethnic makeup of the Soviet Republics. Russian. Predominant nationality. Other. Percentage earning less than 75 rubles per month.
U.S. News Nov 19, 1990 p. 36

Drop in Soviet GDP. Average monthly wage in U.S.S.R. Official ruble exchange rate. Percentage who say they are dissatisfied with their life. Percentage of marriages that end in divorce: U.S.S.R. U.S.A.
U.S. News Nov 12, 1990 p. 12

How *perestroika* ruined the Soviet economy. GNP growth. Inflation. Official ruble rate. Black market rate. 1985-1990.
Business Week Oct 1, 1990 p. 140

Attention, comrade shoppers. Raisa Gorbachev may shop for furs and fancy clothes, but ordinary Soviet citizens are keen on household products. Here are some purchases they hope to make in the next two years.
U.S. News May 7, 1990 p. 75

Production line. % from each republic. Refrigerators. Wheat. Motors. Cotton. Gas. Beef. Poultry. Televisions. Corn. Coal. Map.
Time Mar 12, 1990 p. 30

Empire blues. Tensions between rich and poor complicate the Soviet Union's already difficult nationalities problem. Economic output per capital [by country].
U.S. News Jan 29, 1990 p. 34

Lagging output. Per capita GNP (1988). U.S. Japan. West Germany. East Germany. Singapore. U.S.S.R. Hungary. Bulgaria.
U.S. News Nov 20, 1989 p. 32

Bitter harvest. Soviet farmers still cannot produce enough food to feed the nation. Value of agricultural output. Fixed-capital stock. 1970-1988.
U.S. News Nov 20, 1989 p. 32

Redder ink. Enormous debts hamper Gorbachev's reforms. Budget deficit (in rubles). 1980-1989.
U.S. News Nov 20, 1989 p. 32

A Third World export pattern. U.S. Malaysia. G-7 trading partners. U.S.S.R. Composition of exports (1985).
U.S. News Nov 20, 1989 p. 32

Slow growth. Despite *perestroika's* promises, Soviet production is stagnant. Average annual rate of GNP growth. 1961-1989.
U.S. News Nov 20, 1989 p. 32

More equal than others. People. Average per capita monthly income. Rubles per month. 1988.
U.S. News Nov 20, 1989 p. 35

Last among equals. When it comes to basic measures of the quality of life, the Soviet Union lags behind its Eastern European satellites. Annual meat production. Automobiles registered. Life expectancy.
U.S. News Nov 20, 1989 p. 35

Shopping centers. The U.S.S.R. imports mostly from Eastern Europe and the Third World. Goods. Value (rubles). Percentage from West. Primary source.
U.S. News Nov 20, 1989 p. 35

Lofty goals, harsh realities. Consumer goods. Recent output and new targets for selected goods. 1986-88. 1990 plan.
U.S. News Nov 20, 1989 p. 35

High-tech, low budget. Personal computers in the Soviet Union. Actual and planned production of personal computers in the Soviet Union. 1986-1990.
U.S. News Nov 20, 1989 p. 35

Half a loaf, on a good day. Grain spilled by railroad cars. Percent of each Soviet wheat harvest used as seed for the next crop. Grain lost before it gets to market. Diagram.
U.S. News Nov 20, 1989 p. 36

The sinking ruble. Dollars per ruble. Official. Black market. 1986-1989.
Business Week May 1, 1989 p. 41

The economy. Soviet budget deficit. 1981-1988. Planned and actual GNP growth. Total GNP. 1966-1990.
U.S. News Apr 3, 1989 p. 43

Food on the table. Consumption (pounds per person). 1913, 1917, 1987. Hours of work required to purchase 5 pounds of [various products]. 1927, 1960, 1985.
U.S. News Apr 3, 1989 p. 44

Agriculture. Fixed capital investment in agriculture. Metric tons of grain produced. 1970-1988. What market prices would mean. Estimated market price. State price.
U.S. News Apr 3, 1989 p. 44

Growing pains and a budget in the red. Soviet economic growth. 1980-1988. Soviet budget deficit. 1980-1989.
Newsweek Mar 13, 1989 p. 29

Indices of development of the Soviet economy. National income. Capital. Capital productivity. Labor productivity. Investment.
Bull Atomic Sci Dec 1988 p. 23

SOVIET UNION - FOREIGN INVESTMENTS

Falling investment. Percent change in Soviet fixed investment. 1983-1991.
U.S. News Sep 9, 1991 p. 38

Soviet business deals in the works. Company [and country]. Project.
Business Week Sep 2, 1991 p. 29

Partners in progress. Soviet joint ventures with the West. Value. Biggest project. [Selected countries.]
U.S. News Jul 31, 1989 p. 40

The surge in joint ventures. In quantity and in investment. Total ventures registered. Operational. Millions of dollars. 1987-1989.
Business Week Jun 5, 1989 p. 64

SOVIET UNION - HISTORY

Growth of the Russian Empire [11th - 20th centuries]. Early Russia. U.S.S.R. 1917-1947. Map.
Time Mar 12, 1990 p. 47

Russia under the Czars covers two continents. Until 1689. Annexed 1690-1796. Annexed 1797-1917. Map.
U.S. News Jan 15, 1990 p. 32

The Soviet Union: Rebuilding the empire. Annexed since World War II. Map.
U.S. News Jan 15, 1990 p. 33

SOVIET UNION - IMMIGRATION AND EMIGRATION

Open gates. Jewish emigration from the U.S.S.R. 1974-1989.
Newsweek Sep 25, 1989 p. 52

SOVIET UNION - MILITARY STRENGTH

On their way home. Soviet and American bases face each other along the Pacific Rim. Map.
Newsweek Jun 18, 1990 p. 24

Military balance in the Pacific and Asia. U.S. Pacific forces. U.S.S.R. Pacific forces. Map.
U.S. News Apr 23, 1990 p. 33

Soviet military might in Europe. Moscow's global reach: A blue-water navy and growing political ties buttress foreign policy. Maps.
U.S. News Mar 13, 1989 pp. 20–21

Big brother's occupying forces. Soviet military strength in Warsaw Pact countries. Active-duty personnel. Armored divisions. Helicopters. Combat aircraft.
U.S. News May 16, 1988 p. 32

SOVIET UNION - NATIONALISM

State of disunion. Core state. Affiliated states. States aiming at independence. Independent states. Map.
Business Week Sep 9, 1991 pp. 28–29

A multinational empire. The Baltics. Western frontier. Transcaucasia. Russia. Central Asia. Map.
Newsweek Sep 9, 1991 pp. 20–21

Estonia. Latvia. Lithuania. Population. Per capita GNP. [Ethnic make-up.] Map.
Time Sep 9, 1991 p. 36

Coming apart. All of the republics but Russia are dependent on one another for basic necessities, and their defenses also are integrated. [Various statistics.]
U.S. News Sep 9, 1991 p. 24

[Soviet Union]. Republic. Type of unrest. Troops deployed during past three years. Map.
Time Jan 28, 1991 p. 82

Map/databank. USSR. [The republics.] Area. Population. Ethnic groups. Map. Chart.
Scholastic Update Dec 7, 1990 pp. 6–7

A disintegrating empire. Ethnic makeup of the Soviet Republics. Russian. Predominant nationality. Other. Percentage earning less than 75 rubles per month.
U.S. News Nov 19, 1990 p. 36

Empire blues. Tensions between rich and poor complicate the Soviet Union's already difficult nationalities problem. Economic output per capital [by country].
U.S. News Jan 29, 1990 p. 34

The Soviet Union's unruly republics. Lithuania. Latvia. Estonia. Moldavia. Georgia. Azerbaijan. Map.
Time Jan 22, 1990 p. 32

A house divided. Restless nationalities have grievances against Moscow—and each other. Georgia. Armenia. The Crimea. Baltic States. Moldavia. Uzbekistan.
Newsweek Apr 24, 1989 p. 52

Soviet mosaic. Population of ethnic groups in millions, based on 1979 census.
Time Nov 28, 1988 p. 46

Balance of power. While ethnic Russians are losing majority status, they remain the dominant group. Population [by ethnic group] in year 2000. Where they are. Map.
Newsweek Mar 14, 1988 p. 26

SOVIET UNION - NATURAL RESOURCES

Raw materials. Gold. Diamonds. Hard coal. Oil. Natural gas. Uranium. Map.
U.S. News Nov 19, 1990 p. 37

SOVIET UNION - POLITICS AND GOVERNMENT

The Soviet Union's competing centers of power. Communist party. Soviet government. Chart.
Business Week Feb 19, 1990 p. 32

Gorbachev (General Secretary, President). The Party. Politburo. Central committees. Party Congress. The Government. Supreme Soviet. Council of Ministers. Congress of People's Deputies. Chart.
Time Feb 19, 1990 p. 34

How the election works. Supreme Soviet. Congress of People's Deputies. District seats. Territorial seats. Organization seats. Chart.
Time Feb 6, 1989 p. 48

SOVIET UNION - RELIGION

Faith and doubt. Breakdown of Soviet population by affiliation.
Time Oct 15, 1990 p. 71

SOVIET UNION, AID TO

Who's sending care packages to Moscow. Government. Type of aid. Value (billions).
Business Week Oct 28, 1991 p. 42

The rain of grain. [Donors of grain to the Soviet Union]. Country. Donated food (tons).
Newsweek Dec 24, 1990 p. 37

Doing well, doing good. How West Germany compares with the U.S. on Soviet trade—and aid. Imports. Exports. Aid.
Time Jul 16, 1990 p. 17

SOVIET UNION AND CHINA

Trading the China card. Total Soviet exports. Total Soviet imports. Soviet exports to China. Soviet imports from China. 1980-1990.
U.S. News Aug 19, 1991 p. 30

Two giants. Soviet Union and the People's Republic of China [vital statistics]. Map.
Scholastic Update May 5, 1989 pp. 4–5

U.S.S.R. China. [Locations of armed forces.] Strategic missile sites. Military headquarters. Map.
Time Jul 18, 1988 p. 32

SOVIET UNION AND EASTERN EUROPE

Moscow's red ink. Annual Soviet trade with Eastern Europe. 1980-1988.
U.S. News Mar 27, 1989 p. 36

SOVIET UNION AND THE MIDDLE EAST

The Soviet stake in the Middle East. Country. Arm sales. Advisers. Base access.
Business Week Feb 25, 1991 p. 33

SOVIET UNION AND THE UNITED STATES

The mushrooming race. Three decades of agreements. [Timeline of nuclear weapons treaties.] United States. Soviet Union. 1945-1990.
U.S. News Oct 7, 1991 pp. 24–25

U.S. private citizens living in the Soviet Union. U.S. high-school students enrolled in Russian-language classes. [Other selected statistics.]
U.S. News Sep 2, 1991 p. 13

On their way home. Soviet and American bases face each other along the Pacific Rim. Map.
Newsweek Jun 18, 1990 p. 24

The U.S. vs the Soviets. Civilian research. Scientists and engineers. Nobel prizes in science. Man-hours in space. Computer production.
Business Week Nov 7, 1988 p. 83

Fickle trade winds. U.S.-U.S.S.R. bilateral trade. 1970-1987.
Newsweek Jun 13, 1988 p. 47

Soviet tortoise, American hare. U.S.S.R. and U.S. manned space achievements. 1961-1990. Chart.
U.S. News May 16, 1988 pp. 48–49

Staying aloft. Periods when the U.S.S.R. and the U.S. have had a man in orbit. 1961-1988.
U.S. News May 16, 1988 pp. 52–53

Here today, gone tomorrow. U.S. exports. U.S. imports from U.S.S.R. 1976-1987.
U.S. News Apr 4, 1988 p. 51

Comparing two worlds. U.S. U.S.S.R. Geography. The people. Quality of life. The economy. Military might. Map of Soviet Republics.
Scholastic Update Mar 25, 1988 pp. 8–9

A timeline of U.S.-Soviet ties. 1689-1987. Chart.
Scholastic Update Mar 25, 1988 p. 19

SOYBEANS
See GRAIN

SPACE FLIGHT
Soviet tortoise, American hare. U.S.S.R. and U.S. manned space achievements. 1961-1990. Chart.
U.S. News May 16, 1988 pp. 48–49

Staying aloft. Periods when the U.S.S.R. and the U.S. have had a man in orbit. 1961-1988.
U.S. News May 16, 1988 pp. 52–53

SPACE FLIGHT - UNITED STATES
Cost of space program. NASA employees. Manned space flights. Satellites now in space.
U.S. News Jun 3, 1991 p. 10

The long road to Jupiter [Galileo mission timetable]. Diagram.
U.S. News May 15, 1989 p. 58

An all new bird. Booster rocket. Orbiter. Main engines. External tank. Diagram.
Time Oct 10, 1988 p. 24

SPACE POLLUTION
Neither bird nor plane. The uncertain fate of Cosmos 1900 underscores the growing threat from radioactive space debris. Total launches 1957-1988. Total nuclear launches 1961-1988.
Newsweek Aug 29, 1988 p. 53

SPACE SHUTTLE MISSIONS
Getting a scientific backlog into orbit. Some shuttle projects slated for the next 10 years.
Business Week Oct 17, 1988 pp. 50–51

An all new bird. Booster rocket. Orbiter. Main engines. External tank. Diagram.
Time Oct 10, 1988 p. 24

SPAIN - ECONOMIC CONDITIONS
An economic portrait of Eastern Europe. East Germany. Poland. Czechoslovakia. Hungary. Spain [used as a comparison]. Per capita GNP. Hourly wages. Education level. Population. Industrialization. Debt. Hard currency assets. 1987, 1988.
Business Week Nov 27, 1989 pp. 62–63

SPELLING ABILITY
Spelled by the champs. The winning words [in the National Spelling Bee] in previous years. 1965-1988.
U.S. News Feb 19, 1990 p. 70

SPORTS
Killer sports. Here are the death rates per 100,000 participants in deadlier sports.
U.S. News Jan 15, 1990 p. 67

The price of fun. Americans parted with more than $13 billion for sports clothing last year. Here is how much the average participant spent. [By activity.]
U.S. News Oct 9, 1989 p. 81

On a roll. Americans in pursuit of fitness and recreation spent $677 million on these balls last year. Here's where the money went.
U.S. News Aug 21, 1989 p. 66

The $20 billion golf industry is growing, while more active sports are fading. Golf. Tennis. Jogging. 1984-1987.
Business Week Mar 27, 1989 p. 77

Dollars for desk jocks. Sports federations fill tall buildings and fat committees. Federation fund [selected sports]. 1987 budget. Amount given to athletes (% total).
Newsweek Aug 15, 1988 p. 62

The games people play. Percent of women and men who participate in selected sports.
Ms. May 1988 p. 69

SPORTS - INJURIES

A guide for aching athletes. Here's how to tell when your injury warrants a visit to the doctor. Knee. Foot. Back. Shoulder.
U.S. News Jul 31, 1989 p. 57

SPORTS, COLLEGE

NBA revenue. Average NBA salaries. College referee's pay per game. Graduation rate of college players. [Other selected statistics.]
Newsweek Nov 25, 1991 p. 9

See Also PUBLIC SCHOOLS - SPORTS

SPORTS, ECONOMIC ASPECTS

Sponsorship becomes big business. Corporate outlays to sponsor sporting events. 1984-1991.
Business Week Jan 21, 1991 p. 88

ST. PAUL (MINNESOTA)

St. Paul. Population. Median family income. Median home. Unemployment.
Newsweek Feb 6, 1989 p. 43

STANDARD OF LIVING

The top 10: Having it all. Population. Median family income. Median home. Unemployment.
Newsweek Feb 6, 1989 pp. 42–44

Americans' changing fortunes. How Americans compare their lives with five years before. How Americans compare the state of the nation with five years before. Better today. Worse today. No change. 1979-1988.
U.S. News Aug 29, 1988 pp. 102–103

STATE GOVERNMENT - FINANCE

Budget busters. State expenditure growth as a percentage of personal income, 1979-89. Corrections. Medicaid. Education. Highways. Welfare.
U.S. News Dec 23, 1991 p. 52

Harmful deficits. State and local budget balances, excluding social insurance funds. 1982-1991.
U.S. News Jul 8, 1991 p. 51

Dividing up the pie. State and local government expenditures 1988-89. Education. Public safety. Interest on general debt. Transportation. Utility. Social services. Other.
Newsweek Jul 1, 1991 p. 30

Constant growth. Government spending as a percent of GNP. State and local. Federal. 1950, 1970, 1990.
Newsweek Jul 1, 1991 p. 31

Boosting revenues. State and local government budget surplus/deficit. 1980-1991.
U.S. News May 27, 1991 p. 56

A tide of red ink swamps state and local budgets. Operating balance. 1985-1990.
Business Week Apr 15, 1991 p. 22

Where the states get their tax dollars. Tax/percent. Sales. Personal income. Corporate income. Other.
Business Week Mar 4, 1991 p. 24

Who's in the poorhouse? Largest budget deficits as a percentage of budget [by state].
U.S. News Feb 18, 1991 p. 45

Who's got the money? Largest budget surpluses as a percentage of budget [by state].
U.S. News Feb 18, 1991 p. 46

Who's got the best credit. Top bond ratings [by state]. Worst bond ratings [by state].
U.S. News Feb 18, 1991 p. 48

How far off target? Shortfall (deficits in millions of dollars) as a % of total fiscal year 1991 budget. [By state.]
Time Dec 31, 1990 p. 15

State and local budgets are deep in the red. Operating balance excludes social insurance funds. 1983-1990.
Business Week Jul 30, 1990 p. 12

Capital of the states. Here is a comparison of leading *Fortune* 500 company revenues with those of top states.
U.S. News May 7, 1990 p. 75

State and local budgets head into a deeper rut. Operating budget. 1984-1988 (est).
Business Week Dec 26, 1988 p. 40

State and local budgets slide deeper into the red. Operating balance. 1984-1988.
Business Week May 2, 1988 p. 24

STEALTH BOMBER

The B-2: How does it work? For military planners, the Stealth's beauty is its ability to escape detection by conventional radar. The Pentagon will reveal only some of its secrets. Diagram.
Newsweek Dec 5, 1988 p. 19

How the Stealth evades detection. Diagram.
Time Dec 5, 1988 p. 21

In a general nuclear war, B-2 Stealth bombers have the crucial mission of destroying mobile Soviet SS-24 and SS-25 missiles. [How this mission will be accomplished.] Diagram.
U.S. News Nov 28, 1988 pp. 20–21

How Stealth works. Shape. Radar-absorbing material. Traps. How to fight Stealth. Bistatic radar. Carrier-free radar. Diagram.
U.S. News Nov 28, 1988 pp. 24–25

STEEL INDUSTRY

Big steel's big headache. Operating profit per ton. 1986-1991.
Business Week Jul 1, 1991 p. 27

A decline in steel demand. U.S. steelmaker's domestic shipments. 1985-1990.
Business Week Jan 8, 1990 p. 71

Steelmakers are prospering once again but steelworkers are still feeling the pinch. Earnings (losses) from steel operations of the six biggest U.S. steelmakers. 1983-1988. Average hourly earnings.
Business Week Mar 13, 1989 p. 40

Another healthy year for steelmakers. Average operating profit per ton for 'Big Six' steel companies. 1984-1989.
Business Week Jan 9, 1989 p. 79

Demand [for steel] climbs. Total U.S. consumption. 1981-1988.
Business Week Sep 26, 1988 p. 49

Prices soar. Average U.S. steel prices. 1981-1988.
Business Week Sep 26, 1988 p. 49

Big strides at the mini-mills. Percent of total domestic steel shipments. Annual steel shipments. 1975-2000.
Business Week Jun 13, 1988 p. 102

Big steel's profit comeback. Aftertax income of the six biggest U.S. steelmakers. 1983-1988.
Business Week Jan 11, 1988 p. 94

As steel imports drop. Imports as a percentage of U.S. market. 1981-1988.
Business Week Sep 26, 1988 p. 49

STEROIDS

Steroids may give you more than you bargained for. Side effects and adverse reactions.
FDA Consumer Sep 1991 p. 26

Teenagers blasé about steroid use. Chart 1: Most frequent steroid-associated health problems cited by current and former users. Chart 2: Some reasons cited by current users as to why they disagree with experts about risks.
FDA Consumer Dec 1990 p. 2

STOCK MARKET

Stock and bond issues. Mergers and acquisitions. 1987-1991.
Time Sep 16, 1991 p. 46

Recession's signs and portents. Stock Market. S&P 500 index. 1989, 1990.
U.S. News Sep 24, 1990 p. 80

A recessions'-eye view of the Dow. Dow Jones industrial average. 1945-1990. [Chronology including eight recessions.]
U.S. News Sep 24, 1990 pp. 82–83

The case for holding on. Holding period [of stocks]. Your chance of earning. Chance of losing.
Newsweek Jun 4, 1990 p. 63

Vanishing little guy. Net excess of sales over purchases of common stocks by individuals. 1980-1989.
U.S. News Dec 4, 1989 p. 70

Stocks that did best during the 1980s. Company. Primary business. Price appreciation.
U.S. News Dec 4, 1989 p. 70

Hot stocks to watch. 10 investment strategists were asked to pick a single stock for the 1990s.
U.S. News Dec 4, 1989 p. 71

Profiting from market peaks. [Profits of] investor who put $2,000 into the stock market every year for the last 20 years right at its peak. Portfolio value. Cash invested. 1969-1989.
U.S. News Nov 13, 1989 p. 89

Investors love leveraged companies. "Stub" stock index. S&P 500 index. 1986-1988
Business Week Nov 7, 1988 p. 141

[More junk bonds are being used.] Resulting in a shrinking stock market. Equity issues. 1980-1988.
U.S. News Nov 7, 1988 p. 63

Dow Jones picks the President. Percentage change in Dow Jones industrial average. Year incumbent party won. Year incumbent party lost. 1900-1984.
U.S. News Sep 19, 1988 p. 72

As Wall Street's capital has grown. Total invested capital of New York stock exchange firms, excluding specialists. 1982-1988. The rate of return has fallen. Annual pretax return on average capital. 1982-1988.
Business Week Aug 1, 1988 p. 80

Profiting by bad news. How short selling works.
Newsweek Apr 4, 1988 p. 44

The January barometer. Is January a stock-market bellwether for the rest of the year? January change. Yearly change. 1950-1988.
U.S. News Feb 15, 1988 p. 88

Setting up a low-risk portfolio. Common stocks. Bonds. Money market. Gold stocks.
Black Enterprise Feb 1988 p. 63

STOCK MARKET - WORLD

Taking stock. The top foreign stock markets [by country].
U.S. News Sep 10, 1990 p. 92

Share of the world's stock market money. Other nations. Europe. Japan. U.S. 1980, 1989.
Scholastic Update Sep 7, 1990 p. 21

STRATEGIC ARMS LIMITATION TALKS

See ARMS CONTROL

STRESS

Rough day at the office. Stressful events for managers.
U.S. News Jul 25, 1988 p. 66

STRIKES

Where strikers lost jobs. Company. Year of walkout. Strikers. Back at work.
Business Week Aug 5, 1991 p. 27

Striking out. Number of workdays lost to labor disputes (in thousands). U.S. Japan. West Germany.
U.S. News Jul 16, 1990 p. 26

More power to the picketers. Americans may be more sympathetic to striking workers, but fewer are walking off the job in labor disputes. [Various occupations.] 1978, 1989.
U.S. News Feb 26, 1990 p. 63

STUDENT LOAN PROGRAMS

A class of deadbeats. These four-year colleges and universities have the highest percentage of defaults on student loans made under the Guaranteed Student Loan program. 1986, 1987.
U.S. News Jul 3, 1989 p. 62

Deadbeats. Default rates on student loans by type of school.
U.S. News May 22, 1989 p. 55

STUDENTS

No pain, no gain. Time [spent] on homework [by] seventeen-year-old students. 1988. Chart.
Newsweek May 27, 1991 p. 62

It's arithmetic, in a landslide. Favorite subjects of children in school. Girls. Boys. Math. English. Art. Reading. Science.
U.S. News Sep 12, 1988 p. 71

STUDENTS - EMPLOYMENT

Hardworking high-schoolers. Percentages of 16 and 17-year-old students who work. Male. Female. 1953-1988.
U.S. News Jun 26, 1989 p. 74

SUB-SAHARAN AFRICA - ECONOMIC CONDITIONS

The economics of despair. Per capita GDP. Change in per capita GDP 1980-85. Total external debt as share of GNP. Change in investment 1980-85.
U.S. News Jun 27, 1988 p. 29

The debt bomb. Total external debt of sub-Saharan Africa. Official development assistance to sub-Saharan Africa. 1981-1986.
U.S. News Jun 27, 1988 p. 31

SUBMARINE WARFARE

Superpower hide-and-seek. U.S. and allied navies track enemy submarines with sonar and underwater listening devices. But new, quieter Soviet subs are harder to detect. Diagram.
Newsweek Sep 12, 1988 p. 28

SUBURBS

More people live in the suburbs. Share of metropolitan population outside core city. 1970, 1988.
Business Week Nov 26, 1990 p. 81

[Suburban voters] traditionally vote Republican. Share voted Republican, 1988 presidential election. City. Suburb.
Business Week Nov 26, 1990 p. 81

Suburbs outearn cities. Suburbs. Central cities. Real family income. 1959-1989.
Business Week Sep 25, 1989 p. 95

SUBWAYS

Take the 'A' train. Here are the world's busiest [subway] systems. Annual ridership. Base fare.
U.S. News Jun 25, 1990 p. 66

SUICIDE

Choosing death. Elderly suicides. People 65 years or older. National average. 1977-1987.
Newsweek Jun 18, 1990 p. 46

Deaths by firearms. Accidental deaths. Homicides. Suicides. 1950-1986.
Time Jul 17, 1989 p. 61

The risks. AIDS [deaths]. Homicides. Suicides. Motor-vehicle deaths.
Newsweek Apr 11, 1988 p. 67

SUN

Easier described than understood. The unexplained. Temperature differences. Sound waves. Rotation. Sunspots. Diagram.
Newsweek Jul 15, 1991 p. 60

Dynamo in the sky. Core. Intermediate zone. Convective zone. Photosphere. Sunspots. Sunspot number. Diagram.
Time Jul 3, 1989 p. 49

SUNSPOTS

Making the connection. Colder-than-normal winter temperatures track closely with high solar activity. 1952-1980.
U.S. News Mar 6, 1989 p. 53

SUNTAN

Sales of sunscreens. 1988, 1992.
FDA Consumer Jun 1989 p. 12

SUPERCOMPUTERS - MARKET SHARE

Control Data: Still back in the pack. Total supercomputers installed worldwide, as of June 30, 1988.
Business Week Oct 10, 1988 p. 43

SUPERCONDUCTORS AND SUPERCONDUCTIVITY

Cern's Large Electron-Positron Collider. Stanford Linear Collider. Fermilab's Trevatron. Superconducting Supercollider. Diagram.
Time Apr 16, 1990 p. 52

Superconductors. Japan grabs the lead in research. Japan. U.S. Launches work on applications. Area of superconductor development. Percent of Japanese companies involved.
Business Week Sep 19, 1988 p. 76

SUPERFUND

The DOE's [Department of Energy] expensive studies. Costs of RIFS work at 12 DOE sites. DOE site. Original estimate. Cost to date.
U.S. News Feb 19, 1990 p. 28

SUPERMARKETS

Pies like mom's? Here is the percentage of in-store bakeries that prepare various baked goods from: Scratch. A mix. Prepared dough.
U.S. News Jun 25, 1990 p. 66

Grocery gripes. In a survey of 400 supermarket customers, shoppers said these things irritate them.
U.S. News Oct 9, 1989 p. 81

Why shopping takes forever. Average number of items in a supermarket. 1978-1987. 1987's new products.
U.S. News Jun 20, 1988 p. 78

Comparing grocery bills. Weekly food costs at home [by size of family].
U.S. News May 2, 1988 p. 77

SUPERNATURAL

Scared silly. Percentage who believe in the supernatural [by belief].
U.S. News Oct 29, 1990 p. 107

SUPREME COURT

A conservative court. Justice. Year sworn in. Age. Appointed by.
Newsweek Aug 6, 1990 p. 18

SURGERY

In the operating womb. First major operation on a fetus. Diagram.
Newsweek Jun 11, 1990 p. 56

Places you'll get cut. Differences in the relative frequency with which [surgeons] perform common surgical procedures. Northeast/Mid-Atlantic. Midwest. South. West. 1987.
U.S. News Apr 30, 1990 p. 56

Surgery's hot list. Most-common inpatient surgical procedures.
U.S. News Apr 30, 1990 p. 57

Where to cut [inappropriate medical treatment]. Angina. Cataracts. Suspected heart disease. Back pain. Treatment. Number performed. Number inappropriate.
U.S. News Jan 30, 1989 p. 70

SWEDEN

Tough choices. Electricity generated from: Nuclear power. Hydropower. Coal. Petroleum. Sweden. 1987.
U.S. News Jul 18, 1988 p. 43

SYRIA

Quality versus quantity. Country. Population in millions. Armed forces: Regular, reserves. Tanks. Combat aircraft.
Newsweek May 1, 1989 p. 44

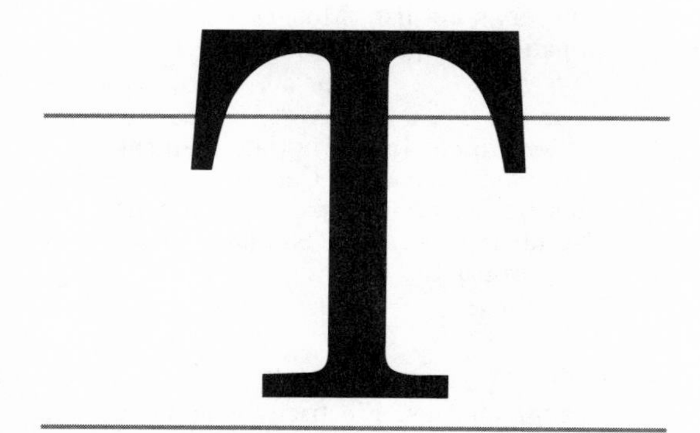

TAX EVASION

Who's got the profits? A corporation could slash its U.S. tax bill by transferring profits to low-tax countries. Typical transaction. Germany. Ireland. United States.
Newsweek Apr 15, 1991 pp. 48–49

TAX REFORM

Closing the loopholes. Number of new registrations for tax shelters including oil and gas and real estate partnerships since the 1986 Tax Reform Act. 1986-1989.
U.S. News Mar 12, 1990 p. 62

TAX SHELTERS

Permissible reasons for 401(k) hardship withdrawals.
Black Enterprise Jun 1990 p. 73

Closing the loopholes. Number of new registrations for tax shelters including oil and gas and real estate partnerships since the 1986 Tax Reform Act. 1986-1989.
U.S. News Mar 12, 1990 p. 62

The 401(k) advantage. Nontaxable 401(k) savings account. Taxable savings account. 5 yrs. 10 yrs. 15 yrs. 20 yrs. 25 yrs.
Time Feb 22, 1988 p. 51

TAXATION

Will rising taxes sink Bush next year? Change in personal tax rate since previous presidential election. Party in White House loses. Party in White House wins. 1956-1988.
Business Week Dec 9, 1991 p. 24

The tax bite just gets bigger. Taxes as a share of GNP. Federal. State & local. 1970-1990.
Business Week Sep 2, 1991 p. 32

A 10% luxury tax is charged on the portion of the purchase price [of:] Furs. Jewelry. Cars. Boats. Airplanes.
Time Jul 1, 1991 p. 52

Taxing payments. Tax receipts as a percentage of GNP. 1970-1991.
U.S. News May 27, 1991 p. 56

Bad timing. Taxes raised for 1991. Income taxes. Gasoline. Alcohol. Tobacco.
U.S. News May 27, 1991 p. 56

Museum pieces take a great fall. Value of objects donated. 1985-1990.
Business Week Feb 4, 1991 p. 102

U.S. Almanac: Comparing state stats. Per capita income. Poverty rate. Job growth. U.S. taxes paid per person. U.S. spending per capita. [By state.]
Scholastic Update Dec 2, 1988 pp. 18–19

Where the living is taxing. Total taxes collected per person in [12 countries].
U.S. News Jan 11, 1988 p. 63

See Also INCOME TAX

TAXATION, STATE

Tax bites. New York's tax increases have been smaller than the revenue hikes in most large states. Percentage increase in tax collections, 1990 versus 1992. [By selected states.]
U.S. News Dec 23, 1991 p. 53

Boosting revenues. State and local government budget surplus/deficit. 1980-1991.
U.S. News May 27, 1991 p. 56

Where the cap pinches. In 1987, residents of these states paid the highest average state and local taxes per household [state and amount paid].
U.S. News Aug 13, 1990 p. 50

The state tax man cometh. The average American spends 2 hours and 45 minutes of every 8-hour workday earning enough to pay taxes. The time spent each day working for the state and local tax man. 1980-1989.
U.S. News Jun 11, 1990 p. 38

Cities of the big taxes. Estimated state and local taxes owed by families of four with a household income of $50,000 and living in the biggest city in each state.
U.S. News Jan 9, 1989 p. 65

Taxing vacations. States with heaviest taxes in three categories. Gasoline. Sales. Cigarettes.
U.S. News Aug 15, 1988 p. 78

TAXES
See INCOME TAX

TEACHERS
Tired teachers. Public-school teachers [who] hold another job during the school year. A profile.
U.S. News Sep 10, 1990 p. 92

TEACHERS - SALARIES
Paying to learn. Average salary of public school teachers. Spending per pupil. 1979-1991. [Other selected statistics.]
U.S. News Sep 16, 1991 p. 13

Up with teachers. The average salary for teachers for the 1990-91 school year rose 5.4 percent. Here are the states that upped pay the most.
U.S. News May 27, 1991 p. 79

More money, more interest. Annual salaries. Freshmen interested in teaching careers. Percentage by sex. 1970-1987.
Newsweek Oct 17, 1988 p. 74

Where it pays to be a teacher. Starting pay. Average pay. The best. The worst. [By state.]
Newsweek Oct 17, 1988 p. 76

Teachers' salaries have gone nowhere. Average salaries. 1971-1988.
Business Week Sep 19, 1988 p. 132

Dollars, not apples. Average salaries of public school teachers [by state].
U.S. News Sep 12, 1988 p. 71

Rochester's investment. Average teacher pay in Rochester city schools. Average big-city teacher salaries.
U.S. News Jan 18, 1988 p. 64

TEACHING
More money, more interest. Annual salaries. Freshmen interested in teaching careers. Percentage by sex. 1970-1987.
Newsweek Oct 17, 1988 p. 74

Few want a teaching career. Entering college students who say they'd like to teach. 1971-1987.
Business Week Sep 19, 1988 p. 132

TECHNOLOGY
A small boom. Personal computers. Fax machines. Cellular phones. Home copiers.
Newsweek Apr 24, 1989 p. 59

TEENAGE PREGNANCY
Young and pregnant. Pregnancy rates per 1,000 women. By age. By race.
Newsweek Jan 8, 1990 p. 33

TEENAGERS
See YOUTH

TEETH
Look, ma, no cavities! Percentage of children [by age] who have never had a cavity. 1979-80. 1986-87.
U.S. News Dec 26, 1988 p. 120

A toothy smile. Percentage of children free of cavities and decay. 1986-1987.
U.S. News Jul 4, 1988 p. 12

TELECOMMUNICATIONS
What it costs to log on. Compuserve. Genie. Prodigy. Start-up cost. Monthly/hourly cost. Extras.
Business Week Jun 17, 1991 p. 113

The shortest commute. Americans who work at home. 1980-2000.
U.S. News Dec 25, 1989 p. 67

A squeeze on equipment makers and big opportunities in services. Worldwide switch R&D. Switch lines shipped. Average price per line in the U.S. Revenues from toll-free and data services. Total long-distance revenues.
Business Week Mar 13, 1989 pp. 138–139

Fax crazy. Facsimile machines sold in U.S. 1983-1988.
Business Week Mar 21, 1988 p. 136

A scramble for global networks. Worldwide communications systems.
Business Week Mar 21, 1988 p. 140

TELEPHONE - LONG DISTANCE SERVICE

Reach out and hold, please. Only one in 10 calls made to the Soviet Union actually gets through. Phone circuits from U.S. Calls from U.S. (yearly).
U.S. News Sep 23, 1991 p. 15

AT&T's shrinking share. Share of long-distance revenues. AT&T. MCI. U.S. Sprint. Others. 1985, 1990.
Business Week Nov 12, 1990 p. 55

Market share. Others. US Sprint. MCI.
Newsweek Jan 29, 1990 p. 71

Share of long-distance market. MCI revenues in billions of dollars. 1982-1989.
Time Sep 18, 1989 p. 61

Long-distance carriers gain on the Baby Bells. Operating margins. Local phone companies. Long-distance companies. 1984-1988.
Business Week Jan 11, 1988 p. 113

TELEPHONE INDUSTRY

The global giants. 1989 phone equipment sales. [Top six companies.]
Business Week Nov 26, 1990 p. 74

How the Baby Bells stack up overseas. Estimated market value of existing or approved foreign operations. Pacific Telesis. U.S. West. Bell South. Nynex. Bell Atlantic. Southwestern Bell. Ameritech.
Business Week Jun 25, 1990 pp. 104–105

While dominance in local service gives the Bells healthy returns, the companies say regulatory limits will handicap the U.S. phone network.
Business Week Mar 12, 1990 pp. 118–119

A squeeze on equipment makers and big opportunities in services. Worldwide switch R&D. Switch lines shipped. Average price per line in the U.S. Revenues from toll-free and data services. Total long-distance revenues.
Business Week Mar 13, 1989 pp. 138–139

TELEPHONE RATES

Ringing up 800 tolls. Direct charge for 15-minute call from New York to Los Angeles-based 800 number at 8 p.m. on a weekday [by carrier].
Business Week Dec 10, 1990 p. 225

Comparing toll-free service to pay-per-call. 800 number. 900 number.
Business Week Nov 5, 1990 p. 162

Business rings up lower phone charges. Index of telecommunications prices for commercial users. 1989-1990.
Business Week Jun 11, 1990 p. 14

Will long-distance prices keep falling? AT&T average per-minute charge for domestic direct-dial calls. 1984-1989.
Business Week Apr 23, 1990 p. 30

What's this call costing me? Evening weekday rate for 12 minutes between Washington, D.C., and Chicago. AT&T. MCI. Sprint.
Business Week Mar 26, 1990 p. 106

Cheap talk no more. Basic monthly telephone charge before and after AT&T breakup [by U.S. city]. And abroad. 1983, 1987.
U.S. News Feb 29, 1988 p. 75

TELEPHONES

The 900 club. 900 numbers are plentiful, solid, up-to-date information and a reasonable price. 31 services listed.
U.S. News Jun 17, 1991 pp. 61–63

Sorry, no number. Unlisted phone numbers [by selected cities].
U.S. News Jun 10, 1991 p. 75

No number, please. Percentage of phone numbers that are unlisted. 1984-1989
U.S. News Sep 18, 1989 p. 77

TELEPHONES, CELLULAR

Cellular's top 10. Company. Millions of potential customers.
Business Week Dec 5, 1988 p. 143

TELESCOPES

Eye on the sky. Keck telescope in Hawaii. Diagram.
Newsweek Jun 3, 1991 p. 50

TELEVISION - AUDIENCE

But average ratings shrink. Average number of television households tuned into sports telecasts on ABC, CBS, and NBC. Millions of households. 1980-1990.
Business Week Dec 10, 1990 p. 221

Network nightmare. % drop in TV ratings (1990 compared with 1989). Jan. Feb. March. Adults 18-49, prime time. Women 18-49, daytime.
Time Jun 4, 1990 p. 80

Shrinking audiences for TV's big three [ABC, CBS, NBC]. The three major networks' prime-time audience share. 1979-1989.
Business Week Apr 9, 1990 p. 26

Average hours of household TV use per day. 1950-1989.
Scholastic Update Jan 12, 1990 p. 24

A little less time for TV. Market. Time [spent watching television weekly by household].
U.S. News Jun 19, 1989 p. 74

Small screen, giant step. Television shows that got the largest audience share. Audience size in millions. Portion of all viewers.
U.S. News Jun 27, 1988 p. 64

Tuning out. Average percentage of homes tuned to major TV networks. 1984-1988.
Business Week Jun 13, 1988 p. 35

TELEVISION - NEWS

TV evening-news ratings. ABC. CBS. NBC. 1985-1990.
Time Sep 3, 1990 p. 69

Broadcast blues. Network news stories filed by women. 1974-1989. News stories about women. February, 1989. ABC. NBC. CBS.
Ms. Sep 1990 p. 89

Where we get our news. TV is now America's number one source of news. Primary news sources: 1959-1988. (Percent of Americans who named each medium as a top source). Television. Newspapers. Radio. Magazines.
Scholastic Update Sep 8, 1989 p. 6

TV is the news source Americans trust most. Most trusted media: 1959-1988. (Percent of Americans who found each source the most credible). Television. Newspapers. Radio. Magazines.
Scholastic Update Sep 8, 1989 p. 6

The ratings: Gloomy news for NBC. Average percentage of TV households tuning in the evening news in the fall season's first eight weeks. 1987, 1988.
Business Week Dec 5, 1988 p. 137

TELEVISION - RECEIVERS AND RECEPTION

Pies in the sky. Satellite dishes. Installed. 1980-1990.
Newsweek May 21, 1990 p. 68

TELEVISION - SPORTS

Growth in sports programs. Major sports telecasts. Number of national telecasts. 1980-1990.
Business Week Dec 10, 1990 p. 221

N.F.L. [television] contract. Increase from 1987.
Time Mar 26, 1990 p. 66

N.B.A. [television] contract. Increase from 1986.
Time Mar 26, 1990 p. 67

Baseball [television] contract. Increase from 1984.
Time Mar 26, 1990 p. 67

A growing number of fans can watch regional sports. Millions of homes receiving [regional sports programs]. 1985-1989.
Business Week Aug 22, 1988 p. 67

Cable networks now offer big-league attractions. Network/location. Major sports. Subscribers.
Business Week Aug 22, 1988 p. 67

TELEVISION ADVERTISING
See ADVERTISING, TELEVISION

TELEVISION, CABLE
Pay-per-view's knockout numbers. Top eight U.S. pay-per-view telecasts. Event. Date. Telecast revenues.
Business Week Sep 2, 1991 p. 56

American households hooked to a cable system in 1990. Share of viewers. Network TV. Cable TV. Cable services with the most U.S. subscribers.
U.S. News Mar 18, 1991 p. 17

Tuning in, paying up. Percent of TV households with cable. Increase in CPI since 1972. Increase in cable costs since 1972. 1971-1990.
Time May 28, 1990 p. 47

Hooked on cable. Number of basic-cable subscribers. Monthly cable fees. 1976-1989.
U.S. News Apr 30, 1990 p. 46

The people's choices. This percentage of subscribers watches these community-access programs.
U.S. News Oct 16, 1989 p. 112

Cable rates: The lid is off. Average monthly fees for basic cable. 1980-1988.
Business Week Jun 5, 1989 p. 136

Competing for cable. Basic subscribers in millions. Pay-TV services. % of total cable audience.
Time May 29, 1989 p. 29

Cable television's box-office bonanza. Cable network revenues. 1984-1989.
Business Week Jan 9, 1989 p. 81

Who's on top in cable guides.
Business Week Oct 31, 1988 p. 34

The cabling of a nation. Program services available on cable. Basic services. Subscribers. No. of cable affiliates.
U.S. News Oct 17, 1988 p. 82

Cable networks now offer big-league attractions. Network/location. Major sports. Subscribers.
Business Week Aug 22, 1988 p. 67

[Cable television]. Ad revenues of the cable industry in billions of dollars. 1980-1988. Cable share of total TV household viewing. 1983-1988. U.S. cable penetration into total TV household market. 1980-1988.
Time May 30, 1988 p. 52

TELEVISION, HIGH DEFINITION
The shape of things to come. HDTV. Conventional TV. Aspect ratio (width vs. height). Competing media. Four ways to transmit HDTV.
U.S. News Sep 10, 1990 p. 75

How to deliver an HDTS signal. Broadcast. Direct-broadcast satellite. Fiber-optic cable. VCR. Diagram.
U.S. News Jan 23, 1989 p. 48

TELEVISION INDUSTRY
Money in the air. Average sale price for radio and TV station. 1960s, 1970s, 1980s.
Business Week Jul 23, 1990 p. 52

A growing appetite for new shows. Syndicated sales of TV programs and movies to TV stations and cable networks. 1985-1990.
Business Week Jan 8, 1990 p. 93

Plugging into the airwaves. Minority ownership of radio and television stations, 1979-1986.
Black Enterprise May 1988 p. 36

TELEVISION PROGRAMS
Boob-tube diet. The menu on prime-time TV. Number of times these foods were displayed or referred to in a 14-hour sampling of shows.
U.S. News Sep 24, 1990 p. 99

Never-ending TV programs. The oldest scheduled network TV shows still running.
U.S. News Feb 26, 1990 p. 63

Still laughing after all these years. These TV comedy series have the most syndicated episodes.
U.S. News Aug 7, 1989 p. 62

Wham! Pow! Bam! Violent acts per hour on cartoon shows.
U.S. News Jun 13, 1988 p. 71

TEMPERATURE

The doggiest days ever? Highest temperature on record in selected cities. Temperature. Date occurred.
U.S. News Aug 1, 1988 p. 62

TEMPORARY EMPLOYMENT

See JOBS, TEMPORARY

TENNIS, PROFESSIONAL

Men's pro tennis: How the prize pool has grown. Millions of dollars. 1975-1988.
Business Week Oct 24, 1988 p. 46

TERRITORIES - UNITED STATES

Capital district and other territories. Origin of name. Date acquired. Area (sq. mi.). Population. Capital. Head of govt.
Scholastic Update Jan 15, 1988 p. 15

TERRORISTS, IRANIAN

Decade of terror. Bombings, murders, kidnappings and hijackings.
U.S. News Mar 6, 1989 pp. 22–23

TEXAS

Back on the job. Unemployment rate. Texas. U.S. 1984-1988.
Newsweek Jan 23, 1989 p. 40

Texas' jobless rate still tops the nation's. Texas. U.S. Civilian unemployment rate, quarterly average. 1986-1988.
Business Week Aug 1, 1988 p. 50

THANKSGIVING DINNER

Tackling turkey day. Thanksgiving dinner. Saturated fat (grams). Cholesterol (milligrams). Calories.
U.S. News Nov 27, 1989 p. 89

THATCHER, MARGARET

Thatcher's report card. Accomplishments. Failures. 1979-1990. Chart.
Business Week Dec 10, 1990 p. 26

Maggie's long run [highlights of Thatcher's career as Prime Minister]. 1975-1990. Chart.
Time Dec 3, 1990 p. 62

Maggie's economic report card. Unemployment. Inflation. Economic growth. Prime lending rate. 1987-1990.
Time Nov 26, 1990 p. 43

THEATER

Way off Broadway. Plays from the 1980s are edging into high-school repertoires. Here are the current favorites.
U.S. News Oct 8, 1990 p. 83

The role's the thing. Shakespeare's longest roles. Role (play). Number of lines.
U.S. News Jul 2, 1990 p. 62

Born on Broadway. Here are the top-grossing movies of the '80s that were based on stage plays. U.S. box-office receipts.
U.S. News Jun 4, 1990 p. 78

On the great white way. Broadway run and performances. Andrew Lloyd Webber. Stephen Sondheim.
U.S. News Feb 1, 1988 p. 53

THIRD WORLD COUNTRIES

See DEVELOPING COUNTRIES

TIMESHARING (REAL ESTATE)

A lease-a-week boom. Growth of U.S. time-sharing resorts. Sales. Number of resorts. 1973-1987.
U.S. News May 23, 1988 p. 70

TIPPING

Tips on tipping. For travelers in the U.S. who get flustered over gratuities, these guidelines may provide some help. Service. Suggested tip.
U.S. News Aug 21, 1989 p. 66

TIRE INDUSTRY

Tires. Key acquisitions. 1988. 1990. Worldwide sales. Market share. 1985 ranking. 1990 ranking.
Business Week Oct 14, 1991 p. 87

How the tiremakers stack up. U.S. market share.
Business Week Mar 21, 1988 p. 62

TITHES

Untithed. Religious denomination. % of income members give to charity.
Newsweek Jan 22, 1990 p. 8

TOBACCO INDUSTRY

Up in smokes. U.S. cigarette exports to East Asia. 1985-1990.
Business Week Feb 25, 1991 p. 66

Puffing more U.S. smokes abroad. Cigarette and cigar exports. 1984-1988.
Business Week Oct 9, 1989 p. 61

TOBACCO INDUSTRY - MARKET SHARE

Leaders of the pack. Share of the U.S. cigarette market held [by company]. 1985-1989.
U.S. News Mar 5, 1990 p. 58

The top cigarette brands of 1988. Brand. Company. Share. Sales. Billions of cigarettes. Growth rate. 1986-1988.
Business Week Jan 23, 1989 p. 58

How the producers rank. Share. Billions of cigarettes. Sales. Growth rate.
Business Week Jan 23, 1989 p. 59

TOKYO (JAPAN)

We'll take Manhattan. Goods cost more in Tokyo than in New York City. Product. Maker. Tokyo. New York.
U.S. News Nov 20, 1989 p. 13

TORTURE

Torture: A guide to human rights abuses around the world. Nations where torture is practiced. Map.
Scholastic Update Oct 21, 1988 pp. 8–9

TOURIST TRADE

Promotion pays off. Money budgeted 1988-1989 to promote tourism. Amount visitors spent in 1987. [Selected cities.]
U.S. News Jul 10, 1989 p. 62

Caribbean tourism. How much they spent. Caribbean tourist arrivals. Tourist arrivals by origin. U.S. tourists main destinations. 1986.
Black Enterprise May 1988 p. 61

The rewards of rest and relaxation. [Caribbean country.] Travel receipts. Other exports. Total. Travel receipts as % of total.
Black Enterprise May 1988 p. 66

Cruising toward increased revenue. Estimates of visitor expenditure: 1982-1986 [by Caribbean country].
Black Enterprise May 1988 p. 68

More tourists are opting for cruises. Caribbean tourism has been booming but a capacity glut may be on the way. 1984-1988.
Business Week Feb 29, 1988 p. 55

See Also TRAVEL

TOWNS

See CITIES AND TOWNS

TOXIC SHOCK SYNDROME

Decrease in tampon-associated TSS. The peak in 1980 reflects an epidemic associated with tampon use. 1978-1990.
FDA Consumer Oct 1991 p. 32

U.S. reports of Toxic Shock Syndrome. January 1979 through March 1990.
FDA Consumer Oct 1990 p. 4

TOXIC WASTE

See HAZARDOUS WASTE

TOY INDUSTRY

Less jumping for joysticks. Total U.S. toy sales. Video game sales. 1987-1992.
Business Week Nov 19, 1990 p. 52

Tops in toyland. Best selling toys of 1988.
U.S. News Apr 24, 1989 p. 80

Toyland's top sellers. 1988, 1985, 1978.
Business Week Dec 19, 1988 p. 59

TRADE BALANCE

See INTERNATIONAL TRADE

TRAFFIC ACCIDENTS

See ACCIDENTS - MOTOR VEHICLE

TRANSPLANTATION OF ORGANS, TISSUES, ETC.

Number of transplants in the U.S. Pancreas. Lung. Heart. Liver. Kidney. 1982-1990.
Time Jun 17, 1991 p. 61

The organ crisis. Kidneys. Hearts. Livers. Others. Transplants performed in 1988. Patients on the waiting list October, 1989.
Business Week Nov 27, 1989 p. 96

Two ways the body rejects organ transplants. Antibody rejection. Cell-mediated rejection. Diagrams.
Business Week Feb 27, 1989 p. 65

Transplant scoreboard. Organ. Transplants, 1987. People on waiting list. Cost. Success rate.
Newsweek Sep 12, 1988 p. 63

TRANSPORTATION

How maglev works. Japanese and German versions of magnetically levitated high-speed vehicles. Magnetic repulsion (Japanese maglev). Magnetic attraction (German maglev).
U.S. News Jul 23, 1990 p. 52

Take the 'A' train. Here are the world's busiest [subway] systems. Annual ridership. Base fare.
U.S. News Jun 25, 1990 p. 66

Schools and transportation get less. Elementary and secondary education. Transportation. 1987-1989.
Business Week Jun 5, 1989 p. 88

TRAVEL

Where the tourists go. The 15 most popular destinations were [various countries].
U.S. News Feb 26, 1990 p. 63

A tale of nine cities. What visitors can expect to pay in some popular tourist centers. First-class hotel. Breakfast, lunch, dinner. Cup of coffee. Taxi ride.
U.S. News May 8, 1989 p. 83

I get around. The time and cost, door to door, for a selection of trips. Plane. Train. Car. Map.
U.S. News Mar 7, 1988 p. 77

More tourists are opting for cruises. Caribbean tourism has been booming but a capacity glut may be on the way. 1984-1988.
Business Week Feb 29, 1988 p. 55

See Also TOURIST TRADE

TREES

Trees that turn first. Approximate number of trees in the nation's parks, yards and streets. Trees used annually for Sunday editions of the New York Times. [Other selected statistics.]
U.S. News Oct 14, 1991 p. 15

What old-growth trees do for the ecosystem and for the economy. Diagram.
Time Jun 25, 1990 p. 62

Biggest trees around. The stoutest trees in the U.S., by species.
U.S. News Apr 24, 1989 p. 80

TRITIUM

How tritium works in a boosted fission explosion. Diagram.
Bull Atomic Sci Jan 1988 p. 41

TRUCKS

Truck sales in a skid. U.S. unit retail sales of heavy trucks. 1986-1990.
Business Week Jul 30, 1990 p. 40

TRUCKS, FOREIGN

Japan's light-truck sales: No pickup. U.S. market share held by imported Japanese light trucks. 1987, 1988.
Business Week Jun 13, 1988 p. 31

TUBERCULOSIS

Reported cases of tuberculosis in the U.S. 1960-1990.
Time Dec 2, 1991 p. 85

Tuberculosis increases. Reported U.S. TB cases. Earlier CDC [Centers for Disease Control] projections. 1970-1989.
FDA Consumer Mar 1991 p. 23

U.N.

See UNITED NATIONS

UKRAINE

Russia's breadbasket. Ukraine. Population. Languages. Religion. State productivity.
U.S. News Sep 16, 1991 p. 40

ULTRAVIOLET RAYS

Assault on the skin. Ultraviolet A. Ultraviolet B. Melanoma. Basal-cell carcinoma. Squamous-cell carcinoma.
Time Jul 23, 1990 p. 69

UNDERCLASS

See POVERTY

UNEMPLOYMENT

Big layoffs. Recently announced plans for staff reductions [by company].
Business Week Dec 23, 1991 p. 25

Pink slips. Layoffs announced by major U.S. corporations in 1991. Company. Number of jobs. Percent of work force.
Time Dec 23, 1991 p. 61

Bicoastal pain. This recession has hit New York and California harder than the rest of the country. Civilian unemployment rate. 1989, 1990, 1991.
U.S. News Dec 23, 1991 p. 52

More people out of work. The unemployed, in millions. 1988-1991.
Time Dec 9, 1991 p. 22

Chronic unemployment. Unemployment rate. Black. Total. 1970-1991.
U.S. News Nov 4, 1991 p. 63

Bad karma. 1987-1991.
Newsweek Oct 28, 1991 p. 50

How job-losers fare after being laid off.
Business Week Sep 16, 1991 p. 24

Nagging unemployment. U.S. unemployment rate. 1978-1991.
U.S. News Aug 12, 1991 p. 39

How bad was it? Real gross national product. Civilian unemployment rate. Industrial production. Change during recession. 1990-91. Average.
Newsweek Aug 5, 1991 p. 64

Mounting unemployment. Unemployment rate. U.S. Mass. 1980-1991.
U.S. News May 29, 1991 p. 64

Increase in unemployed persons, June 1990 to March 1991. Total number of staff cuts by U.S. companies. Quarterly unemployment rate. 1987-1991.
U.S. News Apr 15, 1991 p. 16

Unemployment is climbing. Civilian unemployment rate. Jan. 90 - Jan. 91. So are business failures. 1990-1991.
Business Week Mar 18, 1991 p. 31

A sick economy. National unemployment rate January-December 1990.
Scholastic Update Mar 8, 1991 p. 3

Jobless in the U.S.A. [unemployment rates in Nov. 1989 and Nov. 1990 by region].
Scholastic Update Mar 8, 1991 pp. 10–11

Out of a job. Industry. Total employment. Net number of jobs lost.
Time Feb 11, 1991 p. 66

Unemployment rate. Growth in wages, salaries and benefits. Increase in inflation. Average real disposable personal income. 1990.
U.S. News Feb 11, 1991 p. 12

Unemployment marches lower. Change in unemployment rate. World War I. World War II. Korean War. Vietnam War.
U.S. News Jan 28, 1991 p. 51

How bad will it get? New unemployment claims. 1990.
Newsweek Jan 14, 1991 p. 33

Civilian labor force and employment/ unemployment, by race, 1989-1991.
Black Enterprise Jan 1991 p. 48

Employment status of civilian labor force by age, sex and race. Employed—age 45-64. Unemployed—age 45-64. Black males. White males. Black females. White females. 1989.
Black Enterprise Dec 1990 p. 41

U.S. joblessness: A spotty performance. Areas where unemployment rate rose by 1% or more. Areas where unemployment rate fell by 1% or more. July-August, 1989, to July-August, 1990. Map.
Business Week Nov 5, 1990 p. 86

Unemployment: White-collar workers are hit the hardest. Increase in thousands of unemployed workers, Aug.'89 to Aug. '90. [By type of work.]
Business Week Oct 1, 1990 pp. 130–131

Recession's signs and portents. Unemployment claims. New claims for unemployment benefits. 1989, 1990.
U.S. News Sep 24, 1990 p. 81

U.S. population, civilian labor force, employment, and unemployment; By race, 1987-1990.
Black Enterprise Jan 1990 p. 60

Unemployment rate. United States. Delta region. 1987, 1988.
Black Enterprise Dec 1989 p. 20

More of the jobless are in their prime years. Adults 25 to 54 as a percent of the unemployed. 1966-1988.
Business Week Feb 6, 1989 p. 26

Back on the job. Unemployment rate. Texas. U.S. 1984-1988.
Newsweek Jan 23, 1989 p. 40

Unemployment rates for black workers, age 16 to 19 years. Year. Annual average (in percent). 1985-1987.
Black Enterprise Jan 1989 p. 33

Jobless claims head for a new low. Initial unemployment insurance claims. Four-week moving average. Seasonally adjusted. 1983-1988.
Business Week Nov 14, 1988 p. 46

Texas' jobless rate still tops the nation's. Texas. U.S. Civilian unemployment rate, quarterly average. 1986-1988.
Business Week Aug 1, 1988 p. 50

The Bay State vs. the nation. Unemployment rate. Massachusetts. U.S. 1974-1988.
U.S. News Jun 13, 1988 p. 14

The canning business. Days of the week employes get fired. Time of day employes get fired.
U.S. News Apr 18, 1988 p. 83

How education lowers the risk of unemployment. Jobless rates. High school dropouts. High school graduates. College graduates. 1967-1987.
Business Week Feb 29, 1988 p. 20

A permanent condition? U.S. youth unemployment (ages 16-19). Whites. Blacks. Hispanics.
Black Enterprise Feb 1988 p. 55

Jobless claims are trending higher. Initial unemployment insurance claims. Eight-week moving average. 1986-1987.
Business Week Feb 1, 1988 p. 18

The state of jobless pay. 1988 weekly unemployment compensation maximums [by state].
U.S. News Jan 25, 1988 p. 77

UNEMPLOYMENT - WORLD

Foreign job jitters. Unemployment rate. U.S. France. Britain. 1988-1991.
U.S. News Jul 22, 1991 p. 47

UNEMPLOYMENT BENEFITS

See INSURANCE, UNEMPLOYMENT

UNIONS

See LABOR UNIONS

UNITED NATIONS

America and the U.N. Cumulative U.S. debt to the U.N. budget. 1980-1988.
Time Aug 22, 1988 p. 44

Top five debtors. U.S. S. Africa. Brazil. Iran. U.S.S.R. % of 1988 budget assessed. Amount due from prior years. Amount due in 1988. Amount paid in 1988. Total amount outstanding.
Time Aug 22, 1988 p. 44

Paying for peacework. U.N. peacekeeping forces. In the field. Future.
U.S. News Aug 22, 1988 p. 32

In arrears on the wages of peace. Country. Debt to U.N. Assessment as % of U.N. budget.
Newsweek Aug 8, 1988 p. 36

UNITED STATES

A state-by-state almanac. Nickname. Origin of name. Entered union. Capital. Largest city. Area. Population. Chief industries. Senators. Governor. U.S. House.
Scholastic Update Jan 11, 1991 pp. 18–25

Comparing state stats. Per capita income. Poverty rate. Job growth. Federal taxes paid per person. [By state.]
Scholastic Update Jan 11, 1991 pp. 26–27

Comparing state stats. Public school spending per student. Percent graduated from high school. [By state.]
Scholastic Update Jan 11, 1991 pp. 26–27

A state-by-state almanac. Nickname. Origin of name. Entered union. Capital. Largest city. Area. Population. Chief industries. Senators. Governor. U.S. House.
Scholastic Update Jan 12, 1990 pp. 20–27

Comparing state stats. Per capita income. Poverty rate. Job growth. Federal taxes paid per person. [By state.]
Scholastic Update Jan 12, 1990 pp. 30–31

Comparing state stats. Public school spending per student. Percent graduated from high school. [By state.]
Scholastic Update Jan 12, 1990 pp. 30–31

U.S. Almanac: The basic information. Nickname. Origin of name. Entered union. Capital. Largest city. Area. Population. Chief industries. Driving age. Marriage age.
Scholastic Update Dec 2, 1988 pp. 10–15

U.S. Almanac: Comparing state stats. Percent population change. Population in metropolitan areas. Divorces. Marriages. [By state.]
Scholastic Update Dec 2, 1988 pp. 18–19

U.S. Almanac: Comparing state stats. Per capita income. Poverty rate. Job growth. U.S. taxes paid per person. U.S. spending per capita. [By state.]
Scholastic Update Dec 2, 1988 pp. 18–19

U.S. Almanac: Comparing state stats. Public school spending per student. Percent graduated from high school. [By state.]
Scholastic Update Dec 2, 1988 pp. 18–19

U.S. Almanac: Comparing state stats. Violent crimes per 1,000 people. [By state.]
Scholastic Update Dec 2, 1988 pp. 18–19

State and nickname. Origin of name. Entered union. Area (sq. mi.). Population. Capital. Governor. U.S. Sens. U.S. Reps. [numbers, political party].
Scholastic Update Jan 15, 1988 pp. 13–15

UNITED STATES - FOREIGN RELATIONS

U.S. wars and hot spots. 1945-1991. Major armed conflicts. Armed interventions. Other military actions. Use of U.S. military advisers abroad. World map.
Scholastic Update Feb 8, 1991 pp. 10–11

UNITED STATES - MILITARY STRENGTH

On their way home. Soviet and American bases face each other along the Pacific Rim. Map.
Newsweek Jun 18, 1990 p. 24

The price of overcommitment. American forces are stretched thin around the world at a time of declining U.S. economic power. Number of U.S. military personnel. World map.
Newsweek Feb 22, 1988 p. 55

UNITED STATES - SOCIAL CONDITIONS

Social problems are getting worse. Percentage increases. High school dropouts. 1970-88. Violent crime. 1979-88. Poverty rate. 1975-85. Drug abuse cases. 1980-87.
Scholastic Update Feb 23, 1990 p. 3

UNITED STATES - SOCIAL POLICY

Federal aid has dropped. Changes in the federal government's spending on education, job training, and other social programs. 1965-1990.
Scholastic Update Feb 23, 1990 p. 3

How social investment in children pays off. $1 invested in [a social program] saves [this amount in remedial care].
Business Week Sep 19, 1988 p. 123

UNIVERSE

Evolution of the universe. New satellite measurements of "fossil" radiation from the big-bang explosion at the origin of the universe are causing theoreticians to scratch their heads. Diagram.
U.S. News Jan 29, 1990 pp. 48–49

UNMARRIED COUPLES

Unmarried-couple households (with and without children). 1970-1987.
U.S. News Jul 3, 1989 p. 29

UNMARRIED MOTHERS

Birth rates per 1,000 unmarried women age 15-44 per year. Blacks. Whites. 1970-1986.
U.S. News Jul 3, 1989 p. 29

Percentage of women with first child born out of wedlock who eventually married: Blacks. Whites. Total.
U.S. News Jul 3, 1989 p. 29

See Also SINGLE PARENT FAMILIES

URANIUM

Bomb potential for South America. Uranium mining area. Map.
Bull Atomic Sci May 1989 p. 16

Grades of uranium. Natural. Slightly enriched. Low-enriched. Highly enriched. Weapon-grade.
Bull Atomic Sci May 1989 p. 17

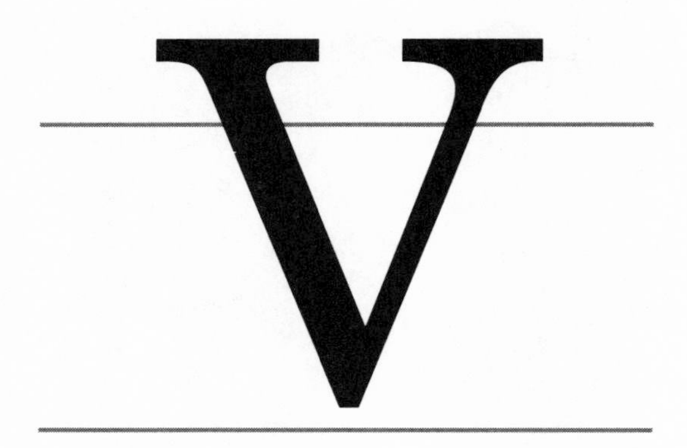

VACATIONS

Work hard and play hard. Average number of working hours and vacation days in major (international) cities. [City.] Hours worked per week. Vacation days per year.
U.S. News Nov 14, 1988 p. 82

Hot cities. Where Americans will spend their vacations this summer.
U.S. News Jun 13, 1988 p. 71

Nation of vacationers. Number of people in U.S. taking vacations. Long trips. Weekend trips. 1980-1987.
U.S. News Jan 18, 1988 p. 75

Executive breaks. Vacation time top managers get. How much they take.
U.S. News Jan 18, 1988 p. 75

See Also TRAVEL

VACCINATION

Recommended immunization schedule. Immunization schedule recommended for infants and children. Recommended age. Vaccines.
FDA Consumer Sep 1990 p. 22

Immunization requirements. Grades K-12. [By state.] Diphtheria. Tetanus. Pertussis. Measles. Mumps. Rubella. Polio.
FDA Consumer Sep 1990 p. 24

VALDEZ

See EXXON VALDEZ

VALUES

See SOCIAL VALUES

VEGETABLE GARDENS AND GARDENING

Shrinking green thumbs. Percentage of households with vegetable gardens. Percentage of gardeners who grow [specific vegetables]. 1976-1987.
U.S. News Feb 22, 1988 p. 91

VEGETABLES

Garden of eating. Fruits and vegetables with the most calories.
U.S. News May 15, 1989 p. 73

Vegetable scoop. Yearly per-person consumption [of various vegetables]. Pounds. Percentage change since 1975.
U.S. News Feb 13, 1989 p. 84

See Also FOOD

VENEREAL DISEASES

See SEXUALLY TRANSMITTED DISEASES

VENTURE CAPITAL

Capital punishment. Venture capital invested in small businesses. 1980-1990.
U.S. News Jun 3, 1991 p. 62

The rise and fall of venture capital funds in the 1980s. 1980-1988.
Black Enterprise Feb 1990 p. 188

More fresh venture capital, fewer juicy deals. New venture capital. Billions of dollars. 1980-1988. Venture-capital-backed initial public offerings. Number of offerings. 1980-1988.
Business Week Dec 12, 1988 p. 37

See Also CORPORATIONS - FINANCE

VERBAL SKILLS

Job skills: A widening gap. Percent of students and jobs available each level of verbal skill. Employers' current needs. Employers' future needs. Students' current level.
Black Enterprise Dec 1989 p. 53

VETERANS

Five decades of V.A. spending. 1940-1990.
Black Enterprise Jul 1991 p. 23

The graying of U.S. veterans. Percentage of veterans by service period. World War II. Korean War. Vietnam War. Peacetime. Average age.
U.S. News Jun 5, 1989 p. 40

VICENNES (WARSHIP) ACCIDENT, 1988

See IRANIAN AIR DOWNING, 1988

VICTIMS OF CRIME

See CRIME, VICTIMS OF

VIDEO DISCS

The picture is in the pits. How it works. Diagram.
Newsweek Jun 5, 1989 p. 68

VIDEO GAMES

Zoom boom. Retail sales of video games. Total industry sales. Nintendo sales. 1985-1989.
Newsweek Mar 6, 1989 p. 67

VIDEOTAPE RECORDERS AND RECORDING

The tape's the thing. Huge hits like "Batman" and "Lethal Weapon" made 1989 a boom year for prerecorded videocassettes, but blanks dropped by 20 million. Here are the figures for the last decade. 1980-1989.
U.S. News Apr 2, 1990 p. 68

Couch-potato country. Major metropolitan areas with the largest percentage of VCR-owning households. Major metropolitan areas with the smallest percentage of VCR-owning households.
U.S. News Apr 10, 1989 p. 72

That's entertainment. Share of VCR owners who each week record TV shows. 1985, 1988. Number of videotape rentals. 1984-1987.
U.S. News Jul 25, 1988 p. 66

Betamax dwindles away. U.S. sales of videocassette recorders and camcorders. 1983-1987.
Business Week Jan 25, 1988 p. 37

VIDEOTAPES

Funny business. For all rental cassettes here are the percentages by type. Comedy. Drama. Action, adventure. Children, family. Science fiction.
U.S. News Oct 16, 1989 p. 125

VIDEOTAPES - RENTING

Videotape saga. Percentage of people who say they make this many trips to a video store to rent a new release [one to five trips].
U.S. News Mar 20, 1989 p. 90

VIETNAM

Bare facts of economic desperation. Vietnam. Cambodia. Land area . Population. Life expectancy at birth. Per capita GNP. Value of exports. Value of imports.
U.S. News Aug 1, 1988 p. 34

VIETNAM WAR, 1957-1975 - CASUALTIES

U.S. battle deaths. Killed in action. Vietnam. Lebanon. Grenada. Panama. Persian Gulf.
Newsweek Mar 11, 1991 p. 28

The toll the war took. Number of U.S. battle deaths. 1957-1975.
Scholastic Update Apr 6, 1990 p. 9

VIETNAM WAR, 1957-1975 - ECONOMIC ASPECTS

The toll the war took. U.S. spending on the Vietnam War. U.S. military aid to South Vietnam. U.S. economic aid to South Vietnam. 1951-1975.
Scholastic Update Apr 6, 1990 p. 9

VIETNAM WAR, 1957-1975 - MISSING IN ACTION

Missing in action. WWII. Korea. Vietnam. MIAs. As a percent of total deaths.
Newsweek Jul 29, 1991 p. 23

Americans still listed as missing in Southeast Asia. POWs who have been returned to the U.S. Remains of U.S. soldiers returned by Vietnam since 1987. [Other selected statistics.]
U.S. News Jul 1, 1991 p. 12

VIETNAM WAR, 1957-1975 - SOLDIERS

The toll the war took. U.S. troops in Vietnam (thousands). 1957-1972.
Scholastic Update Apr 6, 1990 p. 9

VIOLENCE AND TELEVISION

Wham! Pow! Bam! Violent acts per hour on cartoon shows.
U.S. News Jun 13, 1988 p. 71

VIOLENCE, DOMESTIC

Domestic violence: Reasons for assault. [By type of violence.] Maryland, 1988.
Ms. Sep 1990 p. 36

Facts about family violence. Many American women and children find no refuge in their own homes. Figures suggest some form of violence is going on across the street — or even next door.
Newsweek Dec 12, 1988 p. 59

VITAMINS

Vitamins: New RDAs. Nutrient. Male. Female.
Newsweek Nov 6, 1989 p. 84

VOLCANOES

The ring of fire. How plate movement causes volcanoes. Convergent boundaries. Volcanoes active during the past 10,000 years. Other volcanoes.
Time Jun 24, 1991 p. 43

VOLUNTEER SERVICE

What [mandatory and volunteer service] other nations do.
U.S. News Feb 13, 1989 p. 23

VOLUNTEERS

Volunteerism is up. Percent of population doing volunteer work. 1977-1989.
Scholastic Update Feb 23, 1990 p. 3

Who volunteers and why.
Newsweek Jul 10, 1989 p. 37

The nation's unmet needs [and estimated number of workers needed]. Education. Health care. Child care. Environment. Criminal justice.
U.S. News Feb 13, 1989 p. 22

VOTER REGISTRATION

A bigger foothold for the G.O.P. Texas voters in primaries. Florida registered voters. California registered voters. 1980-1990.
Time Apr 23, 1990 p. 22

Registration rules. [Where and when voters can register - by state.] Map.
Scholastic Update Jan 29, 1988 pp. 14–15

VOTING

Who votes for presidents. Percentage of eligible U.S. voters who: Voted. Voted for winning candidate. 1864-1988. Voted in other democracies.
U.S. News Oct 31, 1988 p. 82

Presidential politics in black and white. The gap is narrowing in voter turnout. Whites. Blacks. Percent reported voting. 1968-1984.
Business Week Jul 4, 1988 p. 97

Presidential politics in black and white. Black voters play a bigger role. Percent of total vote. 1968-1984.
Business Week Jul 4, 1988 p. 97

Presidential politics in black and white. Who votes for the Democrats? Percent of total black vote. Percent of total white vote. 1968-1984.
Business Week Jul 4, 1988 p. 97

Women's vote: 1984-1986. Women's vote = margin of victory. Women's vote significantly different.
Ms. Apr 1988 p. 76

Voter turnout by gender, presidential elections. Men. Women. 1972-1984.
Ms. Apr 1988 p. 76

Voter turnout by gender, midterm elections. Men. Women. 1974-1986.
Ms. Apr 1988 p. 76

Races where women's votes won the day. State. Winner. Margin of victory. Difference in women's support. 1984 elections. 1986 elections.
Ms. Apr 1988 p. 77

Voter turnout by gender and age, presidental elections. Women 18-44. Men 18-44. 1972-1984.
Ms. Apr 1988 p. 77

Voter turnout by gender and race, presidential elections. Black women. Black men. 1972-1984.
Ms. Apr 1988 p. 77

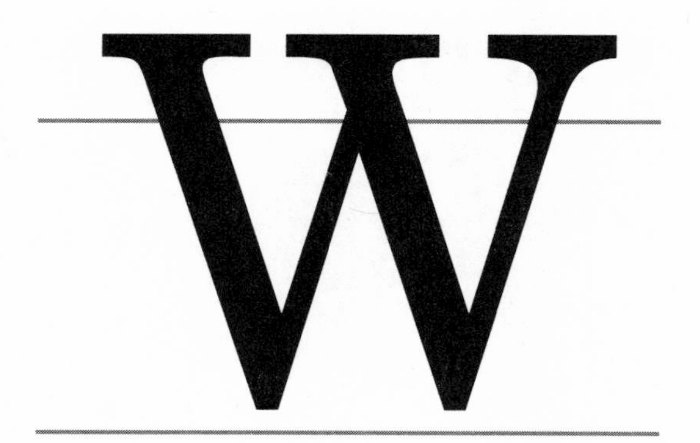

WAGES AND SALARIES

Salary survey. Regional salary survey. Hot tracks in 20 professions.
U.S. News Nov 11, 1991 pp. 88–102

Factory wages overseas are catching up. Hourly compensation. Percent of U.S. level [by country]. 1985, 1990.
Business Week Sep 30, 1991 p. 16

Where pay is low. Mean annual earnings. Men under 30. Services. Retail trade. Manufacturing. 1973, 1989.
Business Week Aug 19, 1991 p. 80

Labor costs are slowing down. Employment cost index. Salary and benefits. 1988-1991.
Business Week Aug 12, 1991 p. 18

Median annual earnings. White males. Black males. White females. Black females. 1969-1989.
Business Week Jul 15, 1991 p. 10

Where the money is. Estimated 1991 starting salaries for graduates of top business schools. Consulting. Investment banking. Marketing.
Business Week May 6, 1991 p. 82

The widening pay gap: CEOs vs. the others. Average total compensation. Chief executive. Engineer. School teacher. Factory worker. 1960-1990.
Business Week May 6, 1991 p. 96

Monthly percentage change in real average weekly earnings. Average weekly earnings. Projected salary increases in 1991. Employers requiring workers to pay partial health insurance costs.
U.S. News Mar 4, 1991 p. 14

Estimated starting salaries for June 1990 graduates with bachelor degrees [by major].
Black Enterprise Feb 1991 p. 48

War of the raises. Federal employes pay hike. Here's how past increases compare with average private-sector raises. 1980-1990.
U.S. News Dec 24, 1990 p. 66

Where you stand in today's economy. If you work in the exporting sector. Percent of private nonagricultural employment. Average weekly wages 1990 dollars. 1979-1990.
Business Week Dec 17, 1990 p. 67

Where you stand in today's economy. If you're in an industry hurt by imports. Percent of private nonagricultural employment. Average weekly wages 1990 dollars. 1979-1990.
Business Week Dec 17, 1990 p. 67

Where you stand in today's economy. If your job is in the domestic sector. Percent of private nonagricultural employment. Average weekly wages 1990 dollars. 1979-1990.
Business Week Dec 17, 1990 p. 67

Managers have done well. Wage and salary index. Managers. Blue-collar workers. 1982-1990. Across the board. Average annual wage and salary increases, 1985-90. Managers. Production workers [by industry].
Business Week Dec 17, 1990 p. 69

Pay hikes stay modest. Average annual pay increases for middle managers at 1,455 companies. Percent. 1981-1991.
Business Week Nov 19, 1990 p. 54

Benefits outpace pay. Annual percentage change. Benefits. Wages. 1981-1990.
U.S. News Nov 5, 1990 p. 58

Union workers lag. Changes in compensation. Nonunion. Union. 1981-1990.
U.S. News Nov 5, 1990 p. 58

The paycheck of the future. Pay tied to performance. Program. Percentage of companies who use this program.
U.S. News Sep 17, 1990 p. 68

The *U.S. News* salary survey. What 20 jobs pay. Hot tracks in 20 professions. Salaries vary considerably region by region.
U.S. News Sep 17, 1990 pp. 81–99

Stationary salaries. Weekly manufacturing wages. Weekly service-sector wages. 1965-2000.
U.S. News Dec 25, 1989 p. 66

Hourly wages of U.S. Hispanic population, age 21-25, by years in the U.S.
U.S. News Sep 25, 1989 p. 31

What jobs are worth around the country. The salaries people earn vary considerably by where they settle. That was the finding of our 10-city survey of 10 jobs at three rungs of the ladder.
U.S. News Sep 25, 1989 pp. 66–67

A few useful pay scales. The advantage of going private. Paths with the biggest payoff. Ground-floor paychecks for '89. What a degree buys.
U.S. News Sep 25, 1989 p. 68

Closing the gap. Here are average earnings of husbands and wives working at full-time jobs, adjusted for inflation in 1987 dollars. 1981, 1987.
U.S. News Aug 28, 1989 p. 102

After-school money. School attended, job. Total cost of degree. Average first-year income.
Newsweek Jun 12, 1989 p. 6

The bottom line. Minimum-wage earners. 1981-1988.
Newsweek Apr 17, 1989 p. 24

Minimum wage, minimal turnout. Hourly workers at or below $3.35 an hour. 1981-1988.
Business Week Mar 27, 1989 p. 35

Steelmakers are prospering once again but steelworkers are still feeling the pinch. Earnings (losses) from steel operations of the six biggest U.S. steelmakers. 1983-1988. Average hourly earnings.
Business Week Mar 13, 1989 p. 40

Average weekly earnings of workers in private companies, by industry. 1987, 1977, 1967.
Black Enterprise Feb 1989 p. 63

Average starting salaries of college graduates, by discipline, 1987-1988.
Black Enterprise Feb 1989 p. 63

Annual average salary of workers in selected fields, March 1988. Occupation. Salary. 1986 salary.
Black Enterprise Feb 1989 p. 63

Paper-chase payoff. Estimated starting salaries for 1988-89 college graduates holding bachelor's degrees. Major and starting salary. Changes from 1987-88.
U.S. News Jan 30, 1989 p. 81

Where parting is sweetest. Percentage of annual salary typically awarded employes who are fired or laid off [by country]. Manufacturing workers. Chief executives.
U.S. News Dec 19, 1988 p. 70

Bigger payoff on payday. Average annual earnings, hourly wage and hours worked per week for manufacturing workers [by state].
U.S. News Oct 31, 1988 p. 82

Time is money. Per-employe cost of various breaks in the workday. Employe's annual income. 10-minute meeting delay. 1-hour meeting. 2-hour lunch.
U.S. News Aug 8, 1988 p. 65

It all adds up. Rise of the minimum wage vs. rise of the average hourly wage. 1938-1987.
Black Enterprise Jul 1988 p. 27

Paychecks for '88. Average starting salary, by degree. Bachelor's degrees. Master's degrees. Doctorates, other advanced degrees.
U.S. News Apr 25, 1988 p. 66

What jobs are worth around the country. How much people make has much to do with where they make it. 10-city survey of 10 jobs at 3 career levels.
U.S. News Apr 25, 1988 p. 68

Drawing power of an M.B.A. Average starting salaries of 1988 M.B.A. graduates, by industry.
U.S. News Apr 25, 1988 p. 71

New engineers. Average starting salaries for 1988 engineering graduates.
U.S. News Apr 25, 1988 p. 73

State of health pay. Average starting salaries in the healing professions.
U.S. News Apr 25, 1988 p. 74

The salary gap. Salaries. Median usual weekly earnings of full-time wage and salary workers, annual averages. White. Black. 1979-1987.
Newsweek Mar 7, 1988 p. 24

Starting salaries of college graduates by discipline, 1986-1987. Discipline. Salary. Change.
Black Enterprise Feb 1988 p. 77

Average weekly earnings of workers in private companies, by industry. 1986, 1976, 1966.
Black Enterprise Feb 1988 p. 77

Annual median salaries of workers in selected fields, March 1986.
Black Enterprise Feb 1988 p. 77

See Also EXECUTIVES - SALARIES, PENSIONS, ETC.

WAGES AND SALARIES - AIRLINE PILOTS

American's pilots' pay: Flying low. American. Northwest. Delta. USAir. Typical monthly pay. Captains. First officers.
Business Week Jul 2, 1990 p. 33

WAGES AND SALARIES - EXECUTIVES

Who earns what. Median weekly earnings for executive, administrative and managerial positions. Men. Women. 1983-1990.
Black Enterprise Aug 1991 p. 42

The 20 highest-paid chief executives and 10 who aren't CEOs. Company. 1990 salary and bonus. Long-term compensation. Total pay.
Business Week May 6, 1991 p. 91

Winners and losers. Sampling of corporate heads who got: Pay raise. Pay cut. 1989, 1990.
Newsweek May 6, 1991 p. 50

The 20 highest-paid chief executives, 10 who aren't CEOs. [Executive.] Company. 1989 salary and bonus. Long-term compensation. Total pay. The 10 largest golden parachutes. Company. Reason. Total package.
Business Week May 7, 1990 p. 57

The decade's biggest CEO money-makers [by name]. Total pay 1980-89.
Business Week May 7, 1990 p. 60

The 25 highest-paid executives. Company. 1987 salary and bonus. Long-term compensation. Total pay.
Business Week May 2, 1988 p. 51

The 10 largest golden parachutes. [Executive.] Company. Reason for payment. Total package.
Business Week May 2, 1988 p. 51

WAGES AND SALARIES - GOVERNMENT EMPLOYEES

State and local workers still get fatter pay hikes. State and local government workers. Private-sector workers. Change in total compensation. 1982-1991.
Business Week Sep 23, 1991 p. 26

The golden rocking chair. Thanks to fat pensions, many retired officials make more than those still on the job. Ex-presidents. Ex-senators. Ex-congressmen.
Time Jun 10, 1991 p. 21

Public servants gain. Changes in compensation. Public workers. Private workers. 1982-1990.
U.S. News Nov 5, 1990 p. 58

A mayoral pay scale. What mayors make in the nation's largest cities.
U.S. News Mar 6, 1989 p. 66

Paying the President. What past Presidents would earn if their salaries were adjusted for inflation in 1988 dollars.
U.S. News Aug 8, 1988 p. 65

Is Uncle Sam a cheapskate? What different jobs pay in the private sector and the federal government.
U.S. News Mar 14, 1988 p. 76

WAGES AND SALARIES - SPORTS

Are salaries soaring out of the ballpark? Salaries as percent of league revenues. Including penalties. As paid. 1986-1990.
Business Week Dec 30, 1991 p. 40

The money is bigger than ever but it doesn't last long. Average salary. Average length of professional-sports career. Baseball. Basketball. Football. Hockey. 1981, 1991.
Business Week Jun 3, 1991 p. 55

Player pay: Gaining yardage. Average annual salary. Increase per year.
Business Week Jun 25, 1990 p. 56

Salary scorecard [selected baseball players].
Newsweek Apr 10, 1989 p. 44

Player salaries are coming on strong. Average player salary. Total player payroll. 1985-1989.
Business Week Apr 3, 1989 p. 87

WAGES AND SALARIES - TEACHERS

Up with teachers. The average salary for teachers for the 1990-91 school year rose 5.4 percent. Here are the states that upped pay the most.
U.S. News May 27, 1991 p. 79

More money, more interest. Annual salaries. Freshmen interested in teaching careers. Percentage by sex. 1970-1987.
Newsweek Oct 17, 1988 p. 74

Where it pays to be a teacher. Starting pay. Average pay. The best. The worst. [By state.]
Newsweek Oct 17, 1988 p. 76

Teachers' salaries have gone nowhere. Average salaries. 1971-1988.
Business Week Sep 19, 1988 p. 132

Dollars, not apples. Average salaries of public school teachers [by state].
U.S. News Sep 12, 1988 p. 71

Rochester's investment. Average teacher pay in Rochester city schools. Average big-city teacher salaries.
U.S. News Jan 18, 1988 p. 64

WAGES AND SALARIES - WOMEN

The biggest gaps. Women's earnings. [As a] percent of men's earnings. [By selected occupations.]
Business Week Oct 28, 1991 p. 35

Trouble at the top. The wage gap. Median weekly earnings. Men. Women. Share of men's wages [by job type].
U.S. News Jun 17, 1991 pp. 40–41

Competing on wages. Hourly compensation costs in manufacturing, U.S. dollars. Germany. U.S. Japan. 1985-1989.
U.S. News Jun 10, 1991 p. 57

The gender gap is narrowing. Women's pay as a percent of men's for full-time workers. 1980, 1988.
Business Week Dec 24, 1990 p. 14

The pay gap between men and women. Annual salary after graduation. Discrepancy in pay scales. By school. By industry.
Business Week Oct 29, 1990 p. 57

The gender gap shrinks. Ratio of women's earnings to men's. Ratio of bachelor's degrees earned by women vs. men. 1960-1988.
U.S. News Apr 2, 1990 p. 45

The pay gap is wide and has gotten wider. Average 1988 pay for six key jobs with comparable responsibility and duties.
Business Week Jan 23, 1989 p. 61

Earnings gap between women and men in selected occupations, 1979, 1986.
Black Enterprise Aug 1988 p. 43

Pink-collar wage gap. Women as percent of occupation. Women's wages as percent of men's.
Ms. Mar 1988 p. 69

While women have made some progress. Ratio of female-to-male earnings. 1979-1986.
Business Week Feb 29, 1988 p. 48

While women have made some progress, future gains may come more slowly. Occupation. Number of new jobs. Percent female. Average pay.
Business Week Feb 29, 1988 p. 48

WAGES AND SALARIES - WORLD

How paychecks compare. Median annual salary and bonus for middle managers. Germany. Japan. France. U.S. Britain.
Business Week Nov 26, 1990 p. 71

Wages are rock bottom. Average hourly wage plus benefits, 1989. Mexico. Singapore. South Korea. Taiwan. U.S. Dollars.
Business Week Nov 12, 1990 p. 105

The big gap in German wages. Factory hourly wage. Hours per week. Holidays per year. West Germany. Japan. Italy. U.S. France. Britain. East Germany.
Business Week May 21, 1990 p. 60

Worker paradise. Hourly industrial wage 1988 [by country].
Newsweek Dec 18, 1989 p. 50

Wages in China: Still the lowest. 1988 average monthly earnings in dollars [by Asian country].
Business Week Jun 26, 1989 p. 78

The explosion in Asian labor costs. Average industrial monthly wage. 1984, 1988. South Korea. Taiwan. Hong Kong. Singapore.
Business Week May 15, 1989 p. 46

Fat paychecks for German workers. Gross hourly earnings 1988 [by country].
Business Week Mar 13, 1989 p. 62

U.S. wages look a lot more competitive. Hourly compensation in manufacturing. Japan. U.S. West Germany. 1985-1987.
Business Week Mar 7, 1988 p. 20

WALL STREET

See STOCK MARKET

WARFARE, ELECTRONIC

The games radars play [electronic warfare]. "Deception" jammers. "Noise" jammers. Decoys. Chaff. Diagram.
U.S. News Sep 10, 1990 p. 44

WARS

U.S. wars and hot spots. 1945-1991. Major armed conflicts. Armed interventions. Other military actions. Use of U.S. military advisers abroad. World map.
Scholastic Update Feb 8, 1991 pp. 10–11

War & peace around the world. [Nations] headed toward peace. [Nations with] major conflicts. Map.
Scholastic Update Mar 24, 1989 pp. 3–4

WARS - COSTS

Still cheap, relatively. Gulf War (estimate). World War II. Vietnam War. World War I. Korean War. Costs in 1991 dollars.
Newsweek Feb 4, 1991 pp. 63–65

WARS - ECONOMIC ASPECTS

War triggers growth. Average annual growth rate in real GNP. World War I. World War II. Korean War. Vietnam War.
U.S. News Jan 28, 1991 p. 51

Unemployment marches lower. Change in unemployment rate. World War I. World War II. Korean War. Vietnam War.
U.S. News Jan 28, 1991 p. 51

WARSAW PACT COUNTRIES - MILITARY STRENGTH

Setting limits. Overall treaty limits. NATO. Warsaw Pact.
Time Oct 15, 1990 p. 62

Current strength. NATO. Warsaw Pact. Troops. Tanks. Artillery. Aircraft.
Time May 14, 1990 p. 27

Force levels. Troops. Tanks. Artillery. Aircraft. NATO. Warsaw Pact.
Time Dec 11, 1989 p. 38

Counting down. Existing forces. NATO's count of its own forces. NATO's count of Warsaw Pact. Warsaw Pact's count of NATO. Warsaw Pact's count of its own forces.
Time Jun 12, 1989 pp. 30–31

The balance of power in Europe. Current NATO forces. Current Warsaw Pact forces. NATO's proposal. Warsaw Pact's proposal.
U.S. News Jun 12, 1989 p. 27

A mean bean count. Mean of estimates [military power]. Warsaw Pact. NATO. Warsaw Pact advantage.
Bull Atomic Sci Mar 1989 p. 31

In-place and immediate reinforcing divisions in central Europe. NATO country. Warsaw Pact country. Total divisions. Total armor division equivalents.
Bull Atomic Sci Mar 1989 p. 32

NATO vs. Warsaw Pact in the central front. Troops. Tanks. Map.
Time Dec 19, 1988 p. 25

So who's counting [conventional forces in Europe]. Troops. Tanks. Combat aircraft. Warsaw Pact. NATO.
Bull Atomic Sci Sep 1988 p. 17

Big brother's occupying forces. Soviet military strength in Warsaw Pact countries. Active-duty personnel. Armored divisions. Helicopters. Combat aircraft.
U.S. News May 16, 1988 p. 32

WASHINGTON, D.C.

Puerto Ricans favoring statehood. 1989 per capita income. Percentage below U.S. poverty line. U.S. adults who favor making Washington, D.C. a state.
U.S. News Feb 18, 1991 p. 10

The case for a 51st state. An economic profile of "New Columbia." [Reasons for] passage of H.R. 51, the bill that would admit the State of New Columbia to the United States.
Black Enterprise Dec 1989 p. 33

City within a city: An insider's guide to CongressWorld. The Capitol has its own peculiar places to eat, meet and carry out the business of the day. Map.
Newsweek Apr 24, 1989 p. 31

Tale of two cities. The northwest corner of D.C. is white and well off. The rest of the city is poorer and mostly black. Average household income. Location of homicides.
Newsweek Mar 13, 1989 p. 17

The statistics: Separate and unequal [blacks and whites compared].
Newsweek Mar 13, 1989 p. 19

Capitol Hill. Map.
Scholastic Update Feb 24, 1989 p. 4

Capital district and other territories. Origin of name. Date acquired. Area (sq. mi.). Population. Capital. Head of govt.
Scholastic Update Jan 15, 1988 p. 15

WATER CONTAMINATION

The chlorination quandary. Solutions. Diagram.
U.S. News Jul 29, 1991 p. 50

The nitrate threat. Solution. Diagram.
U.S. News Jul 29, 1991 p. 51

Lead poisoning. Solutions. Diagram.
U.S. News Jul 29, 1991 p. 52

Microbial contaminants. Solutions. Diagram.
U.S. News Jul 29, 1991 p. 53

Chemical menaces. Solution. Diagram.
U.S. News Jul 29, 1991 p. 54

Toxins from the tap. Five types of contamination in the U.S. water supply: Substance. Source. Risk.
Time Mar 27, 1989 p. 38

WATER SUPPLY

Unequal access. % of population that can get safe drinking water. Per capita availability of water annually, in thousands of cubic meters [various countries].
Time Aug 20, 1990 p. 59

WEALTH

The rich got richer. [Selected statistics.]
U.S. News Nov 18, 1991 p. 34

Tax changes widened the gap. [Selected statistics.]
U.S. News Nov 18, 1991 p. 35

Who the rich are; how they live. [Selected statistics.]
U.S. News Nov 18, 1991 p. 36

Where they get their wealth. [Selected statistics.]
U.S. News Nov 18, 1991 p. 37

The richest counties in America. Top U.S. counties in average household wealth. [Other selected statistics.]
U.S. News Nov 18, 1991 p. 39

The rich get richer. The number of millionaires in the U.S. has grown from about 600,000 in 1980 to an estimated 1.5 million today. 1980-1990.
U.S. News Jun 25, 1990 p. 52

Why the inheritance boom is for real. Who has the wealth. Where the wealth is.
U.S. News May 7, 1990 p. 33

Billionaires on America's Main Street. Where this year's *U.S. News* 100 live or work.
U.S. News Aug 1, 1988 pp. 38–39

The *U.S. News* 100. Individuals or families [who] own at least 5 percent of the shares in publicly traded companies. Rank. Name. Holdings. Value of holdings. % of shares owned. 1987 rank.
U.S. News Aug 1, 1988 pp. 40–41

WEAPON SALES

See ARMS SALES

WEAPONS

U.S. arsenal: Planned obsolescence. U.S. weapons used in the Gulf conflict. Weapon system. Last year of procurement. Prewar inventory. Years until replacement on line.
Newsweek Mar 18, 1991 p. 43

Exotic weapons—and their limitations. Reconnaissance. Tactical weapons.
Business Week Feb 4, 1991 pp. 38–39

Patriot vs. Scud [how Patriot missiles track and destroy Scud missiles].
Newsweek Jan 28, 1991 p. 20

The Tomahawk. Diagram.
Newsweek Jan 28, 1991 p. 22

'Smart' bombs at work. GBU-15(V)2 glide bomb. Weight. Length. Guidance.
Newsweek Jan 28, 1991 p. 23

Before and after [CFE - Treaty on Conventional Armed Forces in Europe]. Weapon. Alliance/country. Current holdings. Holdings after CFE.
Bull Atomic Sci Jan 1991 pp. 33–34

Concentration of forces. Seven of the world's 20 largest military establishments are in the Mideast. Country. Manpower. Tanks. Artillery. Combat aircraft.
Newsweek Jul 2, 1990 p. 29

Weapons: How and what the Third World produced, 1982-87. On the ground. In the air. At sea. [By country].
Bull Atomic Sci May 1990 p. 21

First A-bomb. First H-bomb. United States. Soviet Union. Britain. France. China.
Bull Atomic Sci Jan 1990 p. 29

What delivers the real bang for the buck. ICBMs. Bombers.
Newsweek Jan 23, 1989 p. 20

Biological weapons: The future of war? If they ever make it out of the laboratory, biological weapons could be the next weapons of choice for some armies, or even for terrorists.
Newsweek Jan 16, 1989 p. 23

The tools of chemical war. Poison gas comes in many forms.
Newsweek Jan 16, 1989 p. 23

So who's counting [conventional forces in Europe]. Troops. Tanks. Combat aircraft. Warsaw Pact. NATO.
Bull Atomic Sci Sep 1988 p. 17

Third World ballistic missiles. Country. Designation. Range. First flight. Comments.
Bull Atomic Sci Jun 1988 p. 19

How top 10 stack up. Arms suppliers to the Third World [by country].
U.S. News Apr 11, 1988 p. 45

See Also ARMS DEALERS

WEATHER
See CLIMATE AND WEATHER

WEIGHT (PHYSIOLOGY)
Broader figures. Revised government figures for healthy weights for men and women. Height. 19 to 34 years. 35 years and over.
Time Jul 8, 1991 p. 51

Suggested weights for adults. Height. Weight. 19 to 34. 35 and over.
FDA Consumer Jan 1991 p. 12

WEST BANK
Strategic terrain, precious water. The West Bank [gives] the region great military importance and is crucial to Israel's water supply. The struggle for land. Map.
U.S. News Dec 16, 1991 p. 60

WEST GERMANY
See GERMANY (WEST)

WESTERN STATES
Firing lines. The West. Environmentalists and others challenge business interests for control of the land. Key battlegrounds. Map.
Newsweek Sep 30, 1991 p. 29

Alaska, Northwest. California, Southwest. Employment. Population. Median home price. Annual rate of change from previous year. 1989-1990.
U.S. News Nov 13, 1989 p. 56

WETLANDS
Bogged down. State. Acres of wetlands. Percent of total state acreage. [By state.]
Newsweek Aug 26, 1991 p. 48

WHEAT
See GRAIN

WHITE HOUSE
The White House: Nerve center of the Presidency. Map.
Scholastic Update Jan 13, 1989 pp. 8–9

WILDERNESS
See NATIONAL PARKS AND RESERVES

WILDLIFE
The hunting toll on U.S. wildlife. Animals taken by licensed hunters in the 1988-89 season. Mammals. Birds.
U.S. News Feb 5, 1990 p. 33

The government's kill. The wildlife taken by federal Animal Damage Control program in 1988. Mammals. Intentionally killed. Inadvertently killed. Birds. 1988.
U.S. News Feb 5, 1990 p. 37

WINE AND WINEMAKING
A taste of Europe's finest. The '80s produced a wealth of fine wines across Europe. Here is a sampling of the best: France. Germany. Italy. Spain. Portugal. Map.
U.S. News Jul 23, 1990 pp. 62–63

A vintage year. U.S. wine exports. 1980-1988.
Newsweek Jul 11, 1988 p. 49

The squeeze on French wine. Table wine exports to the U.S. 1984-1987.
Business Week Feb 15, 1988 p. 50

WOLVES
Vanishing packs. Range of the gray wolf. Original. 1990. Map of North America.
Newsweek Aug 12, 1991 p. 46

WOMEN
Percentage of women who say more money would make life better. Highest percentages of women executives. Projected date for closing of the male-female wage gap. [Other selected statistics.]
U.S. News Oct 28, 1991 p. 16

Women and children. More women are working—and fewer are having babies. Female workers as a percentage of the total labor force. 1979-1989. Birthrate. Children per family. 1965-1990.
Newsweek Jul 16, 1990 p. 39

WOMEN - EDUCATION

The gender gap shrinks. Ratio of women's earnings to men's. Ratio of bachelor's degrees earned by women vs. men. 1960-1988.
U.S. News Apr 2, 1990 p. 45

Turning away from high tech. Science and engineering degrees awarded to women as a share of all such degrees in 1986 [by discipline and type of degree]. Baccalaureates. Masters. Doctorates.
Business Week Aug 28, 1989 p. 89

Women in business school. 1988-1990.
Business Week Mar 14, 1988 p. 39

Women in science. Percent of bachelor degrees awarded [by field] to women in 1975-1985.
Ms. Mar 1988 p. 73

WOMEN - EMPLOYMENT

The biggest gaps. Women's earnings. [As a] percent of men's earnings. [By selected occupations.]
Business Week Oct 28, 1991 p. 35

Sideline salaries. Male college coaches earn more, on average, than female coaches. [Sport.] Average salary. Male. Female.
U.S. News Oct 28, 1991 p. 90

Women are taking jobs at a slower pace. Labor force participation rate of women age 20-44. 1970-1990.
Business Week Aug 5, 1991 p. 49

Formidable challenges. Barriers to women advancing to top levels of management, identified by CEOs.
Black Enterprise Aug 1991 p. 37

Ethnic breakdown of women managers.
Black Enterprise Aug 1991 p. 37

Breakdown of U.S. employees in executive, administrative and managerial positions, 1990. [Men, women, white, black].
Black Enterprise Aug 1991 p. 42

Who earns what. Median weekly earnings for executive, administrative and managerial positions. Men. Women. 1983-1990.
Black Enterprise Aug 1991 p. 42

Representation of women in corporate management. Professionals. Managers. Senior managers.
Black Enterprise Aug 1991 p. 42

The changing work force. Percent of work force. Women. Men. Hispanics. Asians. Blacks. 1966, 1980, 1990.
Business Week Jul 8, 1991 p. 62

Trouble at the top. The wage gap. Median weekly earnings. Men. Women. Share of men's wages [by job type].
U.S. News Jun 17, 1991 pp. 40–41

Percentage of the nation's 12 million businesses owned by women in 1982. Percentage of women in U.S. Congress in 1975. Median annual earnings of full-time female workers in 1988.
U.S. News Mar 4, 1991 p. 12

The gender gap is narrowing. Women's pay as a percent of men's for full-time workers. 1980, 1988.
Business Week Dec 24, 1990 p. 14

She's the boss. The number of firms owned by women. An industry breakdown.
U.S. News Dec 3, 1990 p. 79

The pay gap between men and women. Annual salary after graduation. Discrepancy in pay scales. By school. By industry.
Business Week Oct 29, 1990 p. 57

Percent of married couples with wives who work. 1960-1988.
Time Sep 3, 1990 p. 83

Women and children. More women are working—and fewer are having babies. Female workers as a percentage of the total labor force. 1979-1989. Birthrate. Children per family. 1965-1990.
Newsweek Jul 16, 1990 p. 39

Engineering women. Scientists. Engineers. Number of women in field. Percentage in field. 1978, 1988.
U.S. News Jul 16, 1990 p. 66

Mothers who are employed. In the labor force. At home.
Newsweek Jun 4, 1990 p. 66

The mommy track. Percentage of women in the labor force with children under age 6. 1960-1988.
U.S. News May 21, 1990 p. 53

Quitting the race. A greater proportion of men are retiring earlier, but women below 65 are hanging in longer. 1948-1989.
U.S. News May 14, 1990 p. 49

A shrinking gender gap. Nonfarm self-employed individuals. Women. Men. 1963-1988.
U.S. News Oct 23, 1989 p. 78

Salary daze. State-government action on pay equity.
Ms. Sep 1989 p. 88

Closing the gap. Here are average earnings of husbands and wives working at full-time jobs, adjusted for inflation in 1987 dollars. 1981, 1987.
U.S. News Aug 28, 1989 p. 102

Percentage of women, blacks and Hispanics in technical occupations. All workers. All technical workers. Technicians and related support.
Black Enterprise Aug 1989 p. 41

Women are landing better jobs but few mothers have reached the top. Female executives. [Percentage who are] married, no children. Married or divorced, with children. Single.
Business Week Mar 20, 1989 p. 134

A long way from "Avon calling." Industries with the most saleswomen. Industries with fewest saleswomen.
U.S. News Feb 6, 1989 p. 42

Income inequality among women is on the rise. Earnings below $10,000. Earnings above $30,000. 1978, 1986.
Business Week Dec 19, 1988 p. 22

As the pool of young workers shrinks, women will fill the gap, and more working mothers will increase the demand for child care. Population 16-24 [years]. 1979-2000. Female share of the work force. 1950-2000.
Business Week Sep 19, 1988 pp. 112–113

Women in the job market. Percentage of women employed in the 50 largest metropolitan areas and percentage of those women in managerial and professional jobs.
U.S. News Aug 22, 1988 p. 65

Working women. More than 53 million women age 16 and older make up 45% of the total labor force.
Black Enterprise Aug 1988 p. 43

Working women. Percentage of women corporate officers of Fortune 1,000 firms, by race - 1986.
Black Enterprise Aug 1988 p. 43

Earnings gap between women and men in selected occupations, 1979, 1986.
Black Enterprise Aug 1988 p. 43

Women are a growing share of the work force. % of all women in the work force. % of work force that is female. 1950-2000.
Ms. Jul 1988 p. 75

Women at work. States with the highest and lowest proportion of women in selected job categories.
Ms. Jun 1988 p. 67

Pink-collar wage gap. Women as percent of occupation. Women's wages as percent of men's.
Ms. Mar 1988 p. 69

While women have made some progress. Ratio of female-to-male earnings. 1979-1986.
Business Week Feb 29, 1988 p. 48

While women have made some progress, future gains may come more slowly. Occupation. Number of new jobs. Percent female. Average pay.
Business Week Feb 29, 1988 p. 48

WOMEN - POLITICAL ACTIVITIES

Women's vote: 1984-1986. Women's vote = margin of victory. Women's vote significantly different.
Ms. Apr 1988 p. 76

Voter turnout by gender, presidential elections. Men. Women. 1972-1984.
Ms. Apr 1988 p. 76

Voter turnout by gender, midterm elections. Men. Women. 1974-1986.
Ms. Apr 1988 p. 76

Races where women's votes won the day. State. Winner. Margin of victory. Difference in women's support. 1984 elections. 1986 elections.
Ms. Apr 1988 p. 77

Voter turnout by gender and age, presidental elections. Women 18-44. Men 18-44. 1972-1984.
Ms. Apr 1988 p. 77

Voter turnout by gender and race, presidential elections. Black women. Black men. 1972-1984.
Ms. Apr 1988 p. 77

WOMEN ENGINEERS

Engineering women. Scientists. Engineers. Number of women in field. Percentage in field. 1978, 1988.
U.S. News Jul 16, 1990 p. 66

Turning away from high tech. Science and engineering degrees awarded to women as a share of all such degrees in 1986 [by discipline and type of degree]. Baccalaureates. Masters. Doctorates.
Business Week Aug 28, 1989 p. 89

Engineering women. Total engineering degrees conferred on women over 10 years, with most recent breakdown of specialties. 1975-1986.
Ms. Feb 1988 p. 33

WOMEN EXECUTIVES

Measuring the glass ceiling. Why white males monopolize the top of the corporate ladder. Men. Women. Minorities. Employees. Managers. Top execs.
U.S. News Aug 19, 1991 p. 14

Representation of women in corporate management. Professionals. Managers. Senior managers.
Black Enterprise Aug 1991 p. 42

Women are landing better jobs but few mothers have reached the top. Female executives. [Percentage who are] married, no children. Married or divorced, with children. Single.
Business Week Mar 20, 1989 p. 134

The path to the executive suite. Black female officials and managers in selected private industries, compared with white and total, 1985.
Black Enterprise Aug 1988 p. 52

WOMEN IN BUSINESS

Who's sitting on corporate boards? Women. Ethnic minorities. 1985-1989.
Black Enterprise Aug 1991 p. 35

Percent of U.S. industries owned by women. All industries [selected industries].
Black Enterprise Jan 1991 p. 18

Women in business school. 1988-1990.
Business Week Mar 14, 1988 p. 39

WOMEN IN POLITICS

Women at the top. State governments with the most women in top executive-branch posts.
U.S. News Apr 10, 1989 p. 72

Number of women in office. Congress. Statewide offices. State legislatures. 1972-1988.
Ms. Apr 1988 p. 79

Percentage of black elected officials who are women [by region].
Black Enterprise Jan 1988 p. 20

WOMEN IN TELEVISION

Broadcast blues. Network news stories filed by women. 1974-1989. News stories about women. February, 1989. ABC. NBC. CBS.
Ms. Sep 1990 p. 89

WOMEN IN THE ARMED SERVICES

Women at war. Operation Desert Storm. Total number of females. Number who died. Number of POWs.
Newsweek Aug 5, 1991 p. 25

The total numbers: A strong beachhead in the armed forces. Women in the military. Enlisted women. Officers. Occupation. Percent. 1973-1990
Newsweek Sep 10, 1990 p. 23

WOMEN MUSICIANS

Women of note. Women musicians in the "Big Five" American orchestras. Percent of women in orchestras. 1972-1988.
Ms. Mar 1989 p. 82

WOMEN SCIENTISTS

Women in science. Percent of bachelor degrees awarded [by field] to women in 1975-1985.
Ms. Mar 1988 p. 73

WOMEN'S MOVEMENT

Dashed expectations. Has the women's movement made things easier or harder for Americans? Respondents. Total. Men. Women.
U.S. News Feb 12, 1990 p. 61

WOOD

Cutting up the woodpile. One cord of wood produces any of the following [products]. Paper consumed per person [by country].
U.S. News Feb 29, 1988 p. 75

WORDS

Scrabble games sold since 1931. Estimated vocabulary of world's smartest people. Languages with the most words. [Other selected statistics.]
U.S. News Sep 30, 1991 p. 15

WORK

Work hard and play hard. Average number of working hours and vacation days in major (international) cities. [City.] Hours worked per week. Vacation days per year.
U.S. News Nov 14, 1988 p. 82

WORK FORCE

Middle-aged men are leaving the work force. Labor force participation of men 45 to 54 years old. 1986-1990.
Business Week Jul 2, 1990 p. 22

Overworked. A survey of managers in eight industralized nations, including the U.S. and Britain, shows why some complain of being overworked.
Newsweek Jun 11, 1990 p. 6

Quitting the race. A greater proportion of men are retiring earlier, but women below 65 are hanging in longer. 1948-1989.
U.S. News May 14, 1990 p. 49

More and more workers are 'professionals.' Professional and technical workers. 1950-1988.
Business Week Apr 2, 1990 p. 23

White males now dominate the job market. Composition of the labor force, 1985. But they will play a smaller role in the future. New entrants to the labor force, 1985-2000.
Business Week Sep 19, 1988 pp. 102–103

The looming mismatch between workers and jobs. Actual skill levels of new workers. Percent of 21-to 25-year-olds entering the labor market from 1985 to 2000. Skill levels needed for new jobs.
Business Week Sep 19, 1988 pp. 104–105

The coming labor shortage. Percent growth in work force. 1950-2000.
Business Week Sep 19, 1988 p. 113

We need every person in the country. While jobs will increase, the number of new workers, aged 16 to 24, will decrease. 1980-1995.
Black Enterprise Sep 1988 p. 15

WORKER PRODUCTIVITY

See PRODUCTIVITY, INDUSTRIAL

WORKERS' COMPENSATION

See INSURANCE, WORKERS' COMPENSATION

WORKING CONDITIONS

Ties that bind. Company policy on ties and jackets for male executives.
U.S. News Apr 11, 1988 p. 68

WORKING HOURS

Time clock. Average number of working hours per year. U.S. Japan. West Germany.
U.S. News Jul 16, 1990 p. 25

WORKING MOTHERS

Percent of employed women [with children] 18 years and older in nonagricultural occupations by age of youngest child. Married. Not married.
Black Enterprise May 1991 p. 39

Labor force participation of mothers, 1988, by race [black, white] and age of child.
Black Enterprise May 1991 p. 39

Mothers who are employed. In the labor force. At home.
Newsweek Jun 4, 1990 p. 66

The mommy track. Percentage of women in the labor force with children under age 6. 1960-1988.
U.S. News May 21, 1990 p. 53

Working mothers. With children age 6-17. With some children under 6. 1950-1985.
Business Week Sep 25, 1989 p. 102

The older the kids, the longer the hours. Mothers of school-age children. Mothers of preschool children. Full-time work. Part-time work. Not employed.
Business Week Mar 20, 1989 p. 132

Women are landing better jobs but few mothers have reached the top. Female executives. [Percentage who are] married, no children. Married or divorced, with children. Single.
Business Week Mar 20, 1989 p. 134

Fewer moms at home. Percentage of families with two or more wage earners. Percentage of all preschool children with working moms. 1980-1987.
U.S. News Aug 29, 1988 p. 102

WORKING MOTHERS - EUROPE

Europe pays to tend its children. Percentage of children in publicly funded child-care facilities. Percentage of mothers with children (age 4 and younger) who are in labor force [by country].
U.S. News Aug 22, 1988 p. 36

WORLD BANK

Diminished clout. World Bank loans per person in developing countries. 1975-1994.
Newsweek Sep 16, 1991 p. 50

WORLD POLITICS

A world of threats. [Map of the world.] Nuclear powers. Major drug sources. Terrorism. Regional and civil conflicts. Strategic minerals. Oil. U.S. bases. Hard-line Communist states.
U.S. News Dec 11, 1989 pp. 24–25

A world of continuing tensions. Regional conflicts. North America. Central America. South America. Western Europe. Africa. Map.
U.S. News Dec 26, 1988 p. 52

WORLD SERIES (BASEBALL)

Stakes in the October [World series] game. World Series winner's share. Actual prize money per player. Adjusted for inflation. 1908-1987.
U.S. News Oct 24, 1988 p. 81

WORLD WAR, 1939-1945

The battle of Britain. History's first great air campaign. Map and chronology.
U.S. News Aug 27, 1990 pp. 72–73

The murderous wages of 'Total War.' Country. Death tolls. Military. Civilian.
Newsweek Sep 4, 1989 p. 64

History's deadliest war. Total deaths (civilian deaths/military deaths). Allies. Axis.
U.S. News Aug 28, 1989 p. 36

How one war ended and another started. 1914. Roots at Versailles. 1920. A weaker Reich.
U.S. News Aug 28, 1989 p. 51

1945. A divided Europe. From World War II's rubble arose two occupied Germanys, two occupied Austrias and an Eastern Europe dominated by Moscow.
U.S. News Aug 28, 1989 p. 72

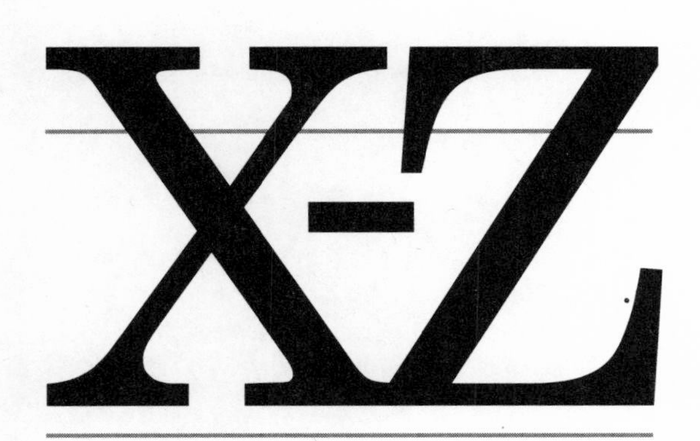

YOUNG MEN

Characteristics of persons ages 14-24. Male. Female.
Black Enterprise Feb 1990 p. 65

YOUNG WOMEN

Characteristics of persons ages 14-24. Male. Female.
Black Enterprise Feb 1990 p. 65

YOUTH

Teens and tobacco. Percentages of ninth to 12th graders who smoke. Male. Female. White. Black. Hispanic.
U.S. News Oct 14, 1991 p. 103

The tuned-out generation [by age group]. % who "read a newspaper yesterday." % who "watched TV news yesterday." % who followed these news events "very closely."
Time Jul 9, 1990 p. 64

Teen tragedies. Teenager's motor-vehicle deaths in 1988. The percentage of teenagers who own a car.
U.S. News Aug 14, 1989 p. 70

Teenage worries. Major concerns of U.S. and Soviet teenagers.
U.S. News Oct 17, 1988 p. 82

A permanent condition? U.S. youth unemployment (ages 16-19). Whites. Blacks. Hispanics.
Black Enterprise Feb 1988 p. 55

Teen power. Where the teens are [by state]. Map.
Scholastic Update Jan 15, 1988 p. 2

The teen population. (U.S. population, aged 13-19, in millions.) 1952-1997.
Scholastic Update Jan 15, 1988 p. 2

Surviving the teen years. Leading causes of death. Ages 15-24. All ages.
Scholastic Update Jan 15, 1988 p. 2

Big spenders. Total spending by U.S. teenagers in billions. 1975, 1987.
Scholastic Update Jan 15, 1988 p. 2

Teen time. How teenagers spend their free time each week, in hours.
Scholastic Update Jan 15, 1988 p. 2

Teen spending power. Number of teenagers. Total value of purchases. Average weekly earnings. Average weekly allowance.
Scholastic Update Jan 15, 1988 p. 9

YOUTH - SOVIET UNION

Teenage worries. Major concerns of U.S. and Soviet teenagers.
U.S. News Oct 17, 1988 p. 82

YOUTH AND CRIME

See CRIME AND YOUTH

YUGOSLAVIA - ECONOMIC CONDITIONS

Troubled economies. Yugoslavia. Hungary. Poland. Change in real wages. Inflation. Gross domestic product per capita. 1984-1988.
Scholastic Update Oct 20, 1989 pp. 16–17

YUGOSLAVIA - NATIONALISM

Unequal economies. % of Yugoslavia's exports. GNP per capita. Average monthly wage. [By republic and province.]
Time Jul 15, 1991 p. 28

Ethnic differences. % of population and main religion. [By ethnic group].
Time Jul 15, 1991 p. 28

Opposing forces. Yugoslav federal forces. Croatian militia. Slovenian militia.
Time Jul 15, 1991 p. 28

Yugoslavia's jigsaw puzzle. As Yugoslavia's religious, cultural and historical tensions explode, a sputtering economy is fueling economic resentment as well. Map.
U.S. News Jul 15, 1991 p. 34

One nation, or many? [Map of republics, ethnic groups, religions.]
U.S. News Jun 18, 1990 p. 32

Yugoslavia. [Republics, provinces and ethnic groups.] Population breakdown. Map.
Time Oct 24, 1988 p. 47

Subject Headings

OF CHARTS, GRAPHS & STATS

1 9 8 8 - 1 9 9 1

Subject Headings

Abortion
Abortion - Laws and Regulations
Accidents
Accidents - Motor Vehicle
Acid Rain
Acupuncture
Addictive Behavior
Adoption
Advanced Placement Examinations
Advertising - Expenditures
Advertising, Direct Mail
Advertising, Magazine
Advertising, Television
Aerobics
Affirmative Action Programs
Afghanistan - Politics and Government
Afghanistan-Russian Invasion, 1979-1989
Africa
Africa - Economic Conditions
Africa - History
Africa - Social Conditions
Aging
Agricultural Subsidies
Agriculture
Agriculture - California
Agriculture - Soviet Union
Agriculture - Soviet Union
Agriculture, Cooperative
AIDS (Disease)
AIDS (Disease) - Africa
AIDS (Disease) - Children
AIDS (Disease) - Research
AIDS (Disease) - Testing
AIDS (Disease) - Vaccines
AIDS (Disease) - Women
AIDS (Disease) - World
Air Transport - Market Share
Air Travel
Air Travel - Accidents
Air Travel - Fares
Air Travel - Safety
Airline Industry
Airline Industry - Market Share
Airplanes, Military
Alaska
Alcohol and Youth
Alcoholic Beverages
Alcoholic Beverages, Drinking of
Alcoholism
Aliens, Illegal
Allergies
Alzheimer's Disease
America - Discovery and Exploration
Americans
Amusement Parks
Andes
Animal Experimentation
Animals
Antarctica
Anthropology
Apartheid
Apportionment (Election Law)
Arab Nations
Armed Forces - United States
Armed Forces - World
Arms Control
Arms Dealers
Arms Sales
Army - United States
Art - Prices
Art Museums
Artificial Body Parts
Asia - Economic Conditions
Asian-Americans

Editor's Note: Subject Headings use Soviet Union prior to December 8, 1991, and Commonwealth of Independent States after that date.

Aspirin
Assault Rifles
Assets, Personal
Associations
Asteroids
Asthma
Astrology
Astronomy
Asylum, Political
Athletic Shoes - Market Share
Atlantic City (New Jersey)
Atlantic States - Economic Conditions
Autographs
Automobile Driving
Automobile Industry
Automobile Industry - Advertising
Automobile Industry - Market Share
Automobile Theft
Automobiles
Automobiles - Air Conditioning
Automobiles - Crashworthiness
Automobiles - Defects
Automobiles - Fuel Efficiency
Automobiles - Imports and Exports
Automobiles - Leasing and Renting
Automobiles - Purchasing
Automobiles - Safety Devices
Automobiles - Sales
Automobiles, Cost of Operation
Automobiles, Experimental
Automobiles in Eastern Europe
Automobiles in Japan
Automobiles, Used
Awards

Baby Boom Generation
Back
Baldness
Baltic States
Bank Failures
Bank of Credit and Commerce International
Bankruptcy
Banks and Banking
Banks and Banking - Market Share
Barbecue Cookery
Baseball Cards
Baseball, Professional
Basketball, College
Basketball, Professional
Batteries
Beaches
Beer Consumption
Beer Industry - Market Share
Beverages
Bicycles and Bicycling
Bilingualism
Billionaires
Biofeedback
Biological Rhythms
Birmingham (Alabama)
Birth Control
Birth Defects
Birth Order
Births and Birth Rate
Black Accountants
Black Athletes
Black Children
Black College Students
Black Colleges and Universities
Black Consumers
Black Families
Black Farmers
Black Teachers
Blacks
Blacks - Economic Conditions
Blacks - Education
Blacks - Employment
Blacks - Health
Blacks - Integration
Blacks - Political Activity
Blacks - Religion
Blacks in Business
Blacks in Newspaper Publishing
Blind
Blood
Blood Banks
Body Fat
Body Image
Bonds
Bonds
Book Publishers and Publishing
Boston (Massachusetts)
Botany - Ecology
Brain
Brazil - Economic Conditions
Breast Examination
Budget - Bush Administration
Budget - United States
Bulgaria - Economic Conditions
Bush, George
Business Ethics
Business Failures
Business Schools

Caffeine
California
California - Economic Conditions
Calories
Cambodia
Campaign Financing
Canada
Canada - Economic Conditions
Canada and the United States
Cancer
Cancer - Causes
Cancer - Nutritional Aspects
Cancer - Research
Cancer - Survival Rates
Cancer (Breast)
Cancer (Cervical)
Cancer (Colon)
Cancer (Lung)
Cancer (Prostate)
Cancer (Skin)
Cancer (Testicular)
Capital Gains
Capital Punishment
Carbon
Career Plateaus
Caribbean Region
Carpal Tunnel Syndrome
Catholic Church
Cellular Phones
Censorship
Census
Central America
Central America - Economic Conditions
Central America - Social Conditions
Central America and the United States
Charity
Chemical and Biological Weapons
Chemical Industries
Chicago (Illinois)
Child Abuse
Child Care
Child Care - Europe
Child Rearing
Childbirth
Children
Children - Diseases
Children - Fitness
Children - Health
Children - Recreation
Children, Gifted
Children of Divorced Parents
Children's Literature
Chile - Economic Conditions
China
China - Economic Conditions
China - Foreign Investments
China - Military Strength
China - Politics and Government
China and the Soviet Union
China and the United States
Chlorination of Water
Chlorofluorocarbons
Chocolate
Cholera
Cholesterol
Cholesterol Testing
Christmas - Economic Aspects
Christmas Cards
Cities and Towns
Civil Rights
Civil Rights - International Aspects
Civilization, Ancient
Cliffs Notes
Climate and Weather
Clothing and Dress
Clothing Industry - Imports and Exports
Coaching
Cocaine
Coffee
Cold (Disease)
College Athletes
College Education, Cost of
College Graduates
College Students
College Students - Foreign
Colleges and Universities
Colleges and Universities - Enrollment
Colleges and Universities - Graduate Work
Colleges and Universities - Research
Comets
Comic Books, Strips, Etc.
Commodity Futures
Commonwealth of Independent States
Communist Party
Compact Discs
Computer Chips
Computer Crimes
Computer Display Terminals - Health Aspects
Computer Industry
Computer Industry - Market Share
Computer Networks

Computer Operating Systems
Computer Printers
Computer Programmers
Computer Software Industry - Market Share
Computer Viruses
Computers
Computers - Educational Use
Computers - Maintenance and Repair
Computers, Personal
Conglomerate Corporations
Congress
Congress - Elections
Consultants
Consumer Price Index
Consumer Protection
Consumer Spending
Consumers - China
Consumers - Japan
Consumers - Soviet Union
Consumers - United States
Convenience Stores
Copyright - International Aspects
Corporate Social Responsibility
Corporations
Corporations - Acquisitions and Mergers
Corporations - Directors
Corporations - Finance
Corporations, International
Cosmetic Industry - Market Share
Cost of Living - United States
Cost of Living - World
Country Life
Crack (Cocaine)
Creativity
Credit
Credit
Credit Bureaus
Credit Cards
Crime
Crime
Crime and Youth
Crime, Corporate
Crime Prevention
Crime, Victims of
Criminal Justice, Administration of
Cuba and the Soviet Union
Czechoslovakia - Economic Conditions

Dade County (Florida)
Damages (Legal)
Death and Death Rates
Debt
Debt, Corporate
Debt, International
Debt, Personal
Decision Making
Defense Contracts
Defense Spending
Defense Spending - Bush Administration
Defense Spending - Europe
Defense Spending - Reagan Administration
Delivery Industry
Democracy
Democratic Party
Denmark - Military Strength
Dentistry
Department of Energy
Depression (Mental)
Desktop Publishing
Developing Countries
Developing Countries - Defenses
Diabetes
Dieting
Dinners and Dining
Dinosaurs
Disasters
Discrimination
Discrimination in Employment
Discrimination in Housing
Diseases
Diseases, Hereditary
Disk Drives (Computers)
Divorce
Dogs
Drexel Burnham Lambert, Inc.
Drinking Water
Dropouts
Drought
Drug Abuse
Drug Enforcement Administration
Drug Trafficking
Drugs and Automobile Drivers
Drugs and Crime
Drugs and Employment
Drugs and Gangs
Drugs and Pregnancy
Drugs and Sports
Drugs and Youth
Drugs (Pharmaceutical)
Drunk Driving
Ducks - North America

Earth - Internal Structure
Earthquakes
Earthquakes - California
Eastern Europe
Eastern Europe - Economic Conditions
Eastern Europe - Nationalism
Eating Disorders
Eclipses, Solar
Economic Assistance
Economic Assistance, American
Economic Assistance, Domestic
Economic Assistance, Japanese
Economic Conditions - Reagan Administration
Economic Conditions - United States
Economic Conditions - United States
Economic Conditions - World
Education
Education - Finance
Education - Tests and Measurements
Education, Elementary
Education, Higher
Education, Preschool
Education, Secondary
Egypt - Military Strength
Elections
Elections - United States
Elections - United States, 1988
Elections - United States, 1990
Electric Power
Electric Power - Consumption
Electric Power - Sweden
Electric Utilities
Electromagnetic Waves
Electronic Mail Systems
Elephants
Employee Benefits
Employee Dismissal
Employee Training
Employees
Employment References
Endangered Species
Energy Conservation
Energy, Cost of
Energy Supply
Engineering Schools
Engineers
Entertainment Industry
Environment
Environmental Organizations
Equal Pay for Equal Work
Ergonomics
Ethiopia
Europe
Europe - Defenses
Europe - Economic Conditions
European Economic Community
Everglades - Ecology
Evolution
Excellence
Executives
Executives - Salaries, Pensions, Etc.
Exercise
Exercise Equipment
Exxon Valdez (Ship) Oil Spill, 1989
Eye
Eye - Diseases and Defects
Eye - Surgery
Eyeglasses

Factories
Fairs
Family
Famines
Farms and Farming
Fast Food Restaurants
Fathers
Fax Machines
Federal Budget
Federal Debt
Federal Deficit
Federal Deposit Insurance Corporation
Federal Drug Administration
Federally Insured Programs
Fertilization *in Vitro*, Human
Fetus - Surgery
Fiber in Diet
Fireworks
Flags - United States
Flu (Disease)
Food
Food - Content
Food - Labeling
Food - Preservation
Food Additives
Food Allergies
Food Contamination
Food, Frozen
Food Industry
Food Preferences
Food Supply - Africa
Food Supply - Soviet Union
Football, College
Football, Professional
Ford Foundation

Forest Products Industry
Forests and Forestry
France - Economic Conditions
Franchises
Fraternities and Sororities
Free Trade and Protection
Fruit
Fruit Juices - Market Share
Furniture
Furs
Fusion
Fusion, Cold

Galileo Project
Gambling
Games
Garbage
Garbage Industry
Gardening
Gas, Natural
Gasoline - Consumption
Gasoline Prices
Gems
General Motors Corporation
Genetic Engineering
Genetic Mapping
Genetics
Geography
Geology
German Reunification
Germany - Economic Conditions
Germany - History
Germany (East) - Economic Conditions
Germany (West) and the United States
Germany (West) - Economic Conditions
Gifts
Gold Mines and Mining
Golf and Golfers
Golf, Professional
Government Contracts
Government Employees
Government Services
Grain
Great Britain - Economic Conditions
Great Britain - Ethnic Groups
Great Britain - Politics and Government
Greenhouse Effect
Grenada-American Invasion, 1983 - Casualities
Gross Domestic Product
Gross Domestic Product - Africa
Gross Domestic Product - France
Gross Domestic Product - Germany
Gross Domestic Product - Latin America
Gross National Product - Asia
Gross National Product - Europe
Gross National Product - Soviet Union
Gross National Product - United States
Gross National Product - World
Gun Control
Guns and Youth
Guns in the United States

Haiti - Economic Assistance
Halloween
Harvard University
Hazardous Waste
Head-wounds and Injuries
Headaches
Health
Health Education
Health Maintenance Organizations
Heart
Heart - Diseases
Heart Attacks
Hepatitis
Heredity
Heroin
High School Equivalency Examinations
High School Graduates
High School Students
High Schools
Hispanic-Americans
Hispanic-Americans - Economic Conditions
Hispanic-Americans - Education
Hispanic-Americans - Employment
Hobbies
Holidays
Home Heating Oil
Home Shopping
Home-based Businesses
Homeless
Homosexuality
Hormones
Hormones, Female
Hormones, Male
Hospitals
Hospitals, Psychiatric
Hostages - Lebanon
Housework
Housing - Affordability

Housing - Buying and Selling
Housing - Construction
Housing - Costs
Housing - Federal Aid
Housing - Great Britain
Hubble Space Telescope
Human Body
Human Rights
Hungary - Economic Conditions
Hunger
Hunting
Hypnosis

Immigration and Emigration - Europe
Immigration and Emigration - Soviet Union
Immigration and Emigration - United States
Immune System
Imports and Exports
Imports and Exports - Canada
Imports and Exports - Caribbean Region
Imports and Exports - China
Imports and Exports - Europe
Imports and Exports - Germany
Imports and Exports - Iran
Imports and Exports - Iraq
Imports and Exports - Japan
Imports and Exports - Korea
Imports and Exports - Korea
Imports and Exports - Mexico
Imports and Exports - North America
Imports and Exports - Soviet Union
Imports and Exports - United States
Impotence
Incentives in Industry
Income
Income - Disposable
Income - Family
Income - Household
Income - Median
Income - Per Capita
Income - Personal
Income Distribution
Income Tax
Income Tax - Auditing
Income Tax - Corporate
Income Tax - Deductions
India
Industry
Infant Mortality
Infants, Premature
Inflation (Financial)
Inflation (Financial) - China
Inflation (Financial) - Europe
Inflation (Financial) - United States
Inheritance
Injuries, Occupational
Insect Control
Insurance
Insurance, Automobile
Insurance, Health
Insurance Industry
Insurance, Liability
Insurance, Life
Insurance, Unemployment
Insurance, Workers' Compensation
Intelligence
Intelligence Service
Interest (Economics)
Interest Rates
Internal Revenue Service
International Business Machines Corporation
International Trade
International Trade - Europe
International Trade - Japan
International Trade - North America
International Trade - Soviet Union
International Trade - United States
International Trade with Canada
International Trade with China
International Trade with Europe
International Trade with Japan
International Trade with Mexico
International Trade with Pacific Rim Countries
International Trade with the Soviet Union
Interviewing
Investment Trusts
Investments
Investments, American
Investments, Foreign
Investments, Japanese
Iran-Iraqi War, 1979-1988
Iranian Air Downing, 1988
Iraq
Iraq - Military Strength
Irish-Americans
Israel
Israel - Military Strength
Israel - Politics and Government
Israeli-Arab Relations
Israeli-Arab Relations - Territorial Question

Italy - Economic Conditions
Ivory Trading

Japan
Japan - Armed Forces
Japan - Economic Conditions
Japan - Employment
Japan - Politics and Government
Japan - Social Conditions
Japan and the United States
Japanese Language
Jesuits
Jews
Job Hunting
Job Satisfaction
Jobs
Jobs - Europe
Jobs - Future
Jobs, Temporary
Joint Ventures
Jordan
Judges
Jumping
Junk Bonds

Kazakhstan
Kidnapping
Kidney Stones
Korea
Korea - Economic Conditions
Kurds
Kuwait - Economic Conditions
Kuwait - Oil Well Fires
Kuwait and the United States
Kuwait-Iraqi Invasion, 1990
Kuwait-Iraqi Invasion, 1990 - American Involvement
Kuwait-Iraqi Invasion, 1990 - Chemical Warfare
Kuwait-Iraqi Invasion, 1990 - Hostages
Kuwait-Iraqi Invasion, 1990 - Military Strength

Labor
Labor Supply - China
Labor Supply - United States
Labor Unions
Laboratory Animals
Language and Languages
Las Vegas (Nevada)
Latin America
Latin America - Economic Conditions
Latin America and the United States
Law Schools
Lawsuits
Lawyers
Layoffs
Lead Poisoning
Learning Disabilities
Lease and Rental Services
Leisure
Leveraged Buyouts
Libraries
Life Expectancy
Lightning
Literacy
Lithuania
Loans
Lotteries
Lyme Disease

Machine Industry
Magazines
Maglev Trains
Mail Order Business
Mammogram
Man - Origin
Man, Prehistoric
Management
Manufacturers and Manufacturing
Marijuana
Marine Corps
Marriage
Mass Media
Massachusetts
Master of Business AdministrationDegree
Mathematics
Maxwell, Robert
Mayors
Meat
Medicaid
Medical Care
Medical Care - Soviet Union
Medical Care, Cost of
Medical Research
Medical Schools
Medicare
Medicine, Alternative
Memory
Mensa
Mental Health Care
Mentally Handicapped
Methanol
Mexico

Mexico - Economic Conditions
Mexico - History
Miami (Florida)
Microscope
Microwave Cooking
Middle Classes
Middle East Countries
Middle East Countries - Military Strength
Midwestern States - Economic Conditions
Military Assistance, American
Military Assistance, Soviet Union
Military Bases
Military Personnel
Military Service, Compulsory
Milken, Michael
Millionaires
Minimum Wage
Minority Businesses
Missiles
Missing Children
Missing in Action
Mississippi River Valley
Money - International Aspects
Money - Soviet Union
Money Laundering
Money Management
Mortgages
Mortgages - Refinancing
Motion Picture Theaters
Motion Pictures
Motorcycles - Market Share
Moving, Household
Municipal Finance
Municipal Government
Murder and Murder Rates
Music, Popular
Music Recording Industry
Musical Instruments
Musical Pitch
Muslims

Names, Personal
National Aeronautics and Space Administration
National Parks and Reserves
Nations
Native Americans
Native Americans - Education
Native Americans - Employment
Native Americans - Social Conditions
Natural Gas
Navy - United States
Netherlands
New England - Economic Conditions
New York (City)
New York (State)
News Media
Newspaper Publishers and Publishing
Newspapers, Tabloid
Nicaragua - Politics and Government
Nicaragua and the Soviet Union
Nicaragua and the United States
Noise
Nonprofit Organizations
North America
North Atlantic Treaty Organization - Military Strength
Northwestern States - Economic Conditions
Nuclear Power Plants
Nuclear Reactors
Nuclear Ships
Nuclear Weapons
Nuclear Weapons - Accidents
Nuclear Weapons - China
Nuclear Weapons - Commonwealth of Independent States
Nuclear Weapons - Europe
Nuclear Weapons - France
Nuclear Weapons - Great Britain
Nuclear Weapons - India
Nuclear Weapons - Iraq
Nuclear Weapons - Pakistan
Nuclear Weapons - South America
Nuclear Weapons - Soviet Union
Nuclear Weapons - Testing
Nuclear Weapons - United States
Nursing Homes
Nutrition
Nuts

Oceanography
Office Buildings
Office Employees
Office Equipment and Supplies
Oil Prices
Oils and Fats
Older Americans
Older Americans - Home Care
Older Women - Economic Conditions
Olympic Games
Opium
Orange Juice Industry - Market Share
Organization of Petroleum Exporting Countries
Ozone

Pacific Rim Countries
Pain
Pakistan
Palestinian Arabs
Pan American Flight 103 Disaster, 1988
Panama
Panama-American Invasion, 1989-1990
Parental Leave
Patents
Patriot (Missile)
Peace Corps
Pearl Harbor, Attack On, 1941
Pensions
Pentagon Procurement Scandal
Persian Gulf War, 1991
Persian Gulf War, 1991 - Aerial Operations
Persian Gulf War, 1991 - Campaigns and Battles
Persian Gulf War, 1991 - Casualities
Persian Gulf War, 1991 - Cost
Persian Gulf War, 1991 - Defenses
Persian Gulf War, 1991 - Environmental Aspects
Persian Gulf War, 1991 - Equipment and Supplies
Persian Gulf War, 1991 - Journalists
Persian Gulf War, 1991 - Military Strength
Persian Gulf War, 1991 - Personalities
Persian Gulf War, 1991 - Soldiers
Persian Gulf War, 1991 - Technology
Persian Gulf War, 1991 - Weapons
Pesticides
Pesticides - Health Aspects
Petrochemicals
Petroleum - Soviet Union
Petroleum Consumption
Petroleum Industry
Petroleum Industry - Soviet Union
Petroleum Prices
Petroleum Production
Petroleum Prospecting
Petroleum Refineries
Petroleum Supply
Petroleum Tankers
Philadelphia (Pennsylvania)
Photography
Physical Fitness
Physicians
Physics
Pilgrims (New England Colonists)
Poaching
Poisonous Plants
Poland - Economic Conditions
Police
Police Brutality
Political Action Committees
Political Parties
Pollution
Pollution, Air
Pollution, Air - Europe
Pollution, Marine
Pollution, Radioactive
Pollution, Water
Population - Africa
Population - Europe
Population - India
Population - Israel
Population - Mexico
Population - Soviet Union
Population - United States
Population - World
Porter, Cole
Portland (Oregon)
Postal Rates
Postal Service
Potassium (Mineral Supplements)
Poverty
Pregnancy
Presidential Campaigns, 1984
Presidential Campaigns, 1988
Presidential Election, 1984
Presidential Election, 1988
Presidential Elections
Presidents - United States
Price Indexes
Prisons and Prisoners
Privacy
Pro-Democracy Movement
Productivity, Industrial
Profit
Promotions
Property
Property Taxes
Proteins
Protestant Church
Psychiatrists
Psychologists
Psychologists
Public Schools - Chicago (Illinois)
Public Schools - Integration
Public Schools - New York (City)
Public Schools - Sports
Public Schools - Violence
Public Welfare

Public Works
Puerto Rico
Pulitzer Prize

Quality of Life
Quarks

Race Relations - United States
Radar Defense Networks
Radiation
Radio Industry
Radon
Railroads
Rain and Rainfall
Rain Forests
Rape
Reading
Real Estate - Japan
Real Estate, Commercial
Real Estate Industry
Recession (Financial)
Recycling (Waste, Etc.)
Refugees
Refugees, Iraqi
Refugees, Kurdish
Refuse and Refuse Disposal
Regulatory Agencies
Religion
Remarriage
Rent and Renters
Republican Party
Research - Federal Grants
Research and Development
Resolution Trust Corporation
Restaurants
Retail Trade
Retirement
Roads
Robots, Industrial
Rock Concerts
Romania - Economic Conditions

Sales Personnel
Salt in the Body
Saudi Arabia
Saudi Arabia - Economic Conditions
Saving and Savings - United States
Saving and Savings - World
Savings and Loan Associations
Savings and Loan Associations - Failures
Savings and Loan Associations - Federal Aid
Savings and Loan Associations - Fraud
Scandinavia - Economic Conditions
Scholastic Aptitude Test
Science - Soviet Union
Scientists
Scientists
Securities Exchange Commission
Security Guard Services
Self-Employed
Self-Perception
Selling
Semiconductor Industry
Senses
Service Industries
Sex Discrimination in Education
Sexual Harassment
Sexually Transmitted Diseases
Shakespeare, William, 1564-1616
Shoplifting
Shopping
Shopping Centers and Malls
Single Parent Families
Single People
Skiing
Sleep
Small Businesses
Smoking
Smoking - Costs
Smoking - Laws and Regulations
Snack Foods
Soccer
Social Programs
Social Security
Social Values
Soft Drink Industry - Market Share
South Africa
South Africa - Ethnic Groups
South Africa - Race Relations
South America
South America - Antiquities
Southeast Asia
Southern States - Economic Conditions
Soviet Union
Soviet Union - Breakup of the Union
Soviet Union - Economic Conditions
Soviet Union - Foreign Investments
Soviet Union - History
Soviet Union - Immigration and Emigration
Soviet Union - Military Strength
Soviet Union - Nationalism

Soviet Union - Natural Resources
Soviet Union - Politics and Government
Soviet Union - Religion
Soviet Union, Aid to
Soviet Union and China
Soviet Union and Eastern Europe
Soviet Union and the Middle East
Soviet Union and the United States
Space Flight
Space Flight - United States
Space Pollution
Space Shuttle Missions
Spain - Economic Conditions
Spelling Ability
Sports
Sports - Injuries
Sports, College
Sports, Economic Aspects
St. Paul (Minnesota)
Standard of Living
State Government - Finance
Stealth Bomber
Steel Industry
Steroids
Stock Market
Stock Market - World
Stress
Strikes
Student Loan Programs
Students
Students - Employment
Sub-saharan Africa - Economic Conditions
Submarine Warfare
Suburbs
Subways
Suicide
Sun
Sunspots
Suntan
Supercomputers - Market Share
Superconductors and Superconductivity
Superfund
Supermarkets
Supernatural
Supreme Court
Surgery
Sweden
Syria
Tax Evasion
Tax Reform
Tax Shelters
Taxation
Taxation, State
Teachers
Teachers - Salaries
Teaching
Technology
Teenage Pregnancy
Teeth
Telecommunications
Telephone - Long Distance Service
Telephone Industry
Telephone Rates
Telephones
Telephones, Cellular
Telescopes
Television - Audience
Television - News
Television - Receivers and Reception
Television - Sports
Television, Cable
Television, High Definition
Television Industry
Television Programs
Temperature
Tennis, Professional
Territories - United States
Terrorists, Iranian
Texas
Thanksgiving Dinner
Thatcher, Margaret
Theater
Timesharing (Real Estate)
Tipping
Tire Industry
Tithes
Tobacco Industry
Tobacco Industry - Market Share
Tokyo (Japan)
Torture
Tourist Trade
Toxic Shock Syndrome
Toy Industry
Transplantation of Organs,
Transportation
Travel
Trees
Tritium
Trucks
Trucks, Foreign
Tuberculosis

Ukraine
Ultraviolet Rays
Unemployment
Unemployment - World
United Nations
United States
United States - Foreign Relations
United States - Military Strength
United States - Social Conditions
United States - Social Policy
Universe
Unmarried Couples
Unmarried Mothers
Uranium

Vacations
Vaccination
Vegetable Gardens and Gardening
Vegetables
Venture Capital
Verbal Skills
Veterans
Video Discs
Video Games
Videotape Recorders and Recording
Videotapes
Videotapes - Renting
Vietnam
Vietnam War, 1957-1975 - Casualties
Vietnam War, 1957-1975 - Economic Aspects
Vietnam War, 1957-1975 - Missing in Action
Vietnam War, 1957-1975 - Soldiers
Violence and Television
Violence, Domestic
Vitamins
Volcanoes
Volunteer Service
Volunteers
Voter Registration
Voting

Wages and Salaries
Wages and Salaries - Airline Pilots
Wages and Salaries - Executives
Wages and Salaries - Government Employees
Wages and Salaries - Sports
Wages and Salaries - Teachers
Wages and Salaries - Women
Wages and Salaries - World
Warfare, Electronic
Wars
Wars - Costs
Wars - Economic Aspects
Warsaw Pact Countries - Military Strength
Washington, D.C.
Water Contamination
Water Supply
Wealth
Weapons
Weight (Physiology)
West Bank
Western States
Wetlands
White House
Wildlife
Wine and Winemaking
Wolves
Women
Women - Education
Women - Employment
Women - Political Activities
Women Engineers
Women Executives
Women in Business
Women in Politics
Women in Television
Women in the Armed Services
Women Musicians
Women Scientists
Women's Movement
Wood
Words
Work
Work Force
Working Conditions
Working Hours
Working Mothers
Working Mothers - Europe
World Bank
World Politics
World Series (Baseball)
World War, 1939-1945

Young Men
Young Women
Youth
Youth - Soviet Union
Yugoslavia - Economic Conditions
Yugoslavia - Nationalism

About the Editor

Robert Skapura taught math, physics and English before becoming a librarian more than fifteen years ago. He is a frequent speaker at state and national library conferences and for three years chaired the Technology Committee for American Association of School Librarians (AASL). He has designed computer software for both library automation and bibliographic instruction and, published two student term paper manuals, one on researching and writing about history, the other covering literature. In 1990 he edited the first volume of *The Cover Story Index*.

Mr. Skapura is addicted to spy novels and plays fullback on a soccer team. He lives with his wife and children near San Francisco where he is the librarian for Clayton Valley High School.